POLYMER PHOTOPHYSICS

Luminescence, energy migration and molecular motion in synthetic polymers

POLYMER PHOTOPHYSICS

Luminescence, energy migration and molecular motion in synthetic polymers

Edited by

David Phillips BSc PhD FRSC

Wolfson Professor of Natural Philosophy
The Royal Institution
London

LONDON NEW YORK
CHAPMAN AND HALL

First published 1985 by
Chapman and Hall Ltd
11 New Fetter Lane, London EC4P 4EE

Published in the USA by
Chapman and Hall
29 West 35th Street, New York NY10001

Printed in Great Britain at the University Press, Cambridge

ISBN 0 412 16510 4

British Library Cataloguing in Publication Data

Polymer photophysics.
1. Polymers and polymerization
2. Photochemistry
I. Phillips, David, *1939-*
547.84′270455 QD381

ISBN 0-412-16510-4

Library of Congress Cataloging in Publication Data

Main entry under title:

Polymer photophysics.
Bibliography: p.
Includes index.
1. Polymers and polymerization.
2. Luminescence.
3. Phosphorescence. I. Phillips, D. (David)
QD381.8.P65 1985 547.7′0455 84-15517
ISBN 0-412-16510-4

Contents

Contributors

D. Baeyens-Volant	Université Libre de Bruxelles, Faculté des Sciences, Campus Plaine 206/1, 1050 Bruxelles, Belgium.
Emo Chiellini	Centro C.N.R. Macromolecule Stereordinate Otticamente Attive, 56100 Pisa, Italy. and Istituto di Chimica Generale, Facoltà di Ingegneria, Università di Pisa, 56100,Pisa, Italy.
C. David	Université Libre de Bruxelles, Faculté des Sciences, Campus Plaine 206/1, 1050 Bruxelles, Belgium.
Giancarlo Galli	Istituto di Chimica Organica Industriale, Università di Pisa, 56100 Pisa, Italy. and Centro C.N.R. Macromolecule Stereordinate Otticamente Attive, 56100 Pisa, Italy.
K.P. Ghiggino	Department of Chemistry, University of Melbourne, Parkville, Victoria 3052, Australia.
Anthony Ledwith	Department of Inorganic, Physical and Industrial Chemistry, The University of Liverpool, Liverpool L69 3BX, UK.

CONTRIBUTORS

L. Monnerie	Laboratoire Physicochimie Structurale et Macromoleculaire, Ecole Supérieure de Physique et de Chimie Industrielles de Paris, 10, Rue Vauquelin, 75231 Paris, France.
J.H. Nobbs	Department of Colour Chemistry, University of Leeds, LS2 9JT, UK.
David Phillips	The Royal Institution, 21, Albemarle Street, London W1X 4BS, UK.
A.J. Roberts	Davy-Faraday Research Laboratories, The Royal Institution, 21, Albemarle Street, London W1X 4BS, UK.
Roberto Solaro	Instituto di Chimica Organica Generale, Università di Pisa, 56100 Pisa, Italy. and Centro C.N.R. Macromolecule Stereordinate Otticamente Attive, 56100 Pisa, Italy.
I. Soutar	Chemistry Department, Heriot-Watt University, Riccarton, Currie, Edinburgh EH14 4AS, UK.
K.L. Tan	Department of Chemistry, University of Melbourne, Parkville, Victoria 3052, Australia.
I.M. Ward	Department of Physics, University of Leeds, Leeds LS2 9JT, UK.
Stephen E. Webber	Department of Chemistry and Center for Polymer Research, University of Texas at Austin, Austin, Texas 78712, USA.

Foreword

Luminescence spectroscopy, including latterly time-resolved techniques, has become of some importance as a means of studying molecular motion, order, and excitation diffusion and energy transfer in synthetic polymers. In this volume, a selection of the many eminent practitioners of these experimental studies have contributed chapters relating their own particular expertise, and it is hoped that the resulting volume will prove of great value to those with interests in the structure and dynamics of polymer chains, and also to photophysicists who might be seeking complex molecular systems upon which to practice their ever more sophisticated techniques.

The field is moving rapidly, but it is to be hoped that the volume will remain for some time a timely statement of achievements to date.

David Phillips
March 1985

CHAPTER ONE

Introduction to photophysics of organic molecules

This volume is devoted to the study of the photoluminescence of synthetic polymers in which the absorbing and emitting chromophores are organic molecules that may be expected in principle to behave in photophysical terms like their small molecule analogues. We introduce the subject of luminescence in polymers therefore by a detailed summary of the photophysical processes that occur in small molecules [1]. These can be introduced in the conventional way with reference to a state diagram, after Jablonskii. The lowest energy (ground) state of almost all organic molecules is a singlet state in which electrons with opposing spin angular momenta are paired in molecular orbitals. In a simple nomenclature, this is shown as S_0 in Fig. 1.1. Absorption of light by the molecule results in promotion of an electron from the highest occupied molecular orbital (HOMO) into an orbital normally unoccupied in the ground state, the lowest unoccupied orbital (LUMO). However, this process does not through the electric dipole mechanism change the spin of the electron making the transition, and thus the excited state produced is also a singlet state, designated S_1 in Fig. 1.1. Since in the excited state the pair of electrons from the original orbital are now in different orbitals, there are no constraints of spin angular momentum upon them, and they may adopt parallel spins, resulting in a state, shown in Fig. 1.1 as T_1. For the same electron configuration, that is, occupancy of orbitals, the triplet state is always lower in energy than the corresponding singlet state, owing to lower electron repulsion

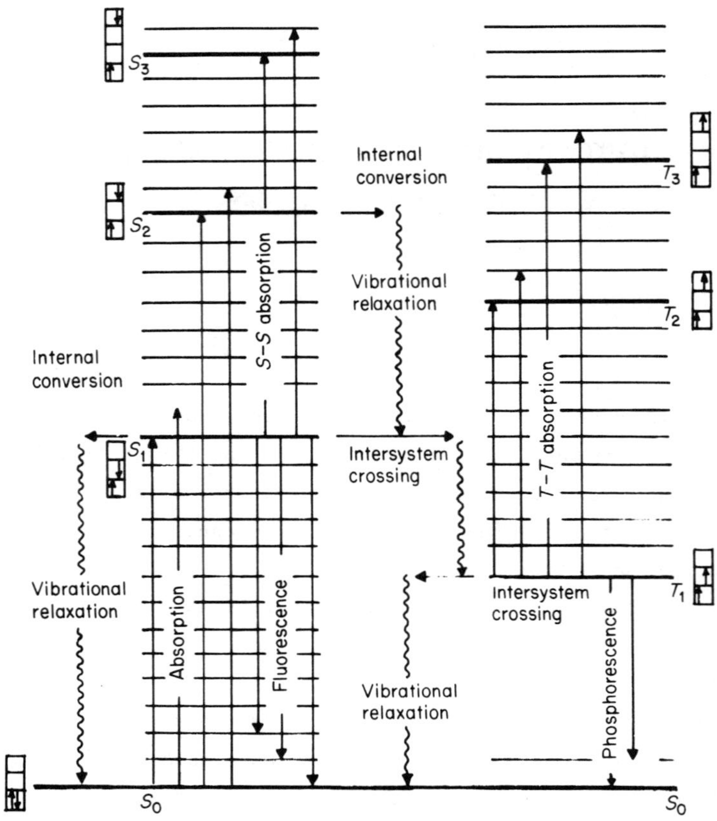

Fig. 1.1 Jablonskii diagram showing fates of polyatomic molecules upon photoexcitation.

in the latter. Promotion of electrons from the HOMO to other than LUMO must produce a higher energy state than S_1. Examples of such higher energy state are shown in Fig. 1.1 as S_2 and S_3 with the triplet states corresponding shown as T_2, T_3. In principle, any vibrational level of the excited state can be formed from the ground state upon absorption, and thus a whole distribution of wavelengths will suffice to populate excited electronic state; an intensity profile of this distribution is of course the absorption spectrum. The possible fates of a molecule excited to a specific vibrational level of the first excited singlet state can now be considered. For convenience, this is assumed to be the zero-point level of the S_1 state, and that the molecule exists in isolation.

1.1 UNIMOLECULAR FATES

The first, obvious fate is that which results from the reverse process to absorption, that is, the removal of the electron from the LUMO back to the HOMO, with corresponding emission of a photon. This process, occuring between states of the same multiplicity (spin) is defined as fluorescence. Since in principle, emission may produce any vibrational level in the ground electronic state, a distribution of wavelengths corresponding to the fluorescence transition would be observed, this being the fluorescence spectrum. In addition to this radiative fate of the excited singlet state, non-radiative decay processes may occur. The first and obvious of these is chemistry, resulting in the case of the isolated molecule in fragmentation, or isomerization. There are two non-radiative fates, however, that result in degradation of all or part of the electronic excitation energy by the formation of an electronic state of the molecule lower in energy. In the first, termed internal conversion, a lower state of the same multiplicity is produced. In the specific case being considered, this would be the ground electronic state, and the transition would be denoted $S_1 \rightarrow S_0$ internal conversion. It should be noted that this process produces an isoenergetic level of the the lower state, and that collisional processes are then required to take away the excess vibrational energy so produced in the ground-state. This collisional vibrational relaxation process is clearly shown in Fig. 1.1. Since the vibrational level of the S_0 state produced initially is very 'hot' thermal reaction sometimes occurs before vibrational relaxation can take place.

A much more common fate of the excited singlet state would be a spin inversion of one of the original electron pair, producing the excited triplet state, T_1. This process is given the somewhat cumbersome name intersystem crossing. Note again that the transition from S_1 to T_1 must be considered as a two-step process, that of electronic relaxation to produce an isoenergetic vibrational level of the lower electronic state, followed by collisional vibrational relaxation. Since the electronic energy of a triplet state of the same electronic configuration is always lower than that of the corresponding singlet state, intersystem crossing must initially produce a vibrationally excited triplet state, which may therefore be reactive.

The triplet state formed by S_1 - T_1 intersystem crossing may itself undergo two unimolecular fates, both analagous to fates of the singlet excited state. The first is the radiative $T_1 \rightarrow S_0$ process, which is termed phosphorescence. A distribution of wavelengths results from this process, giving rise to a phosphorescence spectrum. A typical energy relationship between absorption, fluorescence and phosphorescence spectra is shown in Fig. 1.2. On an energy scale, fluorescence is often, but not invariably, a mirror image of absorption, whereas phosphorescence is of lower-energy (red-shifted)

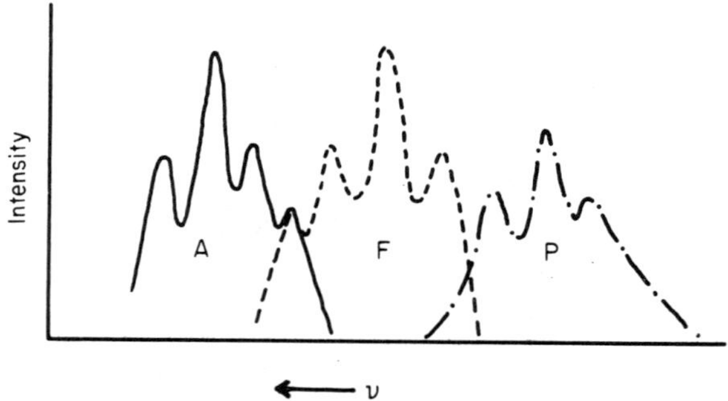

Fig. 1.2 Energy relationship between absorption, A, fluorescence, F, and phosphorescence, P.

with respect to the fluorescence reflecting the lower electronic energy level of the triplet state.

All fates discussed above, with the exception of simple vibrational relaxation, are unimolecular. Before identifying bimolecular fates, it would be as well to point out the competitive nature of the processes referred to. An excited state produced upon photoexcitation might be depopulated by many, if not all of these processes. This is illustrated with a simple kinetic scheme. Suppose a molecule M is raised to the excited singlet state referred to above, namely, the zero-point level, which we can denote ${}^1M_0{}^*$ using a common notation. Here the superscript 1 refers to multiplicity, * identifies excited states, and subscript 0 refers to excess vibrational energy content of the state, here zero. Reactions (1.2) - (1.4) simply show the various fates of the excited states:

	Process	Rate	
$M + h\nu \rightarrow {}^1M_0{}^*$	Absorption	I_a	(1.1)
${}^1M_0{}^* \rightarrow M + h\nu_F$	Fluorescence	k_F	(1.2)
${}^1M_0{}^* \rightarrow {}^3M_m{}^*$	Intersystem crossing	k_{ISC}	(1.3)
${}^1M_0{}^* \rightarrow M_m$	Internal conversion	k_{IC}	(1.4)
${}^1M_0{}^* \rightarrow$ products	Reaction	k_D	(1.5)

If we assume continuous illumination of the molecular system, that is I_a, the number of photons absorbed per second is constant and not time-dependent, then the steady-state approximation can be used to derive useful relationships.

Thus the rate of formation of the excited state ${}^1M_0{}^*$ can be equated to the rate of loss:

$$d[{}^1M_0{}^*]/dt = I_a - (k_F + k_{ISC} + k_{IC} + k_D)[{}^1M_0{}^*] = 0 \qquad (1.6)$$

therefore

$$[{}^1M_0{}^*]_{SS} = I_a/(k_F + k_{ISC} + k_{IC} + k_D). \qquad (1.7)$$

A useful quantity in photochemistry is termed the primary quantum yield, Φ, which is defined as the rate of any process relative to the rate of photon absorption. Taking fluorescence as an example:

$$\Phi_F = k_F[{}^1M_0{}^*]/I_a \qquad (1.8)$$

Substituting Equation (1.7) in Equation (1.8) we obtain:

$$\Phi_F = k_F/(k_F + k_{ISC} + k_{IC} + k_D) \qquad (1.9)$$

It can thus be seen that a quantum yield is merely a ratio of rate constants. However, it is very clear that in order to rationalize, or better, to predict the photophysical and photochemical behaviour of molecules it is necessary to know the magnitude of each and every rate constant for all processes that deplete the excited-state population. It is pertinent at this point to consider also the expression for the rate at which a photochemical process would occur given steady-state illumination, by simple consideration of Equation (1.8). The rate of fluorescence is given simply by $k_F[^1M_0^*]$, which is obviously equal to $\Phi_F I_a$. The rate at which light is absorbed in the system is thus a critical factor in determining the rate. The Beer-Lambert law relates light transmitted (I_t) through an absorbing medium to that incident (I_o) in Equation (1.10):

$$A = \log_{10}(I_o/I_t) = \varepsilon c l \qquad (1.10)$$

where A is the optical density, or absorbance, c is the concentration of absorbing molecules, l is the path length and ε is the molar decadic extinction coefficient.

Neglecting reflection from surfaces, we can say:

$$I_a = I_o - I_t. \qquad (1.11)$$

Simple manipulation then gives

$$\log_{10}\{1 - (I_a/I_o)\} = -\varepsilon c l \qquad (1.12)$$

and, noting that $\log_e (1-x) = -x$ for small values of x yields the commonly used Equation (1.13), which however, is only valid for small absorbances. The purpose of this laboured derivation is:

$$I_a = I_o\, \varepsilon c l \times 2.303 \qquad (1.13)$$

merely to demonstrate that the rate of any photochemical process is related to ε, the molar decadic extinction coefficient, and it is thus necessary to devote some consideration to factors determining the magnitude of this molecular property (see below).

The consequences of pulsed excitation of the simple molecular system should now be considered. Clearly, for a δδ-pulse excitation, the excited state population decays according to the differential equation:

$$-\mathrm{d}[{}^1\mathrm{M}_0{}^*]/\mathrm{d}t = (k_\mathrm{F} + k_\mathrm{ISC} + k_\mathrm{IC} + k_\mathrm{D})[{}^1\mathrm{M}_0{}^*]. \tag{1.14}$$

Simple integration yields the result that the excited state population should decline exponentially, with a lifetime (τ), defined as time taken for the concentration to drop to 1/e of the original, given by Equation (1.15). This lifetime is the one that can be measured experimentally:

$$\tau^{-1} = k_\mathrm{F} + k_\mathrm{ISC} + k_\mathrm{IC} + k_\mathrm{D} \tag{1.15}$$

and since fluorescence intensity is a convenient monitor of excited state concentration, the fluorescence decay time τ_F is often referred to. However, where it is possible to monitor non-radiative decay, or chemistry, in a time-dependent fashion, the same result would be obtained, since τ as defined in Equation (1.15) depends upon the rate constants for all decay processes.

The fluorescence lifetime, τ_F, should not be confused with the quantity known as the radiative or natural lifetime of any state. This is the lifetime the state would have if it decayed exclusively by the radiative process, and here we will use the notation τ_R to denote this hypothetical lifetime, which clearly has the relationship to the rate constant for fluorescence defined in Equation (1.16):

$$\tau_\mathrm{R}{}^{-1} = k_\mathrm{F}. \tag{1.16}$$

Thus

$$\tau_\mathrm{F} = \tau_\mathrm{R}\Phi_\mathrm{F} \tag{1.17}$$

or

$$k_\mathrm{F} = \Phi_\mathrm{F}/\tau_\mathrm{F}. \tag{1.18}$$

Before considering bimolecular fates of excited states, the triplet state should be considered further. Simple manipulation of Equations

(1.1) - (1.5) will show that the quantum yield of triplet state formation is given by Equation (1.19):

$$\Phi_{ISC} = k_{ISC}/(k_F + k_{ISC} + k_{IC} + k_D). \quad (1.19)$$

Consideration of the fate of the triplet state under steady state, illumination reactions (1.20) - (1.23) yields an expression for, say, the phosphorescence efficiency, Φ'_P Equation (1.24):

	Process	Rate	
$^3M_m^* \xrightarrow{(M)} {^3M_o^*}$	Vibn. relaxation	$k_C[^3M_m^*][M]$	(1.20)
$^3M_o^* \rightarrow M + h\nu_P$	Phosphorescence	$k'_P[^3M_o^*]$	(1.21)
$^3M_o^* \rightarrow M_n$	Intersystem crossing	$k'_{ISC}[^3M_o^*]$	(1.22)
$^3M_o^* \rightarrow$ Products	Reaction	$k'_D[^3M_o^*]$.	(1.23)

In this text primed quantities denote triplet state parameters. The term quantum efficiency refers to the relative importance:

$$\Phi'_P = k'_P/(k'_P + k'_{ISC} + k'_D) \quad (1.24)$$

of any decay process of an excited state, whereas yield relates rate to rate of light absorption. Of course, for a state excited by light absorption directly, the terms yield and efficiency are interchangeable, but for the case here, of triplet state decay, the phosphorescence quantum yield, Φ_P is a composite, of formation and decay, given by Equation (1.25):

$$\Phi'_P = \Phi_{ISC}\ \Phi'_P \quad (1.25)$$

$$= \{k_{ISC}/(k_F + k_{ISC} + k_{IC} + k_D)\} \times k'_P/(k'_P + k'_{ISC} + k'_D) \quad (1.26)$$

The phosphorescence decay time, τ'_P, is given simply by Equation (1.27):

$$(\tau'_P)^{-1} = k'_P + k'_{ISC} + k'_D. \quad (1.27)$$

Because spin inversion is involved in the radiative and non-radiative fates of triplet states, the rate constants are much smaller than those for singlet-state decay. Thus the phosphorescence decay time is usually many orders of magnitude longer than the fluorescence decay time.

There is one fate of the triplet state that should be included here, since the rate constant describing it is unimolecular in form (Equation (1.28)):

$$^{3}M_{o}^{*} \xrightarrow{\Delta H} {}^{1}M_{o}^{*}. \tag{1.28}$$

The process is the thermal reactivation of the triplet state to reform the singlet state, which may then fluoresce. In such a system, an emission would be observed that had the same spectral distribution as fluorescence, but a decay time that corresponded to the triplet state. This emission is referred to as delayed fluorescence, and can be of two types. That defined here is known as E-type, and is relatively uncommon in aromatic molecules. E-type delayed fluorescence can only be of importance if there is enough thermal energy available in a system to repopulate the singlet state from the triplet, and the phenomenon is thus confined to these molecules where the singlet-triplet energy gap is of the order of kT.

1.2 BIMOLECULAR PATHWAYS

All of the discussion to date has assumed that the molecule under consideration in excited singlet or triplet states did not encounter any other molecule during the excited state lifetime, other than for the purpose of removal of vibrational energy. This is a very artificial condition, wholly inappropriate in polymers where bimolecular encounters will dominate, leading for example to electronic relaxation of the system, termed quenching. Such fates may be accomodated by incorporating into Reactions (1.1) - (1.5) a typical bimolecular interaction between the excited singlet state $^{1}M_{o}^{*}$ and a ground-state quencher molecule, Q.

Assuming:

$$\begin{array}{lll} & & \text{Rate} \\ {}^1M_o^* + Q \rightarrow & \text{Quenching} & k_Q[{}^1M_o^*][Q] \end{array} \quad (1.29)$$

in a fluid medium, such that the quenching is a diffusion-controlled process, the rate will be as given, leading to a modification of the expressions for quantum yield, etc., to those shown in Equation (1.30).

$$\Phi_F = k_F/(k_F + k_{ISC} + k_{IC} + k_D + k_Q[Q]) \quad (1.30)$$

$$\tau_F^{-1} = k_F + k_{ISC} + k_{IC} + k_D + k_Q[Q]. \quad (1.31)$$

Clearly Equation (1.18) still holds, but a convenient expression much used to quantify such bimolecular interacitons can be obtained by re-defining quantities $(\Phi_F)_o$ and $(\tau_F)_o$ for the conditions $[Q] = 0$, i.e.:

$$(\Phi_F)_o = k_F/(\tau_F)_o \ . \quad (1.32)$$

Division of Equation (1.32) by Equation (1.30) yields Equation (1.33), which is one form of the Stern-Volmer relationship. Evidently, measurement of relative yield of:

$$(\Phi_F)_o/\Phi_F = 1 + k_Q(\tau_F)_o[Q] \quad (1.33)$$

any process (here fluorescence, but it could be reaction, for example) and a single decay time as a function of concentration of added quencher results in quantitative estimation of the bimolecular rate constant, k_Q. The possible bimolecular quenching processes are listed below.

1. Chemical reaction
2. Enhancement of non-radiative decay
3. Electronic energy transfer
4. Complex formation

Chemical reaction is, of course, widespread, but will not be discussed further. Enhancement of non-radiative decay needs some explanation. It has been found that in the proximity of a molecule containing an atom of high nuclear charge (and therefore high mass), spin-selection rules break down, leading to the acceleration of the rate of all processes, both radiative and non-radiative, involving spin inversion. Thus intersystem crossing rates from singlet states can be enhanced, and it can be seen from Equations (1.9) or (1.30) that this would lead to a diminution in quantum yield of singlet-state processes, such as fluorescence. The effect on triplet states is less obvious, since both k'_P and k'_{ISC} are subject to the so-called 'heavy-atom' effect. Paramagnetic species, such as ground-state molecular oxygen can induce the same effect.

Electronic energy transfer is very commonly encountered. The condition for it to be observed is the obvious one that the transfer should be exothermic. There are several mechanisms by which the process could occur, one radiative, the others non-radiative. The latter type will be discussed fully later, but radiative transfer, sometimes referred to as the 'trivial' process, consists of sequential emission of fluorescence or phosphorescence from the donor molecule D^*, (Reaction (1.34)) followed by absorption of the photons by the acceptor molecule A (Reaction (1.35)):

$$D^* \rightarrow D + h\nu \tag{1.34}$$

$$A + h\nu \rightarrow A^*. \tag{1.35}$$

It can be seen from this simple scheme that no diffusion of the donor and acceptor molecules is necessary for this transfer to occur, but the spin-selection rules for radiative transitions must be obeyed. The same two conditions apply in the dipole-induced dipole non-radiative mechanism, sometimes referred to as Forster or long-range transfer, but for the electron exchange mechanism, the close proximity of donor-acceptor pairs is a necessity, and in fluid media, this type of transfer usually dominates, whereas the former are observed in addition in solid state media.

It should be recognized that collisional processes are governed by spin-conversion laws due to Wigner, which differ from those for radiative processes. Stated briefly, the Wigner conservation law states that for the collisional reaction between A and B to give products C and D, with the molecules having spins S_A, S_B, S_C and S_D, respectively, a complex between A and B may have any spin given by the series I. Likewise, the products correlate with spins in series II.

$$A + B \rightarrow C + D \tag{1.36}$$

Series I $S_A + S_B, ((S_A + S_B) - 1), \ldots |S_A - S_B|$

Series II $S_C + S_D, (S_C + S_D) - 1), \ldots |S_C - S_D|$.

The following rules then apply:

1. Spin-allowed process: All terms in series I are found in series II.
2. Spin-forbidden process: No terms in series I are found in series II.
3. Spin-restricted process: Some terms in series I are found in series II.

In the case of spin-restricted process, a spin statistical factor may be easily calculated from the probabilities of forming a reactant complex in a particular spin state. Thus in the formation of $O_2(^1\Delta_g)$ sensitized by triplet states of many molecules, (Reaction(1.37)), the product:

$$^3M + O_2(^3\Sigma_g^-) \rightarrow M + O_2(^1\Delta_g) \tag{1.37}$$

spin (series II) must be singlet, whereas the reactant complex spins (series I) may be quintet, triplet or singlet. Only the singlet reactant complex may proceed to products, and thus the reaction can have a maximum rate given only by the collision frequency of ^3M and $O_2(^3\Sigma_g^-)$ modified by the probability of producing the reactant complex in a singlet state, in this case 1/9.

1.3 COMPLEX FORMATION

Complex formation is a very important subject, since this type of bimolecular interaction may precede any of the first three, 1., 2., or 3. We can distinguish between complexes formed between an excited state of a molecule and the ground state of the same molecule, which is termed an excimer, this being the coalescence of the two words excited and dimer. In this case resonance interactions lead to a weak intermolecular force which binds the two species (Fig. 1.3). Aromatic molecules form excimers easily, and very often such excimers exhibit strong fluorescence that is characterized by being red-shifted with respect to the uncomplexed fluorescence, and broad and featureless.

Emission similar in features can be seen from excited complexes formed between excited molecules and ground states of different molecules. Such complexes are termed exciplexes, and are usually substantially charge transfer in nature. Exciplexes may often appear as intermediates in photochemical reactions, but are not invariably fluorescent. Other techniques have thus to be devised to verify their participation.

The bimolecular processes outlined above involve interactions between an electronically excited molecule and a ground-state partner. It is of course possible to observe interactions between two excited-state species. This is particularly important when high concentrations of intermediates are produced, and when these transients are relatively long-lived, as in the case with triplet state molecules. The annihilation of two triplet states can produce an excited singlet state (Reaction (1.38)):

$$^{3}M_{o}^{*} + {}^{3}M_{o}^{*} \rightarrow {}^{1}M_{m}^{*} + M \qquad (1.38)$$

which if it fluoresces, is another form of delayed fluorescence, termed in this case, P-type delayed fluorescence. Following laser

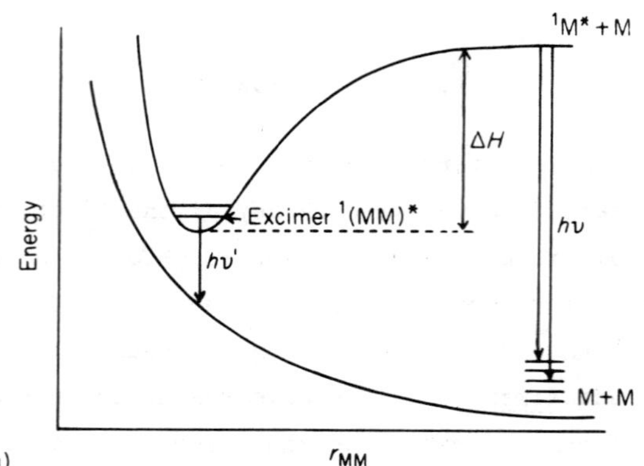

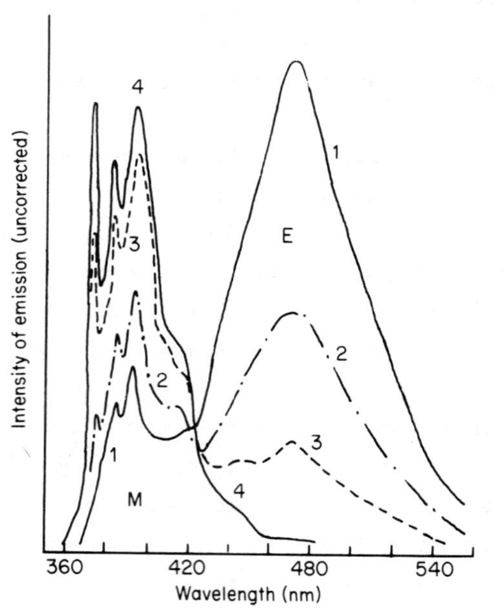

Fig. 1.3 (a) Energy diagram for eximer formation.

(b) Fluorescence spectra of typical aromatic molecule: concentration increases from curve 1 to curve 4. M, fluorescence of monomer; E, fluorescence of excimer.

excitation, singlet-singlet annihilation is not unkown.

This thumb-nail sketch of photophysics and photochemistry can be completed by considering processes that are biphotonic. The most obviously important of these are the sequential absorption of a second photon by a transient species such as the S_1 or T_1 states depicted in Fig. 1.1. The spectral distribution of this absorbed light may be used to identify the transients, and the decay of the transient absorption may be used to monitor the kinetics of the photochemical fates of the species. These commonly used techniques of flash photolysis and laser flash photolysis are based upon sequential absorptions that should not be confused with simultaneous multiphoton excitations, which form a recent and useful brand of spectroscopy, and often lead to ionization of the molecule. For further reading see [1].

1.4 QUANTITATIVE CONSIDERATIONS

Since this volume is concerned with luminescence in synthetic polymers, some attention will be paid now to quantitative aspects of radiative processes, and other fundamental photophysical aspects of importance.

1.4.1 *Transition probabilities*

Fluorescence has been defined above simply as the electric dipole transition from an excited electronic state to a lower state, usually the ground state, of the same multiplicity. Mathematically, the probability of an electric-dipole induced electronic transition is proportional to R_{if}^2 where R_{if}, the transition moment integral between initial state i and final state f is given by Equation (1.39). In this Equation ψ_e represents the electronic wavefunction, ψ_n the vibrational wavefunction, $\hat{M}$ is the electronic dipole moment operator, and the Born-Oppenheimer principle of separability of electronic and vibrational wavefunctions has been invoked. The first integral involves only the electronic wavefunctions of the system, and the second term, when squared, is the familiar Franck-Condon factor:

$$R_{if} = \int_{-\infty}^{+\infty} \psi_{ef} \hat{M}\psi_{ei} \, d\tau_e \int_{-\infty}^{+\infty} \psi_{nf} \, \psi_{ni} \, d\tau_n. \qquad (1.39)$$

To a good approximation the electronic integral is zero unless the states i and f are of the same spin: hence electronic absorption in most organic molecules results in the formation of an excited singlet state from the ground singlet state. There are also symmetry restrictions on transitions between states i and f imposed by the necessity of the electronic integral being totally symmetric if it is not to become of zero value upon integration. For a transition that satisfies this requirement (symmetry allowed), the molar decadic absorption coefficient will have a maximum value of the order of 10^5 dm^3 cm^{-1} mol^{-1}. The corresponding value for the rate constant for radiative decay, k_R, of the excited state via the spontaneous fluorescence process is given approximately by Equation (1.40) and more exactly by Equation (1.41):

$$k_R \, (s^{-1}) \approx 10^4 \, \varepsilon_{max} \, (dm^3 cm^{-1} mol^{-1}) \qquad (1.40)$$

$$k_R = 2.88 \times 10^{-9} n^2 \, \langle \bar{\nu}_F^{-3} \rangle_{av}^{-1} \int \varepsilon \, d\bar{\nu}/\bar{\nu}. \qquad (1.41)$$

where $\langle \bar{\nu}_F^{-3} \rangle_{av}^{-1}$ is a measure of the average frequency of the fluorescence, $\int \varepsilon \, d\bar{\nu}/\bar{\nu}$ is the area under the absorption curve, and n is the refractive index of the medium in which the experiment is carried out. Use of Equation (1.41) for a symmetry-allowed transition gives a value of k_R of $10^9 s^{-1}$, or a radiative lifetime of 1 ns. For a transition that does not satisfy the symmetry restrictions imposed by Equation (1.39), the transition moment integral can be non-zero if a second-order mechanism is invoked which necessitates the excitation of a non-totally symmetric vibration in one or other of the electronic states. This has the effect of reducing the magnitude of the transition moment integral compared with that for a symmetry-allowed transition, and also leads to the absence of the absorption and emission spectral features corresponding to transitions between the vibrationless ground and excited electronic states.

In addition to symmetry restrictions on electronic transitions there are restrictions caused by the necessity of overlap in space of molecular orbitals for the electron in its initial and final states. Where this is small, for example in $n \rightarrow \pi^*$ transitions in carbonyl compounds, the electronic integral in Equation (1.39) is diminished. The 'allowedness' of a transition, F, expressed as an oscillator strength can be summarized as in Equation (1.42) by a series of factors, f, that relate in turn to spin, (s), overlap (o), p (parity), sy (symmetry), and that have the following approximate values $f_s = 10^{-5}$, $f_o = 10^{-2}$, $f_p = 10^{-1}$, $f_{sy} = 10^{-1} - 10^{-3}$:

$$F = f_s f_o f_p f_{sy} F_A \qquad (1.42)$$

F_A is the oscillator strength of a fully allowed transition. For fluorescence then ($f_s = 1$), values of k_R extend from 10^9 s^{-1} for a fully allowed transition, typical say of dyestuffs, through 10^7 s^{-1} for symmetry-forbidden transitions (typical say of aromatic molecules) down to 10^5 s^{-1} and less (for say carbonyl compounds). It should be noted that the radiative rate constant referred to above depends upon the refractive index of the medium in which it is measured.

The transition dipole moment operator $\hat{\boldsymbol{M}}$ contained in Equation (1.43) is a vector quantity:

$$\hat{M}^2 = \hat{M}_x^{\,2} + \hat{M}_y^{\,2} + \hat{M}_z^{\,2} \qquad (1.43)$$

in other words the change in position of the electron in the transition is in a fixed direction with respect to some system of coordinates in which the molecular frame is fixed. The dipole moment operator is resolvable in three directions such that, for any transition, only one component will be finite. Hence if a perfectly ordered system such as a fixed molecular single crystal absorbs plane-polarized light, there will generally be one orientation of the crystal axes with respect to the plane of the polarization that maximizes the absorption probability. Normally, fluorescence involves the transition between the same two states as are observed in absorption (i.e. the light emitted has the same directional properties with respect to the electric vector as has the light absorbed). This is referred to as a parallel transition. Phosphorescence is often polarized perpendicularly to the absorption transition. In a two-dimensional representation the fluorescence intensity component parallel to the plane of polarization of the electric vector of the incident radiation (the reference direction), $I_{||}$ and that perpendicular to the reference direction $I_{\perp}$ varies with θ, the angle between the reference direction and direction of the optical transition moment on a laboratory scale, since the molecule is rigid. The exact magnitude of $I_{||}$ and $I_{\perp}$ as a function of θ can be traced out in the circular diagram shown in Fig. 1.4.

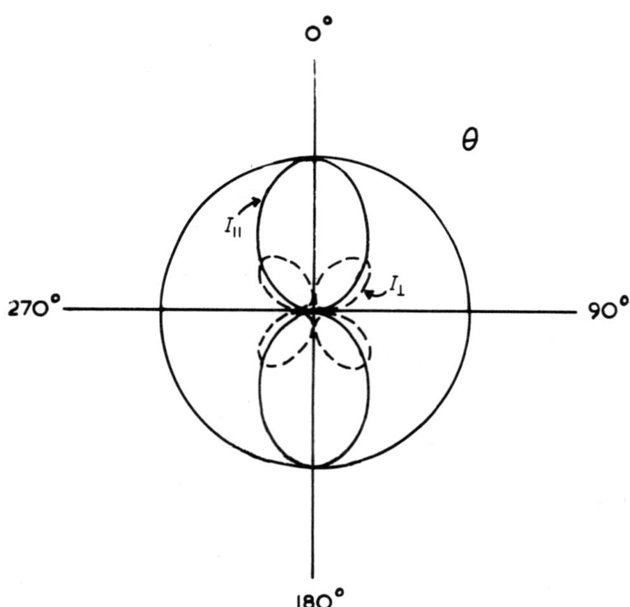

Fig. 1.4 Trace of fluorescence intensity parallel ($I_{||}$), and perpendicular ($I_{\perp}$), to polarization radiation as oriented sample is rotated through 360°.

It should be noted that this diagram has been constructed for a fixed single molecule, but is clearly also appropriate for a perfectly ordered fixed system with transition moments of all absorbing chromophores parallel. The degree of polarization P is defined as:

$$P = (I_{||} - I_{\perp})/(I_{||} + I_{\perp}). \qquad (1.44)$$

it can be seen that it will vary with θ for this system between the limits of +1 and -1 between θ = 0 and θ → 90°. For a perfectly random array of absorbing chromophores, it is easily shown that for a parallel emission, P has the value of +0.5 independent of θ, whereas for a perpendicular transition, P = -0.33.

In fluid media, this analysis is only appropriate if the average rotational relaxation time τ_r is very long compared with the fluorescence decay time, τ_F, ($\tau_r \gg \tau_F$). If the opposite is true, i.e. $\tau_F \gg \tau_r$, all anisotropy introduced in the sample upon excitation is lost through the rotational relaxation before emission. If τ_r and τ_F are comparable, then partial relaxation will occur on the same timescale as emission of fluorescence, and analysis of the time-dependence of the fluorescence anisotropy can yield information concerning the rotational motion. The emission anistropy r is defined as:

$$r(t) = (I_{||}(t) - I_{\perp}(t))/(I_{||}(t) + 2I_{\perp}(t)) \qquad (1.45)$$

and at time zero, r_0 is given by:

$$r_0 = \frac{1}{5}(3\cos^2\delta - 1) \qquad (1.46)$$

where δ is the angle between absorption and emission dipoles, which for fluorescence, is usually zero, yielding and initial anisotropy of 0.4. Development of this treatment of anisotropy of emission and the time-dependence is to be found in subsequent chapters of this volume.

1.4.2 *Medium effects*

The spectral positions of absorption and emission are influenced by the dielectric properties of the medium in which observations are made. Thus, for example, Fig. 1.5 shows that the vapour phase

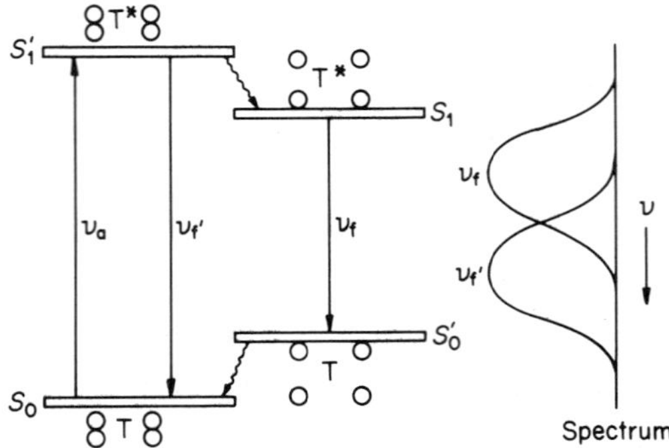

Fig. 1.5 Features of solvent relaxation.

0-0 bands in absorption and fluorescence of a molecule are identical, whereas in solution with solvent of static dielectric constant ε and refractive index n, the bands are no longer coincident. The differences can be rationalized using Onsager theory. A solute molecule of dipole moment μ in a spherical cavity of radius a polarizes the dielectric of the solvent, producing a reaction field. This field, R_o, is given for the ground-state of the solvent molecule (of dipole moment μ_o), by Equation (1.47):

$$R_o = 2\mu_o(\varepsilon - 1)/a^3(2\varepsilon + 1). \tag{1.47}$$

Upon excitation, and invoking the Franck-Condon principle, the electronic excitation is much more rapid than the dielectric relaxation time of the solvent, so the reaction field of the Franck-Condon excited state R_1' is given by Equation (1.48), in which the solvent still experiences the ground-state solvent dipole moment, and the high frequency dielectric constant ($=n^2$) is used in place of ε:

$$R_1' = 2\mu_o(n^2 + 1)/a^3(2n^2 + 1). \tag{1.48}$$

If the decay time of the excited state of the solvent exceeds the dielectric relaxation time, then the equilibrated excited state system of dipole moment μ', has a reaction field given by Equation (1.49):

$$R_1 = 2\mu'(\varepsilon - 1)/a^3(2\varepsilon + 1). \qquad (1.49)$$

Following fluorescence, the non-equilibrium ground-state has a reaction field R_o' (Equation (1.50)):

$$R_o' = 2\mu'(n^2 - 1)/a^3(2n + 1)) \qquad (1.50)$$

Manipulation of these equations shows that the difference in energy of the 0–0 transitions in absorption and emission is given by Equation (1.51):

$$\Delta\bar{\nu} = \frac{2(\mu' - \mu_o)^2}{hca^3}\left[\frac{(\varepsilon - 1)}{(2\varepsilon + 1)} - \frac{(n^2 - 1)}{(2n^2 + 1)}\right]. \qquad (1.51)$$

For non-polar solvents $\varepsilon \simeq n^2$, and no shifts are observed. For polar solvents shifts can be large. Therefore emission spectral positions can be very sensitive to the microscopic environment of the emitting molecule.

There are some specific solvent effects also to be considered. For example in states arising from $n \rightarrow \pi^*$ excitations, the absorption bands undergo blue shifts in polar solvents relative to those observed in non-polar solvents, owing to the binding of the non-bonding electrons to the solvent. Where there are two excited states of differing electronic configuration in close proximity, change of solvent polarity can invert the ordering of the excited states, leading to a change in the nature of the fluorescence, which can clearly influence spectral profile, decay time and yield. In mixed solvents, or in heterogeneous systems, level inversion can lead to time-dependent spectra, and complex fluorescence decays.

Change of pH may have a dramatic effect upon the emission

properties of molecules that undergo acid-base equilibria. Some solvents, particularly chlorinated solvents such as dichloromethane can quench excited states via non-emitting complex formation, or the heavy-atom effect.

1.4.3 *Effects of concentration of emitter, and of quencher*

An obvious consequence of increasing the concentration of emitting molecules or added quencher in any system is to increase the probability that energy can migrate from one molecule to another by a non-radiative process, shown generally in Reaction (1.52):

$$D^* + A \rightarrow D + A^*. \tag{1.52}$$

There are several possible mechanisms through which this might occur namely, the induced dipole interaction, the exchange mechanism, radiative transfer, and excitation diffusion.

(a) Induced dipole interaction

A long-range (15-100 Å), non-radiative, single-step process can arise out of coulombic (dipole-dipole) interactions between stationary excited-state donor (D^*) and ground-state acceptor (A) chromophores. The theory for this mode of energy transfer was developed by Forster[2]. and gives the probability of excitation energy being transferred from D^* to A as, considering only dipole-dipole interactions:

$$k_{D^*\rightarrow A} = \frac{K}{\tau_R \quad R^6} \cdot \int_0^\infty f_D(\bar{\nu})\varepsilon_A(\bar{\nu}) \frac{d\bar{\nu}}{\bar{\nu}^4} \tag{1.53}$$

where $K = (9000\ \kappa^2 c^4 \ln 10)/(128\pi^5 n^4 N)$, where c is the speed of light, n is the refractive index of solvent, N is Avagodro's number, κ is a constant dependent on the mutual orientation of D^* and A, R is the donor-acceptor distance, τ_R is the inherent donor radiative lifetime, and the integral is a measure of the spectral overlap between donor emission and acceptor absorption, measured on a wavenumber scale.

The rate of energy transfer is therefore expected to increase with increasing acceptor extinction coefficient with the proviso that the electronic energy level is greater in energy for the donor than for the acceptor. Hence transfer is more likely to occur between dissimilar molecules. The rate constant as a function of R is given by:

$$k_{D^* \rightarrow A} = \frac{1}{\tau_D}\left(\frac{R^o}{R}\right)^6 \tag{1.54}$$

where τ_D is the actual mean lifetime of D*, and R^o is a critical transfer distance, such that there is an equal probability of energy transfer and spontaneous deactivation of D*. Theoretical values of R^o can be obtained from Equation (1.55):

$$(R^{o6}) = \frac{9000(\ln 10)\kappa^2}{128\,\pi^5 n^4 N} \int_0^{-\infty} f_D(\bar{\nu})\,\varepsilon_A(\bar{\nu})\,\frac{d\bar{\nu}}{\bar{\nu}^4} \tag{1.55}$$

where n is the solvent refractive index; experimental values of R^o are available from Equation (1.56):

$$R^o = (3000/4\pi N\,[A^o])\,\nu^3 \tag{1.56}$$

where $[A^o]$ is that critical concentration of acceptor such that similar conditions defining R^o are met. Comparison of experimental and theoretical values of R^o will give an indication of whether the Forster mechanism is appropriate for the system under consideration.

It is important to note that for this mechanism spin conservation laws apply to donor and acceptor molecules individually, and no change in spin is allowed in either the donor or acceptor during transfer. Thus although singlet-state energy transfer is certainly allowed, triplet-triplet energy transfer is forbidden. However, if the transition is allowed in the acceptor but forbidden in the donor, then there is still a probability of transfer taking place. Thus the process represented by Equation (1.57) may be observed:

$$^3D^* + A \rightarrow D + {}^1A^* \,. \tag{1.57}$$

The Forster process should be independent of the viscosity of either the solvent or medium employed, since this relationship essentially considers the interaction between static donor-acceptor pairs.

(b) Exchange mechanism [3]

The exchange mechanism for energy transfer is similar to the previous interaction in that it is a non-radiative single-step process, but differs substantially in that it operates up to relatively very small distances (15 Å) corresponding to the overlap of molecular orbitals of donor and acceptor. The probability of energy transfer occuring via this mechanism is dependent on the square of an exchange integral, which has extremely low values at distances of over 15 Å separating the acceptor from the donor. Since intermediate complex formation is invoked, the less restrictive Wigner spin conservation laws apply (see above), and thus triplet-triplet energy transfer (Reaction (1.58)) is allowed by the exchange mechanism, whereas it is forbidden by the induced-dipole interaction:

$$^3D^* + A \rightarrow D + {^3A^*} . \qquad (1.58)$$

The probability of energy transfer from D* to A via an exchange mechanism has been given by the empirical expression:

$$k_{D^* \rightarrow A} = K' \; Z^2 \int f_{D^*}(\bar{\nu})\varepsilon_{A'}(\bar{\nu})d\bar{\nu} \qquad (1.59)$$

where K' is an experimental constant, Z is also a constant dependent on the exchange integral, and the integral term is the spectral overlap contribution described in a similar manner to that encountered in the Forster mechanism, although the two are not exactly equivalent, since the acceptor absorption is also normalized in the case of the exchange expression. The Z^2 term contains the distance dependence, and varies as exp $(-2R/L)$, where R is the donor-acceptor distance and L is the effective size of the molecules. An R^0 value may be defined in similar fashion to that for the induced dipole mechanism, and experimental results in many polymer systems (see below) give R^0 values that indicate that the exchange mechanism satisfactorily accounts for the observations, being close to the thoretical limiting R^0 value of ~15 Å.

(c) Excitation diffusion

This results from extended interaction between neighbouring identical chromophores and has long been known to occur in crystals; since a polymer chain can be considered as a one-dimensional crystal it is perhaps not surprising that this form of energy transfer does occur in polymer systems. This mechanism can be considered in terms of the excitation energy not being bound to the original absorbing molecule but able instead to migrate over much larger distances than those associated with the Forster mechanism, either along a chain or across many chains, by a series of coulombic interactions. There are two essential differences between this and the induced dipole-dipole interaction: first, in the former the coupling is considerably stronger and the rate is proportional to $(R^3)^{-1}$ rather than $(R^6)^{-1}$; secondly, transfer between similar molecules is favoured. Two cases of coupling between adjacent chromophores can be distinguished. In the weak coupling case the coupling energy is considerably less than the energy width of the whole vibrational envelope of the particular electronic level associated with the transfer, though it is still considerably greater than that of Forster transfer. The transfer rate (10^{12} s^{-1}) is faster than for the Forster transfer, and transfer can occur from the higher vibrational levels of the appropriate donor excited state. Thus, transfer between similar molecules is favoured although it can also occur between dissimilar molecules if the excited states involved are close in energy. The excitation is essentially spread over a number of molecules and moves initially in a coherent manner with a constant group velocity, being then known as an exciton, although lattice vibrations and irregularities tend to destroy the coherence and make propagation purely diffusive.

In the strong coupling case the coupling energy is considerably greater than the energy width of the vibrational envelope of the appropriate electronic transition, with the result that the excitation transfer rate ($\approx 10^{15}$ s^{-1}) is now much faster than the nuclear vibration rate. The excitation can be considered, in time-independent theory, to be spread out over the whole system of similar molecules. In time-dependent theory the excitation transfer process can, as for weak coupling, be described in terms of a travelling wave packet. In contrast to most other forms of excitation transfer the absorption spectrum of the system will often differ considerably from that of the isolated molecules of which it is composed.

Energy migration may reveal itself as a fast component in the decay of emission, being often in the picosecond region, in singlet state emission of free molecules, but possibly in some synthetic polymer systems on the nanosecond timescale. It gives rise to a non-exponential form of the donor fluorescence decay law, $I(t)$, given by Equation (1.60):

$$I(t) = I_0 \exp\{-t/\tau_0 - 2B\gamma(t/\tau_0)^{\frac{1}{2}}\} \qquad (1.60)$$

where τ_0 is the donor fluorescence decay time in the absence of acceptor and $\gamma = [A]/[A^0]$, the ratio of concentrations of acceptor,

$$B = \left(\frac{1 + 10.87x + 15.5x^2}{1 + 8.743x}\right)$$

$$x = D\alpha^{-\frac{1}{3}} t^{\frac{2}{3}}$$

$\alpha = (R^0/\tau_0)^6$ where R^0 is the value of R such that the rate of energy transfer is equal to the rate of spontaneous decay of the donor.

1.4.4 *Quenching and complex formation*

The interaction of an excited molecule M* with a ground state may in some instances lead to electronic quenching of the excited state (Reaction (1.61)):

$$M^* + M \rightarrow M + M. \qquad (1.61)$$

In many systems concentration quenching of singlet states is accompanied by the appearance of a new emission band to the red of the fluorescence of the uncomplexed molecule $^1M^*$ which is attributable to the formation of an excimer (Reaction (1.62)):

$$^{1}M^{*} + M \rightleftharpoons {}^{1}(MM)^{*}. \quad (1.62)$$

Excimer fluorescence is characteristically broad, as in Fig. 1.3 and decay characteristics are complex, since excimer formation is reversible.

The phenomenon of excited dimer (or excimer) formation was first [4]. recognized by Forster and Kasper while studying the concentration dependence of the fluorescence of pyrene solutions. Subsequently it has become apparent that the effect is general to aromatic moieties and is characterized by the appearance of a broad structureless emission profile at lower energies than that of the corresponding unassociated chromophore as illustrated in Fig. 1.6. Since the formation of excimer occurs at high chromophore concentrations at the expense of excited monomer without distortion of the absorption band of the system from that of the pure unassociated species it has been proposed that the excimer emission occurs from a transient dimeric entity that exists solely in the excited state, being dissociative in the ground state. The mechanism shown has been proposed to describe intermolecular excimer formation [2].

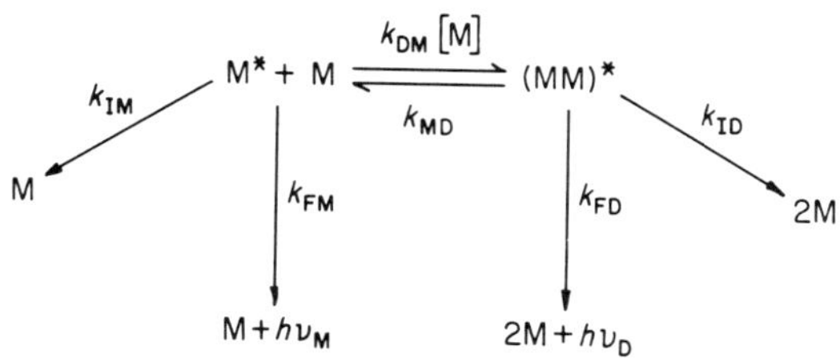

Fig. 1.6 Birks, kinetic scheme for excimer formation [4].

k_{FX}, k_{IX} are the first order rate coefficients governing fluorescence and non-radiative decay and k_{MD} that of excimer dissociation. k_{DM} is the second-order rate coefficient governing the diffusion controlled formation process.

The intermolecular phenomenon is now well understood in terms of its diffusion-controlled nature and resultant well characterized kinetic behaviour. Under conditions of photostationary excited state, the following expression results for the ratio of fluorescence quantum yields ϕ_D and ϕ_M of eximer and monomer respectively;

$$\frac{\phi_D}{\phi_M} = \frac{k_{FD}k_{DM}[M]}{k_{FM}(k_{MD} + k_D)} \tag{1.63}$$

where [M] is the concentration of chromophoric species and $k_D = k_{FD} + k_{ID}$. Under transient excitation more complex behaviour is encountered. The time dependence of the monomer emission $i_M(t)$ following excitation by a δ-pulse is described as the sum of two exponential terms, whereas that of the eximer $i_D(t)$ by the difference of exponentials as detailed below:

$$i_M(t) = A_1 \exp(-\lambda_1 t) + A_2 \exp(-\lambda_2 t) \tag{1.64}$$

$$i_D(t) = A_3\{\exp(-\lambda_1 t) - \exp(-\lambda_2 t)\} \tag{1.65}$$

where $A_1 = k_{FM}(\lambda_2 - X)/(\lambda_2 - \lambda_1)$

$A_2 = k_{FM}(X - \lambda_1)/(\lambda_2 - \lambda_1)$

$A_3 = k_{FD}k_{DM}[M]/(\lambda_2 - \lambda_1)$

and $\lambda_{1,2} = \frac{1}{2}\{(X+Y) \mp \{(X-Y)^2 + 4k_{MD}k_{DM}[M]\}^{\frac{1}{2}}\}$

where $X = k_M + k_{DM}[M]$

$Y = k_D + k_{MD}$

and $k_M = k_{FM} + k_{IM}$.

Exactly analagous kinetics are observed in exciplex systems. Numerous publications have appeared in which the validity of these expressions for intermolecular excimer formation in homogeneous fluid media has been demonstrated.

In a few cases, weak association between a molecule M in the ground state and an added molecule may occur (Reaction 1.66).

$$M + Q \rightleftarrows MQ. \tag{1.66}$$

Excitation of the associated complex MQ then produces an excited state $^1(MQ^*)$ which may resemble closely a true exciplex, produced by diffusional quenching of the uncomplexed excited state. However, the kinetics of fluorescence in such a system are in principle distinguishable, since in the case of the ground-state complex excitation is instantaneous, and the complex fluorescence does not then exhibit a 'growing in' period. Without giving a full analysis, in cases where such static quenching is of importance, measurement of relative quantum yields of fluorescence, $(\phi_F)_o/\Phi_F$, and relative decay times of fluorescence, $(\tau_F)_o/\tau_F$, can be used to obtain a form of the equilibrium constant for ground-state complex formation through Equation (1.67):

$$\left[(\phi_F)_o/\phi_F\right]/\left[(\tau_F)_o/\tau_F)\right] = 1 + K'_{eq}[Q]. \tag{1.67}$$

Since molecular oxygen is an ubiquitous impurity, the processes by which it may quench excited states have been widely studied. Quenching of fluorescent singlet states may occur through Equations (1.68) or (1.69), with very high efficiencies:

$$^1M^* + O_2(^3\Sigma_g^-) \rightarrow {}^3M^* + O_2(^3\Sigma_g^-) \tag{1.68}$$

$$\rightarrow {}^3M^* + O_2(^1\Delta g). \tag{1.69}$$

Triplet-state quenching can occur via Reactions (1.70) and (1.71):

$$^3M^* + O_2(^3\Sigma_g^-) \rightarrow M + O_2(^3\Sigma_g^-) \tag{1.70}$$

$$\rightarrow M + O_2(^1\Delta g) \tag{1.71}$$

The long lifetime of the triplet state makes such quenching processes dominate over unimolecular decay, and hence phosphorescence is not normally observed in fluid media. In fluid media, the various quenching processes discussed above, with the exception of dipole-induced dipole electronic energy transfer, yield decay curves that can be described by an exponential or a sum of exponentials. In some circumstances however, diffusional kinetics can yield an additional, observable decay term that has a $t^{-\frac{1}{2}}$ dependence (Equation 1.72). For a diffusion-controlled quenching reaction the continuum model predicts the fluorescence decay law to be:

$$I(t) = I_o \exp(-at - 2bt^{\frac{1}{2}}) \qquad (1.72)$$

where

$$a = 1 + 4R'D_{AQ}N'[Q] \qquad (1.73)$$

$$b = 4(R')^2(\pi D_{AQ})^{\frac{1}{2}}N[Q]. \qquad (1.74)$$

R' is related to the encounter distance, and D_{AQ} is the mutual diffusion coefficient of A and Q, N' is 6.023×10^{20}, and τ_o is the decay time at [Q] = 0. The ratio of intensities of fluorescence under steady illumination is predicted to be:

$$(I_o/I) = \{1 + 4R'D_{AQ}N'[Q]\}Y^{-1} \qquad (1.75)$$

where $$Y = 1 - (b/a^{\frac{1}{2}})\pi^{\frac{1}{2}} \exp(b^2/a)\,\mathrm{erfc}(b/a^{\frac{1}{2}}) \qquad (1.76)$$

and $$\mathrm{erfc}(x) = 2\pi^{-\frac{1}{2}} \int_x^{\infty} \exp(-u^2)\, du. \qquad (1.77)$$

1.5 EMISSION PROPERTIES OF MOLECULES IN SYNTHETIC POLYMERS

The basis of emission processes of molecules incorporated intrinsically in synthetic polymers will of course be exactly comparable to those of corresponding free molecules described above, save in the following important regards. [2,3,4]

1. Bulk polymers are rigid or viscoelastic solids, and hence kinetic schemes given above which describe diffusional phenomenon

occurring in the fluid phase will be inappropriate.

2. The large molecular size of polymers will make diffusional processes of the whole molecule, e.g. rotational motion, occur in a much slower timescale than that of small free molecules. Restricted relative motions, e.g. of side groups, or segmental motion, may occur in a relatively faster timescale.
3. Any synthetic polymer will in general be characterized by a distribution of molecular weights and may be of different stereoregularity, e.g. atactic, isotactic, or syndiotactic. Many different microenvironments may thus surround an emitting molecule, and this heterogeneity must be taken into account.
4. In polymers in which an emitting molecular species is part of the repeat unit, the local concentration of chromophore will be high. Thus the importance of such bimolecular processes as energy transfer, concentration quenching, excimer formation will be enhanced with respect to the corresponding free molecule.

The remainder of this volume will amplify many of the points above, but for completeness, the kinetics of excited states appropriate to the solid and intermediate physical states ahould be considered.

1.5.1 *The Perrin model*

The simplest model devised for relating the observed donor luminescence as a function of acceptor concentration for those exchange interactions in glasses where diffusion is limited, is that of Perrin [5]; expressed mathematically it is:

$$\Phi_0/\Phi = \exp(VN'[A]) \tag{1.78}$$

where $N' = 6.02 \times 10^{20}$; Φ_0 is the luminescence quantum efficiency in the absence of acceptor, Φ, that in the prescence of a concentration of acceptor [A]; and V is the volume of a quenching sphere (cm^3). This model implies that an excited state is instantaneously deactivated without emitting light if a quenching molecule is within a sphere of quenching action of volume *V*, whereas a molecule of the acceptor located outside this volume has no influence on the donor.

The theory is limited, however, in that the implication of a sharp spherical boundary does not in reality describe the observed more gradual falloff in quenching action as a function of donor-acceptor separation. A further obvious discrepancy is that donor lifetime is in fact a function of acceptor concentration, whereas the Perrin model predicts the complete independence of donor lifetime upon presence of acceptor.

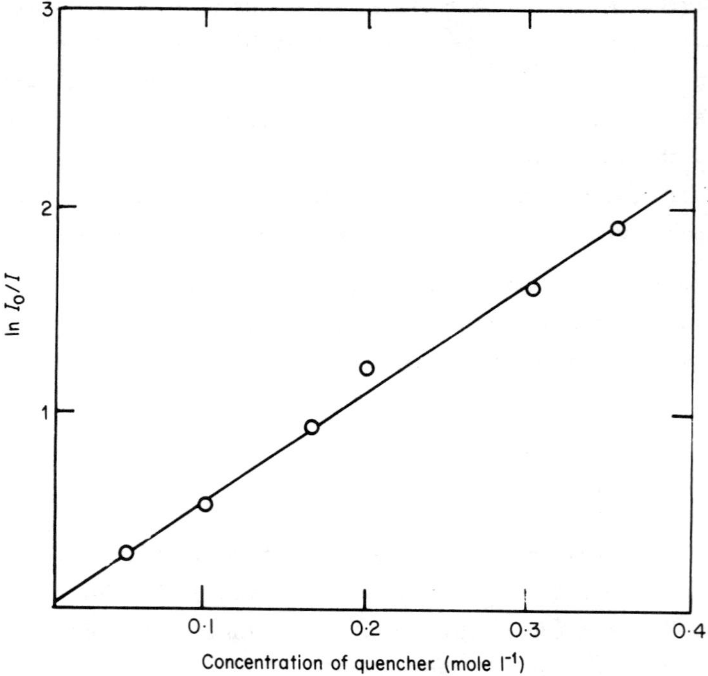

Fig. 1.7. Perrin plot of ln I_0/I against concentration of cyclo-octadiene for thermally oxidized poly(butadiene) in 1:1 tetrahydrofuran-diethyl ether solution at 77K. I_0 is intensity of phosphorescence from an αβ-unsaturated carbonyl in absence of cyclooctadiene; , that at any concentration. (After Fig. in *J. Polm Sci., Polym. Chem. Ed.*, **17**, 1711, (1979).

Nevertheless, the treatment is attractive because of its simplicity, and has been applied to various systems, exemplified by Fig. 1.7, in which the quenching of triplet excited αβ-unsaturated carbonyls by cyclooctadiene at low temperatures in poly(butadiene) is depicted in Perrin form.

1.5.2 *The Inokuti-Hirayama relationship*

A more elaborate method of characterizing the efficiency of triplet energy by an exchange mechanism has been developed by Hirayama and Inokuti [7]: it recognizes that the donor lifetime will be influenced by the acceptor. The efficiency of nonradiative transfer f_{NR} is given by:

$$f_{NR} = 1-(\tau_o^{-1}\int_0^\infty \Phi(t)\mathrm{d}t) \tag{1.79}$$

where $\Phi_t = \exp\{-t/\tau_o-\gamma^{-3}\,[A]/[A^o]g(e^{\gamma t}/\tau_o)\}$ and is a decay function of the excited state, τ_o is the lifetime of the same state in the absence of quencher, $[A^o]_o$ is the critical concentration, γ is a parameter related to R^o, τ_o, and the spectral overlap while:

$$[A^o] = 3/(4\pi R^{o\,3}). \tag{1.80}$$

The method of evaluation involves comparing experimental transfer efficiencies as a function of concentration with the theoretical plot of f_{NR} versus $[A]/[A^o]$ for a certain value of γ. The value of γ is obtained from an experimental determination of the decrease of donor lifetime and luminescence intensity due to increasing acceptor concentration. The best fit will provide a value of $[A^o]$ from which R^o may be calculated. Chapter 2 shows values of R^o when the same experimental data used in the Perrin treatment is treated according to the manner of Hirayama and Inokuti. Another example of data treatment is shown in Fig. 1.8 for the case of phosphorescence quenching of αβ-unsaturated carbonyls in thermally degraded poly(butadiene) at 77K by cyclooctadiene [8].

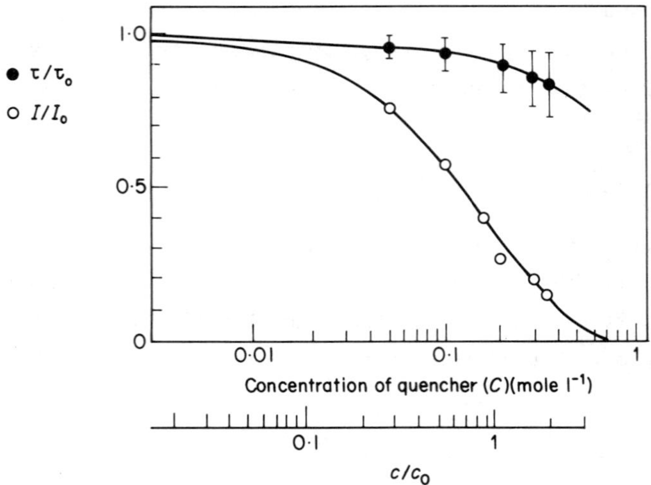

Fig. 1.8 Hirayama-Inokuti plot of relative phosphorescence emission intensities (ɸ) and lifetime (o) under same conditions as in Fig. 1.2. Best fit is obtained for $\gamma = 20$ (see text). (From Fig. in *J. Polym. Sci., Polym. Chem. Ed.)*, 17, 171, (1979).

1.5.3 *Intermediate kinetics*

The two cases of diffusional kinetics and those in rigid media can be defined in terms of a molecular energy transfer distance r where:

$$r = \{2(D + \Lambda)\tau_0\}^{\frac{1}{2}} \tag{1.81}$$

in which D is the relative molecular diffusion coefficient of the donor and acceptor, Λ is the donor excitation energy migration coefficient, and τ_0 is the donor luminescence decay time. If $r \ll R^0$, the molecules remain stationary during transfer, and Forster kinetics will be appropriate. When $r \gg R^0$, Stern-Volmer kinetics apply. For intermediate cases two theories, due to Voltz *et al* [9] and Yokota and Tanimoto [10] have been developed.

1.5.4 *Treatment of Voltz et al.*

These authors applied the diffusion theory to singlet-energy migration and developed a relationship between the rate constant for energy transfer (k_T) and R^o such that

$$k_T = 2\pi N_o (D + \Lambda) R^o \, 10^{-3} l \text{ mole}^{-1} s^{-1}. \quad (1.82)$$

Here N_o is Avogadro's number, D and Λ are as defined above, R^o is either the Forster or Perrin critical transfer distance dependent on whether or not the migration is singlet-singlet or triplet-triplet, and k_T can be obtained from a modified Stern-Volmer plot.

Values of Λ have been obtained for triplet-triplet migration in solid samples of poly(benzophenone)/naphthalene, poly(vinylnaphthalene)/piperylene systems, assuming in all cases immobility triplet state (i.e. $D = 0$). Table 1.1 shows typical values of Λ along with the transfer frequency (ω) obtained from elementary diffusion theory.

Table 1.1 Energy migration from the Voltz theory

Polymer	$\Lambda(cm^2 s^{-1})$	$\omega(s^{-1})$	*Diffusion length* (Å)
Poly(benzophenone)	3.3×10^{11}	0.7×10^5	-
Poly(vinylnaphthalene) Poly(acenaphthene)	4.8×10^{-14}	128	-
Poly(vinyltoluene)	1.7×10^{-4}	-	228
Poly(vinylcarbazole)	-	-	220

Birks [5] compared the experimental data for the naphthalene/9,10-diphenylanthracene system in various solvents with the Voltz theory and found a considerable discrepancy between actual and expected behaviour. This is perhaps due to the incorrect boundary assumptions employed by Voltz *et al.*:

$$p = 0.5 \text{ for } r < R^o$$

$$p = 0 \quad \text{for } r > R^{\text{o}}$$

where p is the transfer probability and r is the energy migration distance, whereas, in fact, $p \propto (R_{\text{o}}/r)$. For this reason the treatment described below is often preferred.

1.5.5 *Yokota and Tanimoto's treatment*

These authors developed a statistical representaion of Forster's theory and gave an expression for dipole-dipole transfer in a fluid medium in which the distribution of donor and acceptor is determined both by diffusion and by the decay and transfer of the donor excited state, such that:

$$I_x(t) = I_x(0)\ \exp\{-t/\tau_{\text{o}} - 2B\gamma(t/\tau_{\text{o}})^{\frac{1}{2}}\} \qquad (1.83)$$

where $I_x(t)$ is the donor fluorescence response function, τ_{o} is the donor fluorescence lifetime in the absence of acceptor, $\alpha = [\text{A}]/[\text{A}^{\text{o}}]$, $x = D\ \alpha^{-1/3}\tau^{2/3}$, $\alpha = (R^{\text{o}}/\tau_{\text{o}})$ and $B = (1{+}10.87x + 15.5x^2/(1{+}8.74x)^{3/4}$. It is then possible to relate the above expression to non-radiative transfer efficiencies.

Comparison of the two intermediate theories, along with that of Forster, for a polymer system was carried out by North and Treadaway [10] using dilute solutions of poly(N-vinylcarbazole) in poly(methylmethacrylate) incorporating anthracene as the quencher. The study of the singlet energy transfer showed similar conclusions as those reached above - that is, that the Yokota-Tanimoto model gave a much better fit to experiment (see Fig. 1.9) than that of Voltz *et al.*, and that singlet energy migration in systems containing homo- or co-polymer donors with small acceptor molecules was better analysed with the former theory because of the finite effect of energy migration along the chain. They also concluded the Forster's model can effectively describe singlet-singlet energy transfer in static small-molecule donor-acceptor systems.

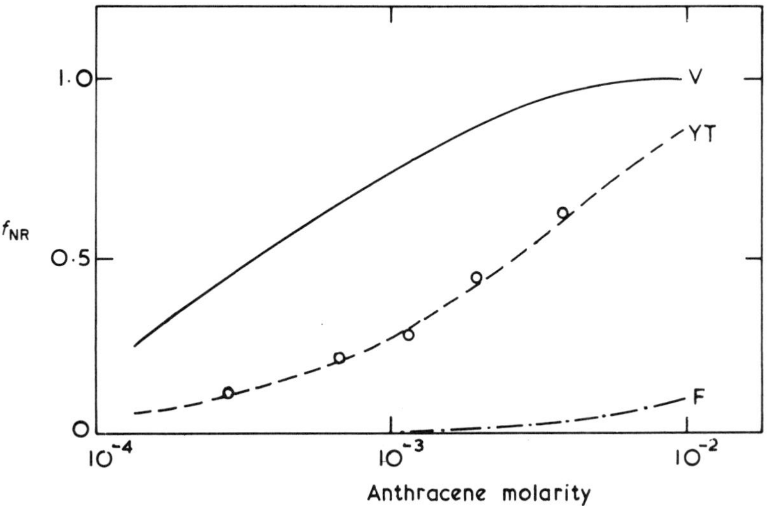

Fig. 1.9 Comparison of theory and experiment for electronic energy transfer from anthracene to methylmethacrylate. Data points, experimental; F = Forster expression; YT = Yokota - Tanimoto expression; V = Voltz expression (see text). (After Fig. in *Eur. Polym. J.* 9, 609 (1973).)

1.6 JUSTIFICATION FOR THE STUDY OF POLYMER LUMINESCENCE

The many studies carried out on luminescence in synthetic polymers have been motivated by a wide spectrum of scientific and technological aims. Some of the more obvious are categorized below.

1.6.1 *Fundamental interests*

Macromolecules exist and despite the complexity arising from the heterogeneous nature, a strong case can be made for increasing fundamental knowledge about such systems. These would include studies on the nature of photoemission from polymers of type B, (see below) in which interchromophoric interactions are of special interest. One could also include here fundamental studies on the photoluminescence of small molecules in polymeric environments.

1.6.2 *Luminescence of probe molecules*

These studies permit evaluation of polymer properties. In particular, measurement of the relative intensities of fluorescence of a probe molecule polarized parallel to and perpendicular to the plane of linearly polarized exciting radiation as a function of orientation of a solid sample yields information concerning the ordering of polymer chains. In solution, similar polarization studies yield information on the rotational relaxation of chains and the viscosity of the microenvironment of the probe molecule. More recently, the study of luminescence intensity of probe molecules as a function of temperature has been used as a method of studying transition temperatures and freeing of subgroup motion in polymers.

1.6.3 *Luminescent species in polymer photooxidation*

The problems associated with establishing a mechanism for the photo-oxidation and weathering of synthetic polymers are great, and any method that provides additional information is useful. In addition to traditional methods, such as product analysis, infrared spectroscopy (both conventional and ATR) and UV-visible absorption spectroscopy, luminescence methods have recently been employed.

1.6.4 *Identification of polymers*

It is a fact of commercial life that there is a frequent call for the rapid identification of synthetic polymers, usually by industrial scientist and technologists interested in a rival product. The growing possibility that recycling of plastic material may become economically attractive compared with disposal would also require identification of different synthetic polymers for sorting purposes. Luminescence spectroscopy could provide a convenient method of rapid identification.

The bulk of the volume will be concerned with aspects 1 and 2 outlined above, each chapter being contributed by authorities in the field. We have thus concentrated on polymers of type B, in the Somersall and Guillet classification [2], in which the chromophore is included in the repeat unit of the macromolecule, rather than in type A where isolated chromophores represent very minor components and are attached to a polymer chain or end-group. This emphasis is deliberate, since in type A polymers, the emitting species is usually an adventitious impurity or oxidation product. Moreover, we have concentrated attention upon singlet state processes, since it is in this area of fluorescence techniques, particularly those of time resolved fluorescence and fluorescence anisotropy, that most recent progress has been made [3], [4].

REFERENCES

1. For further reading see for example

BARLTROP, J.A. and COYLE, J.D. (1978) *Principles of Photochemistry,* Wiley, London.

2. SOMERSALL, A.C., and GUILLET, J.E. (1975) *J. Macromol. Sci. Rev. Markromol. Chem.,* C, 135.
3. BEAVAN, S.W., HARGREAVES, J.S. and PHILLIPS, D. (1979) *Advances in Photochemistry,* 11, 207.
4. GHIGGINO, K.P., PHILLIPS, D. and ROBERTS, A.J. (1981) *Advances in Polymer Science,* 40, 71.
5. PERRIN, F. (1924) *Comptes Rendues Ser. C.,* 178, 1978.
6. HIRAYAMA, F. and INOKUTI, M. (1965) 43, 1978.
7. HARGREAVES, J.S. and PHILLIPS, D. (1979) *J. Polym. Sci. Polymer Chem. Ed.,* 17, 1711.
8. VOLTZ, R., LAUSTRAIT, G. and COCKER, A. (1973) *J. Chim. Physique Phys. Chim. Biol.,* 63, 1255.
9. YOKOTA, M. and TANIMOTO, O. (1967) *J. Phys. Soc. Japan,* 22, 779.
10. NORTH, A.M. TREADAWAY, M.F. (1973) *Eur. Polym. J.,* 9, 609.

CHAPTER TWO

Phosphorescence and other delayed emissions of polymers

2.1 INTRODUCTION

The term 'delayed emission' is inherently subjective since it refers to the time scale on which the emission process occurs. The term is usually applied to differentiate an emission process from fluorescence. Fluorescence is itself, however, a delayed emission process, except that the delay of the emission usually occurs on a time scale of 10^{-11} to 10^{-7} s, whereas phosphorescence or 'delayed emission' typically has a lifetime in the 10^{-6} to 10 s range, as discussed in Chapter 1. These latter emission features obviously must arise from metastable excited states, usually by virtue of spin-selection rules. In the context of polymer spectroscopy, this metastable state is universally an $S = 1$ ('triplet') state, whereas the ground state and fluorescent excited state are both $S = 0$ ('singlet') states. In the absence of spin-orbit coupling (which can be induced by atoms with larger nuclear charge, referred to as the so-called 'heavy atom effect') a transition $S=1 \rightarrow S=0$ is forbidden, and hence the $S = 1$ state is metastable. Kasha pointed out many years ago that the rate of an $S=1 \rightarrow S=0$ transition is often reduced by a factor of about 10^{-6} relative to an analogous spin-allowed transition [1]. Thus phosphorescence and delayed emission in organic systems usually involves the lowest triplet state of the aromatic chromophore. This state will be denoted T_1 or $^3A^*$ (for the A'th species) hereafter. The fluorescence usually involves the lowest excited singlet state of the chromophore (denoted S_1 or $^1A^*$). Both fluorescence and phosphorescence/delayed fluorescence are the result of transitions to the ground state singlet, S_o or A. These features are summarized by the well known Jablonskii diagram discussed in Chapter 1. Different excited state kinetic processes are summarized and defined in Table 2.1.

Table 2.1 Definition of some common photophysical terms

Absorption	Transition induced from lower to higher energy state by absorption of a photon.
T - T absorption S - S absorption	An absorption event between two excited states; two triplet (S = 1) states in the case of T - T absorption and two singlet (S = 0) states in the case of S - S absorption.
Internal conversion	Relaxation of an excited state to a lower state of the same spin multiplicity and without emission of a photon.
Intersystem crossing	Relaxation of an excited state to a lower state of different multiplicity.
Fluorescence	Relaxation of an excited state to a lower state of the same spin multiplicity by emission of a photon; typical time scale 10^{-11} to 10^{-7} s.
Phosphorescence	Relaxation of an excited state to a lower state of different spin multiplicity by emission of a photon; typical time scale 10^{-6} to 10 s.

The two unique features of the T_1 state of typical organic molecules are: †

1. Unpaired electrons (with concomitant paramagnetism).
2. Long lifetime.

Because of the unpaired electron spin, many of the reaction paths available to the triplet state are not observed from the S_1 state. Furthermore, the long lifetime of the T_1 state allows many bimolecular collisions, which further enhances the probability of reaction. Thus, it is not surprising that the photochemistry of many organic molecules is dominated by the T_1 state. From the point of view of the spectroscopist, these features are disadvantageous because the T_1 state is so easily quenched by various molecules, including paramagnetic species (e.g. O_2) or species containing atoms with large nuclear charge (e.g. CH_3Br or CH_3I). In both cases, the quenching mechanism is the enhancement of intersystem crossing (ISC), i.e. $T_1 \rightsquigarrow S_0$. Consequently, observation of phosphorescence usually requires very high purity if either the molecules or the energy can diffuse (see following paragraph). Phosphorescence is observed most easily for solid-state samples, including glasses (typically mixtures of solvents at low temperature), doped crystals or polymer films, because collision-induced quenching is eliminated.

If identical chromophores are sufficiently close together, the phenomena of energy migration (equivalently referred to as 'exciton diffusion') can occur. In disordered systems or crystals well above absolute zero, this energy migration is best visualized as a random walk of the energy from one chromophore to another. In the case of the T_1 state, the probability of an energy transfer step falls off exponentially and is essentially zero over distances greater than 15 Å. ††

---

† A very general discussion of triplet-state processes is given by McGlynn *et al.* [2].

†† This is to be contrasted with Förster energy transfer, in which the energy transfer probability declines with R^{-6} (where R is the

chromophore separation) and there may be a significant probability of energy transfer for distances as large as 50-80 Å in favourable cases.

The theory of triplet exciton energy migration is fairly complex. The quantum mechanical mechanism is analogous to an electron exchange interaction between the pair of molecules. Therefore, intermolecular overlap of molecular orbitals is required. For example, consider the process:

$$^3D + A \rightarrow D + {^3A^*}. \tag{2.1}$$

The rate of energy transfer in Reaction (2.1) is proportional to the exchange matrix element squared:

$$| \langle \phi_A(1)\phi_D{}^*(1)|e^2/r_{12}|\phi_A{}^*(2)\phi_D(2)\rangle|^2. \tag{2.2}$$

The identification of the molecule orbitals involved in Equation (2.2) is given in Fig. 2.1.

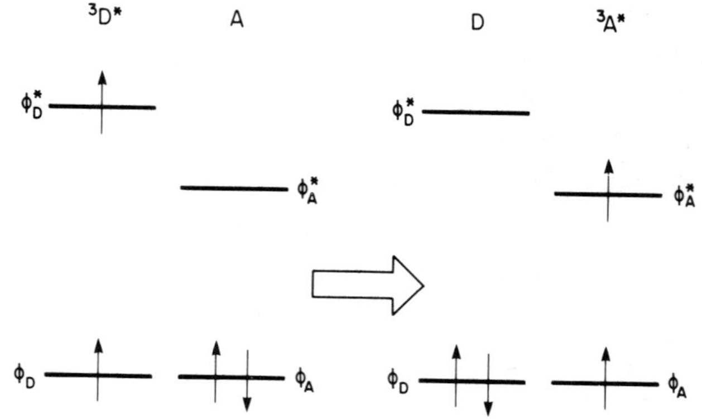

Fig. 2.1 Representation of the 'exchange' or Dexter mechanism (D.L. Dexter, *J. Chem. Phys.* **XXI**, 836 (1953)) of triplet energy transfer.

In Equation (2.2), r_{12} is the separation of a pair of electrons shared between the D and A molecules. In the case of energy migration, the donor and acceptor molecule are chemically equivalent. If for some reason some molecules in a 'lattice' are not equivalent (e.g. at a lattice defect, isotopic substitution, etc.) such that the energy of the T_1 state is slightly depressed at some point, then a shallow triplet exciton trap is produced; if the T_1 energy is elevated, an exciton-reflecting barrier (or 'antitrap') is produced. A chemically distinct species introduced into the lattice may produce a deep triplet trap. Clearly the description 'shallow' or 'deep' trap depends on the ratio $\Delta E(T_m,T_t)/kT$ where $\Delta E(T_m,T_t)$ is the difference in energy of the mobile triplet (i.e. the triplet state energy of the majority, chemically equivalent species in the lattice) and the trap, and kT is the thermal energy (i.e. Boltzmann constant times the absolute temperature). The field of molecular crystal spectroscopy provides many examples of exciton trapping at defects or by dopant species deliberately added to the crystal. As we see in later sections, equivalent phenomena are observed in polymer spectroscopy.

Trapping mechanisms are sometimes referred to as 'static' or 'dynamic', depending on the geometric requirements for formation of a trap. For example, if the excited state energy of the trap is inherently lower than that of the exciton, without geometric rearrangment, then it is a 'static' trap (i.e. no molecular or lattice vibrations are required for trapping). If a significant geometric rearrangement is required for trapping, then the rate of trapping may be very temperature dependent. Thus 'dynamic' trapping may be thermally assisted. In the context of molecular crystals this is referred to as 'phonon assisted'. As a practical matter, static trapping is typically inversely dependent on temperature since thermal energy enhances detrapping, whereas dynamic trapping is enhanced by increasing the temperature over lower temperature ranges, but is ultimately diminished at the higher range because of detrapping. These ideas are illustrated in Fig. 2.2. The best examples of dynamic trapping in the polymer literature are found in the trapping of S_1 excitons at excimer forming sites. A simple kinetic scheme that illustrates static vs dynamic quenching is given in Appendix I.

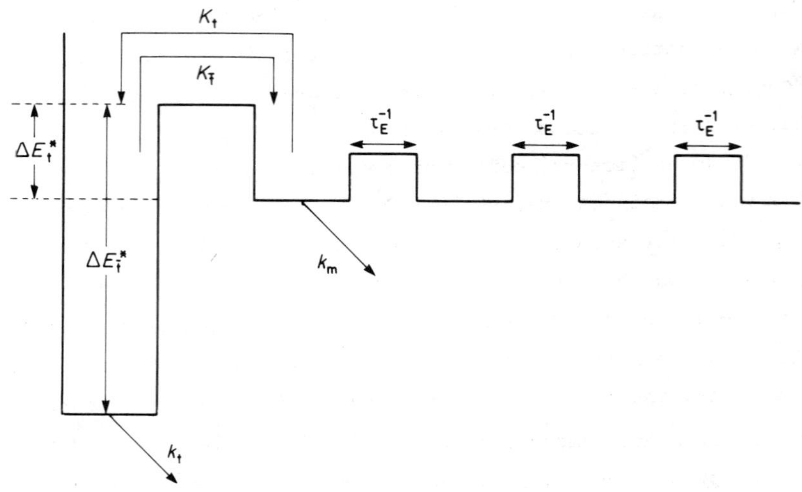

Fig. 2.2 Representation of static and dynamic trapping of energy (see Appendix I). $\tau_{E^{-1}}$ is the rate of energy transfer between equivalent sites.

A schematic representation of the temperature dependence that follows from this kinetic scheme is presented in Fig. 2.3.

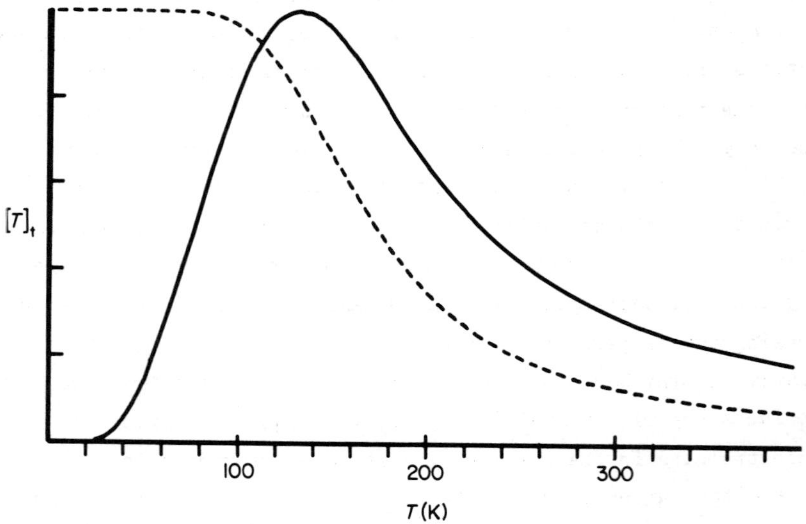

Fig. 2.3 Caption on next page.

A phenomenon that is of special importance to triplet state processes (although not limited to the triplet state) is triplet-triplet annihilation:

$$^{3}A^{*} + {}^{3}B^{*} \rightsquigarrow {}^{2S+1}A^{*} + B \qquad (2{\cdot}3a)$$

$$\rightsquigarrow A + {}^{2S+1}B^{*} \qquad (2{\cdot}3b)$$

where on the right side of Reaction (2.3) S = 2,1,0 are spin allowed. Typically no S = 2 state is energetically accessible, so the annihilation process either:

1. adds a new non-radiative relaxation channel if S = 1

or 2. results in fluorescence from $^{1}A^{*}$ or $^{1}B^{*}$ if S = 0.

Because this fluorescence occurs on the time scale of the lifetime of the triplet states $^{3}A^{*}$, $^{3}B^{*}$ it is referred to as 'delayed' fluorescence. Since this emission feature originates from the metastable triplet state, just as does phosphorescence, and it is often an important feature in polymer delayed emission, we will include a discussion of delayed fluorescence in this review.

In Reaction (2.3) one observes that delayed fluorescence can arise from either the $^{1}A^{*}$ or $^{1}B^{*}$ state, although in general the lower energy state will be formed preferentially (partially because of subsequent singlet energy transfer to the lower energy state). Nevertheless the delayed fluorescence can be expected to be composed of at least two features.

Fluorescence from excimers and exciplexes play an important role in polymer emission spectroscopy (see Chapters 2, 3 and 5). In the annihilation process the $^{3}A^{*}$ and $^{3}B^{*}$ molecules must come within essentially a collision distance of each other before annihilation can occur. Thus one expects that the metastable complex $^{2S+1}(A...B)^{*}$

Fig. 2.3 Plot of the steady-state concentration of T_t (trapped triplet) in arbitrary units as a function of T (see Appendix I), for the case of $\Delta E_t^{*} \neq 0$ (solid) and 0 (dashed).

(excimer if A = B, exciplex if A ≠ B) will tend to be preferentially formed in a sufficiently fluid environment. An analogous phenomenon is known to occur in the fluid phase T-T annihilation of small molecules. However, there are only a few cases of fluid-phase polymer T-T annihilation we are aware of (see Section 2.3.3), and because of experimental limitations it cannot be stated that the singlet excimer is preferentially formed in process 2.3.

2.2 SPECIAL FEATURES OF POLYMER DELAYED EMISSION SPECTROSCOPY

One must distinguish two limiting cases in polymer spectroscopy:

1. Copolymers in which the chromophores are minority species and are essentially isolated from each other.
2. Homopolymers or copolymers in which a chromophoric species is present in sufficient mole fraction that energy transfer or excited state-ground state chromophoric interaction is possible (this case does not exclude a minority luminescent trap species).

In the first case the luminescent molecule is often used as a probe of the polymer dynamics or environment. In the second case a fairly complex set of photophysical phenomena can occur. In the following paragraphs we will present an overview of the special phenomena that occur for homopolymers. Examples will be presented in the later sections.

2.2.1 *Excimer formation in homopolymers*

Because of the high local density of chromophores the following process is often facile:

$$^{2S+1}A^* + A \underset{k_{DM}}{\overset{k_{MD}}{\rightleftharpoons}} {}^{2S+1}(A\ldots A)^* \qquad (2\cdot4)$$

In Reaction (2.4), the Birks notation has been used for excimer formation /dissociation (k_{MD}/k_{DM}) respectively. † If the binding energy of the complex $^{2S+1}$(A...A)* is of the order of kT then thermal dissociation is possible. The metastable complexation in Reaction (2.4) is exactly analogous to the trapping/detrapping process discussed in the Appendix, and has been extensively studied for the singlet (S = 0) state (see Chapters 2, 3 and 6).

The existence of a triplet excimer has been doubted in the case of small molecules, but there seems to be little doubt that such a species exists in polymers. However, there are two important points about triplet excimers in polymers that should be made.

1. A fairly specific geometric configuration is required for maximum excimer stabilisation (although there may be various locally stable excimer configurations) such that changing the way in which chromophore is bound to the polymer backbone may drastically change the likelihood of excimer formation. These factors have been studied in great detail for singlet excited states. Thus one should not suppose that because of the high local density of chromophores in polymer coils that excimer formation is assured.
2. Although triplet excimer formation is much less studied (or even accepted) than singlet excimer formation, it is by no means obvious that the geometric requirements for a stable triplet state excimer are identical to those of the singlet state. Thus it is plausible that a polymer for which singlet excimer formation is relatively improbable triplet excimer formation may be facile. Model compounds have illustrated this effect.

2.2.2 *Energy migration in homopolymers*

The high local density of chromophores in a polymer coil also can lead to extensive intracoil energy migration. This phenomenon is

---

† Perhaps the most influential book in the discussion of molecular photophysics is that of Birks [3].

well documented for both the singlet and triplet state. The net effect of triplet energy migration is that energy trapping (Reaction (2.1)) or T-T annihilation (Reaction (2.3)) is much more important than can be rationalized on the basis of segmental diffusion of two chromophores to within a collision diameter. The bimolecular rate constant for trapping or annihilation can be written:

$$k = (4\pi N_o/10^3)[D(A) + D(B) + \Lambda_E(^3A^*) + \Lambda_E(^3B^*)]PR \quad (2.5)$$

where

$N_o/10^3 = 6.023 \times 10^{20}$ mol^{-1}cm^{-3}

$D(A),D(B)$ = physical diffusion of A,B molecules (segmental diffusion if species is polymer bound) (cm^{2}s^{-1}),

$\Lambda_E(^{2S+1}X^*)$ = energy migration rate of $^{2S+1}X^*$ excitation (cm^{2}s^{-1}),

P = probability of the trapping/annihilation reaction at the collision radius R.

We note that PR may be considered together as a single parameter describing the overall rate of the reaction. The Λ_E terms have the effect of enhancing trapping/annihilation processes relative to what one would predict based on small values of D(X). This enhancement is often referred to as the polymeric 'antenna effect' (by analogy to the photocollection processes of chloroplasts).

The use of a simple rate constant like Equation (2.5) to characterize energy migration in homopolymers is convenient, but misleading. From simple one-dimension walk theory

$$\Lambda_E = (1/2)(\bar{\ell}^2/\tau_E) \quad (2.6)$$

where $\bar{\ell}^2$ is the average of the distance squared for an energy transfer step (i.e. ℓ corresponds to a nearest-neighbour separation), and τ_E is an average transfer time (see Fig. 2.2). Clearly there will be a range of nearest-neighbour separations for a given polymer configuration, with a concomitant range of transfer time.† One would expect the use of an averaged expression like Equation (2.6) to be valid only if thermal averaging of all neighbouring configurations was fast compared to the average energy transfer time, or if the polymer is stereoregular and rigid. In general, energy transfer in a disordered system is difficult to treat theoretically.

Equation (2.5) has also been the basis for estimation of Λ_E by comparing the rate of excited state quenching of a monomeric 'model compound' and the polymer-bound excited state. The argument is as follows: if the intrinsic rate of reaction of a quencher molecule (Q) with the model compound is the same as the polymer chromophore (i.e. $(PR)_{\text{polymer}} = (PR)_{\text{model}}$), then the quenching rate constant may be written:

$$k_Q = (4\pi N_o/10^3)(D(\text{Q})+D(\text{E})+\Lambda_E)(PR) \tag{2.7}$$

For the polymer the physical diffusion of the segment containing E is expected to be slow, such that $D(\text{E})_{\text{polymer}} \simeq 0$. Likewise at low overall concentration one may assume $(\Lambda_E)_{\text{model}} \simeq 0$. Therefore

$$k_{Q(\text{polymer})}/k_{Q(\text{model})} = (\Lambda_E+D(\text{Q}))/(D(\text{Q})+D(\text{E})). \tag{2.8}$$

One expects $D(\text{Q}) \simeq D(\text{E})$ for the monomeric compound, such that

$$2(k_{Q(\text{polymer})}/k_{Q(\text{model})})-1 \simeq \Lambda_E/\bar{D} \tag{2.9}$$

---

† For example, the expression proposed by Inokuti and Hirayama [4] is $\tau_E(\ell) = \tau_o\exp[\gamma(\ell/\ell_o-1)]$ where $\gamma = 2\ell_o/\ell_b$, and τ_o is the unimolecular lifetime for the triplet state, ℓ_b is related to the spatial extent of the MOs of the triplet-state molecule (analogous to an atomic Bohr radius), and ℓ_o is the distance at which $\tau_E(\ell_o) = \tau_o$. Thus using this expression:

$$\Lambda_E(\ell) = (1/2\tau_o)\ell^2\exp[\gamma(1-\ell/\ell_o)].$$

where $\bar{D} = (D(Q)+D(A))/2$. Thus if $k_{Q(polymer)} < (1/2)k_{Q(model)}$ then $\Lambda_E/\bar{D} \simeq 0$. If the $k_{Q(polymer)}/k_{Q(model)}$ ratio is greater than 0.5 this is often taken as 'proof' that energy migration is occurring. However, there can occur 'polymer effects' (e.g. chromophore crowding) that modify the energy of the excited state, such that $(PR)_{polymer} \neq (PR)_{model}$. An example of this situation will be discussed in Section 2.3.3.

2.2.3 *Molecular weight effects and polymer configurations*

There are at least four effects of molecular weight:

1. Coil density, which effects the density of excimer forming sites or the possibility of two segments diffusing together.
2. The mole fraction of chain ends, which are generally expected to be atypical of the remaining polymer and hence may act as energy traps.
3. Increasing the length of the polymer chain increases the average number of excimer forming sites per coil (this is especially important for low molecular weight polymers, since there may be a significant fraction of coils with no excimer forming sites at any instant).
4. Increasing the length of the polymer increases the probability of multiple excitations, which is a critically important factor in T-T annihilation.

There is perhaps no other feature of polymer spectroscopy that is more unique than these molecular weight effects. Thus it is possible for investigators using polymers that are identical in every way except molecular weight to draw somewhat different conclusions from a spectroscopic investigation.

In addition to the molecular weight, polymers may be more or less stereoregular, depending on the method of preparation. Tacticity has been shown to be important in singlet excimer formation, and by extension, to triplet-state processes including triplet energy migration. Also coil density is a function of solvent quality for a given molecular weight, so that one anticipates 'solvent effects' on triplet-state processes (solvent effects have been clearly demonstrated for singlet-state spectroscopy).

For all the above it is clear that, when possible, experiments should be conducted on a range of molecular weights for a given method of polymer preparation, and the choice of solvents for a fluid-phase study should be carefully considered (i.e. there may be fairly drastic changes in going from a good solvent to a θ-solvent). Although not established, it is even possible that solvent effects could carry over to films cast from solution.

2.2.4 *Intrinsic traps and polymer degradation*

As was mentioned in the previous section, end groups can provide energy-trapping sites. This is an example of an intrinsic trap that is part of a polymer and cannot be removed by any purification step once the polymer has been prepared. An exception to this statement is the hydrogenation of residual vinyl groups, but this is generally only partially successful. In addition to the end groups, other species may be incorporated into the main chain if present in the original monomer in a form suitable for copolymerization (e.g. a vinylbenzaldehyde present as an impurity in polystyrene). Quite often these impurity species either show up very easily by producing an unexpected fluorescence or phosphorescence, or they are relatively harmless in that they do not act as a trap or complexation agent for the polymer-bound chromophore (e.g. vinylcyclohexane present in small mole fraction and copolymerized with styrene would produce little or no perturbation to the polystyrene absorption or emission spectrum).

Obviously one must be very careful in the handling and purification of monomers. Since monomeric species are often rather reactive, some 'purification procedures' can actually reduce the quality of the final polymer. Careful characterization of monomer purity by GC or HPLC techniques usually pays, as does fairly careful optical characterization of a given polymer sample before a series of experiments is begun.

One of the most important areas of polymer photochemistry is photodegradation, which has the effect of:

1. chain breaking (decreasing the average molecular weight),
2. cross-linking (decreasing polymer solubility), and
3. photooxidation or photorearrangements (which can be a precursor step to 1 or 2).

Effects 1 and 2 can produce effects on the polymer covered in Section 2.2.3. Effect 3 can produce an intrinsic trap where none existed in the original polymer. Thus, one must not expect a given polymer sample to last forever, and the spectrum often provides a good monitor for degradation (in fact this is one of the potential applications of polymer phosphorescence). As before, once an intrinsic trap has formed it is quite likely that it is irreversibly present on the polymer backbone. Hence the period and method of storage of a polymer can cause differences in the observations for a given polymer system from one investigator to another.

2.3 REVIEW OF EXPERIMENTAL RESULTS FOR POLYMERS IN GLASSY MATRICES, FILMS AND FLUID SOLUTIONS

In general the observation of delayed emissions requires a solid phase, frequently at low temperature, in order to minimize quenching of the long-lived triplet state by O_2 or any other paramagnetic, heavy atom or reactive species. Much less common is the observation of delayed emission from fluid-phase species, primarily because of adventitious quenching.

In the review of experimental observations that follows the organization will be by phase (low-temperature glassy matrix, films and fluid solution), and within each phase, classes of polymer systems will be considered together. The rationale for this organization is that the effect of phase on the polymer spectroscopy often seems to be more important than the precise nature of the chromophore. However, this is not a hard and fast rule, as will be seen.

2.3.1 *Low-temperature glassy matrices*

(a) General considerations

In order to carry out phosphorescence studies at low temperatures (typically at 77K, the boiling point of liquid nitrogen) the glassy matrix must fulfill two criteria:

1. The solvent used to form the matrix must dissolve the polymer.
2. The solvent must form a clear (and hopefully strainfree) glass upon quench-cooling.

Although the clarity of the glass is not absolutely essential, it greatly facilitates experimental work since excitation scattering is minimized and more of the bulk of the sample is excited, thereby enhancing the emission intensity. Suitable solvents that have been used include 1:1 (by volume) of diethylether (Et_2O) and tetrahydrofuran (THF), 2-methyltetrahydrofuran (MTHF) and toluene (with great care, since this glass often will crack and form a 'snow'). It is also possible to cast films of a polymer containing a small mole fraction of the polymer to be studied. The 'host' polymer must absorb shorter wavelength light than the polymer that is the object of the spectroscopic study. After the film has hardened, it may be studied at variable temperatures according to the experimental design. Although this method has been used to advantage in polymer fluorescence [5], it has not been extensively exploited in polymer phosphorescence. There are two problems with using a polymer film as a 'glassy matrix':

1. Quite often the host matrix contains a small concentration of intrinsic luminescent impurity which interferes with the guest polymer emission.
2. Polymer incompatibility may cause aggregation of the guest polymer, compromising the original intent of using a glassy matrix medium.

The low-temperature glassy matrix environment prevents dynamic quenching or trapping phenomena (i.e. excimer formation) and under the conditions of a quench-freeze the polymer coils are isolated if the coil concentration is sufficiently low. Furthermore, the polymer is close to its lowest energy configuration, which serves to minimize excimer formation. These conditions maximize the rate and 'diffusion length' of energy migration, and some of the clearest (and earliest examples of triplet migration were obtained in the glassy matrix phase.

(b) Naphthalenic polymers

The earliest observation of T-T annihilation in polymers was that of Cozzens and Fox [6, 7] in 1969 for the case of poly(1-vinylnaphthalene) (P1VN) and since that time quite a large fraction of what is known about the triplet state of polymers has been centred on the naphthalene chromophore. This is not surprising, since various naphthalene-containing polymers have been synthesized and the naphthalene triplet state is well characterized and convenient for study (excitation at $\lesssim$ 320 nm, τ = /2.3 s at 77K, phosphorescence emission structured in a convenient region (roughly 480-560 nm)). The naphthalenic polymers that have been well studied are summarized in Fig. 2.4.

As stated in the previous paragraph, much of the interest in polymer phosphorescence in glassy matrices can be tied to the initial observation by Cozzens and Fox of T-T annihilation in P1VN. Their original spectrum is reproduced in Fig. 2.5. In addition to T-T annihilation, these workers noted that the quenching of polymer phosphorescence by a small molecule (piperylene) dissolved in the glass was much more effective than for a small molecule model compound (1-ethylnaphthalene). It was argued that this was a manifestation of the migration of the triplet energy along the polymer chain (an intramolecular 'triplet exciton') thereby enhancing the probability that the excitation will encounter a quencher. Their original observations of quenching efficiency are shown in Fig. 2.6. These two observations have remained the primary diagnostics for triplet energy migration in polymers, although as we will see,

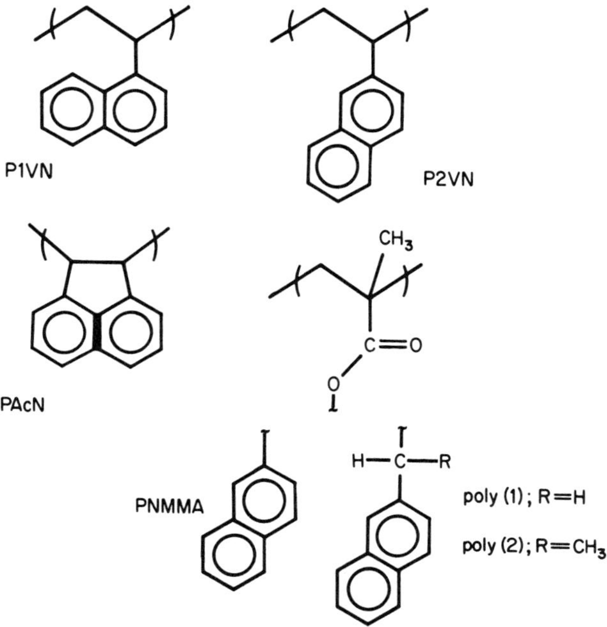

Fig. 2.4 Pendant group structure of naphthalenic polymers (see text for abbreviations).

phosphorescence depolarization has also been used to advantage.

Cozzens and Fox and coworkers carried out a long series of studies of polymer photophysics including more work on naphthalenic polymers. It was shown that in a copolymer of vinylnaphthalene and inert 'spacer groups' (e.g. methylmethacrylate or styrene) energy migration was not significant until the mole fraction of naphthalenes was sufficiently high [8]. In addition the extent of energy migration is often supralinear in naphthalene mole fraction (see Fig. 2.7). Although this is easy to understand qualitatively on the basis of the probability of 'runs' of naphthalene groups of a given length along the polymer chain, it is difficult to deal with this problem quantitatively since the distribution of neighbouring pairs in a copolymer is not necessarily random. To take an extreme example, a 1:1 mole ratio of a 'copolymer' could be two homopolymers of equal length or an alternating copolymer.

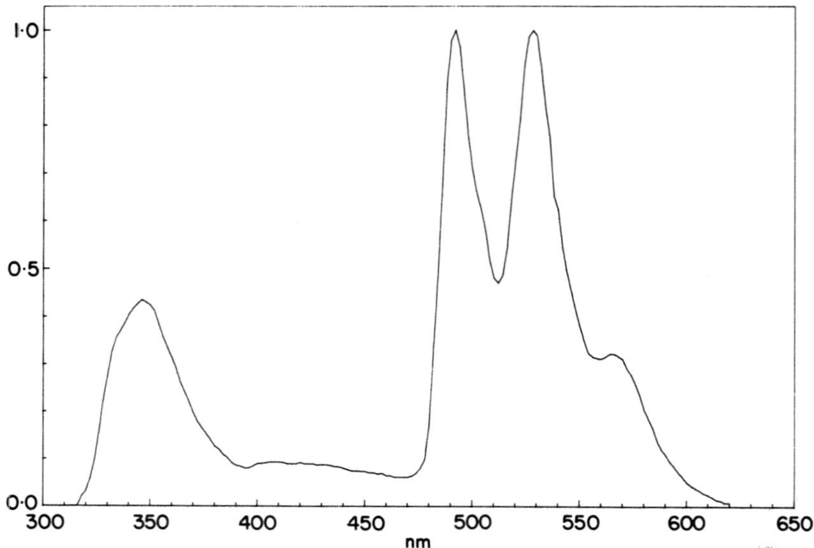

Fig. 2.5 Delayed-emission spectrum of poly(1-vinylnaphthalene) in tetrahydrofuran-diethyl ether at 77K, excitation wavelength 290 nm (Cozzens and Fox [6], reproduced by permission).

Fox *et al* [8] studied poly(2-vinylnaphthalene) (P2VN) as a homopolymer and as a component in various copolymers. As was the case for P1VN, energy migration was strongly dependent on the mole fraction of naphthalene units. In 1976 Pasch and Webber [9] published a study of P2VN in which the molecular weight of the homopolymer was systematically varied. For constant excitation the ratio of delayed fluorescence to phosphorescence increased strongly with molecular weight before levelling off (see Figs. 2.8 and 2.9). No systematic molecular weight effect on extinction coefficient of fluorescence quantum yield was observed (similar studies on poly(*N*-vinylcarbazole) were conducted at about the same time by Klöpffer, see Section 2.3.1.(*c*)). There are at least two plausible explanations for a molecular weight effect on the delayed emission of polymers in rigid glasses (see Section 2.2.2):

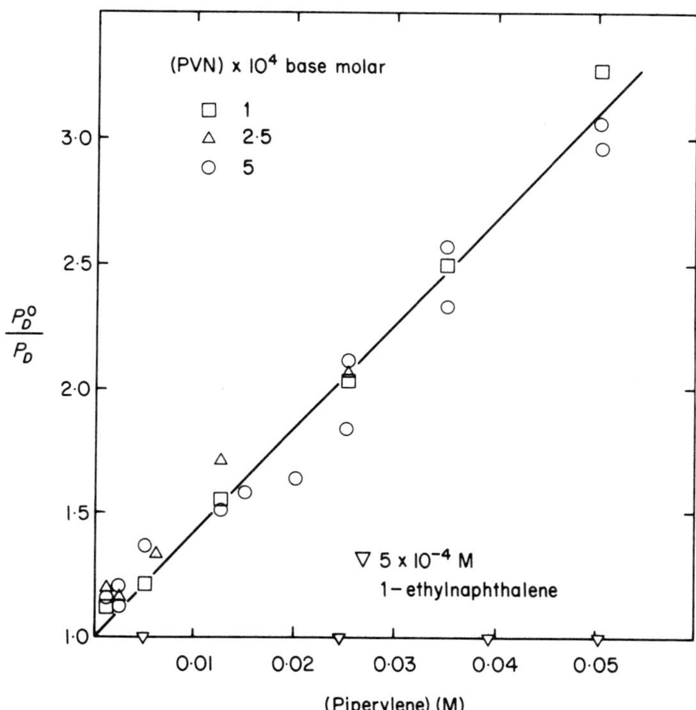

Fig. 2.6 Quenching of poly(1-vinylnaphthalene) (PVN) and 1-ethylnaphthalene phosphorescence at 77K as a function of piperylene molarity (Cozzens and Fox [6], reproduced by permission).

1. Mole fraction of chain ends (if these are intrinsic traps)
2. The probability of multiple excitation.

The paper of Pasch and Webber [9] did not distinguish between these two possibilities, although the multiple excitation mechanism was favoured.

If energy migration is the reason for enhanced excited-state quenching in polymers, then it follows that increasing the molecular weight of the polymer should also enhance the quantum efficiency of quenching. The argument is straightforward: a longer chain length allows a triplet exciton to sample more of the matrix volume, with a concomitant increase in the probability of encountering a quencher molecule. This was verified for P2VN by Pasch et.al. [10] and Webber in two ways:

1. The Stern-Volmer constant for the quenching of phosphorescence by piperylene was found to increase with molecular weight.
2. The efficiency of sensitization of biacetyl phosphorescence via the naphthalene triplet exciton was found to increase with molecular weight.

This latter experiment is interesting because it provides some insight to the period of time during which a triplet exciton is mobile. The situation is illustrated in Fig. 2.10. T_{BiA} (triplet state of biacetyl) has a lifetime ca. 10^{-3} of that of naphthalene, T_{BiA} is a deep trap for the naphthalene exciton at 77K, and the phosphorescence spectrum of biacetyl is distinctly different from that of naphthalene. Also shown in Fig. 2.10 is a T_1' state, considered to be a relatively shallow trap for the triplet exciton such that there is little or no difference in the phosphorescence spectrum of T_1 or T_1'. This corresponds to self-trapping (or end-group trapping). If the phosphorescence spectrum contains a

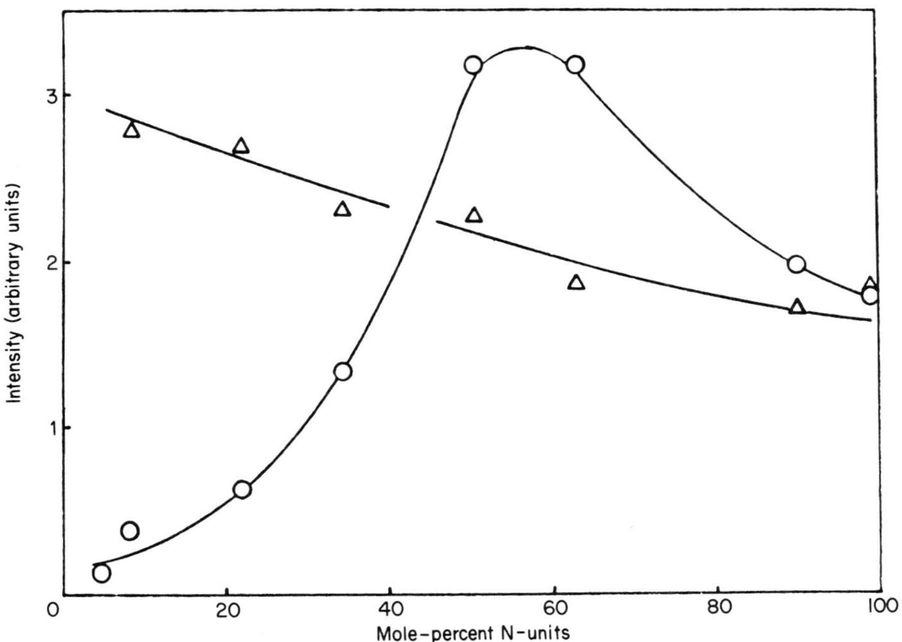

Fig. 2.7 Delayed fluorescence (circles) and phosphorescence (triangles) intensities (normalized to fluorescence intensity) from N-segments in PVN/S as a function of composition; 290 nm, (N-units) = 10^{-2} M (Fox et al [8], reproduced by permission).

significant component of biacetyl at times longer than $\tau(T_{BiA})$ then excitonic sensitization is implicated. If at later times the phosphorescence reverts to that of the pendant chromophore (naphthalene in this case), then it is inferred that the triplet

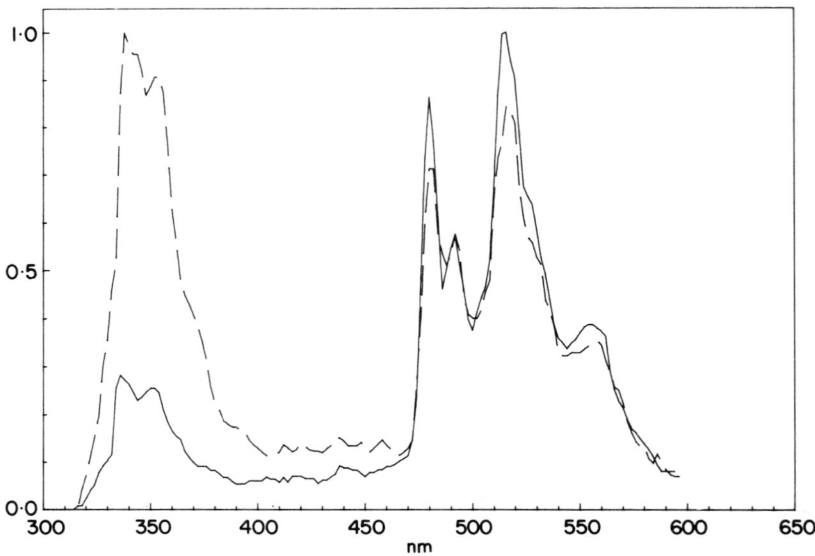

Fig. 2.8 Delayed emission spectra for two samples of P2VN in 1:1 THF:Et_2O at 77K. Dashed curve for sample with MW = 505,000 and solid curve for sample with MW = 39 700 (Pasch and Webber [9], reproduced by permission).

exciton has become immobilized (but see the next paragraph) and hence one has a measure of the period over which the triplet exciton is mobile. This technique is general if it is possible to find an irreversible phosphorescent trap with a lifetime much shorter than the excitonic state studied. The conclusions of the initial study using this technique were that the period of excitonic mobility was greater for P2VN than poly(acenaphthalene) (PAcN) and the efficiency of biacetyl sensitization by the triplet exciton of P2VN increased with molecular weight. The phosphorescence spectra of these species in the presence of 10^{-2} M biacetyl 1.0 s after the cessation of excitation display the following ordering with respect to the relative biacetyl component [12]:P2VN (high MW) > P2VN (low MW) > PAcN (low MW) (see Fig. 2.11).

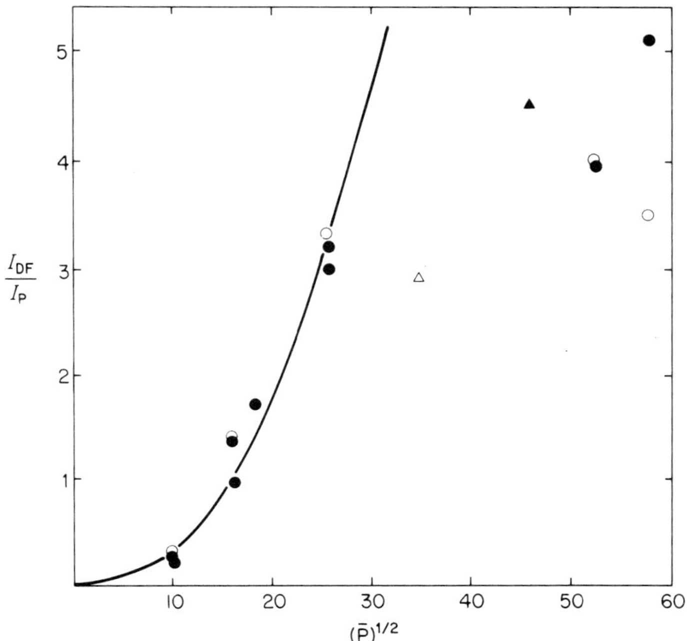

Fig. 2.9 The ratio I_{DF}/I_P vs $(\bar{P})^{\frac{1}{2}}$ (the degree of polymerization) (Pasch and Webber [9], reproduced by permission).

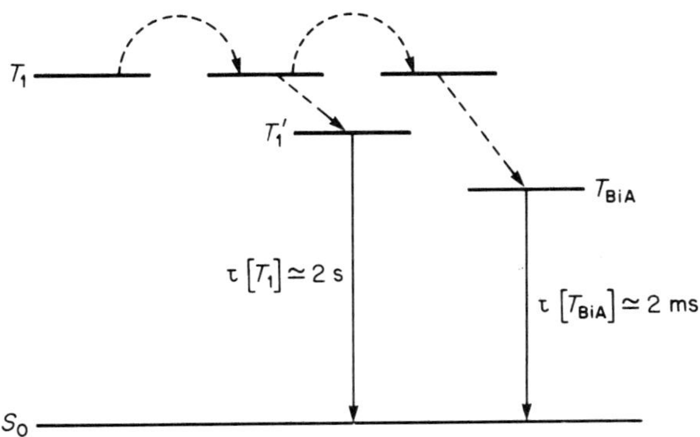

Fig. 2.10 Representation of triplet exciton sensitized biacetyl phosphorescence.

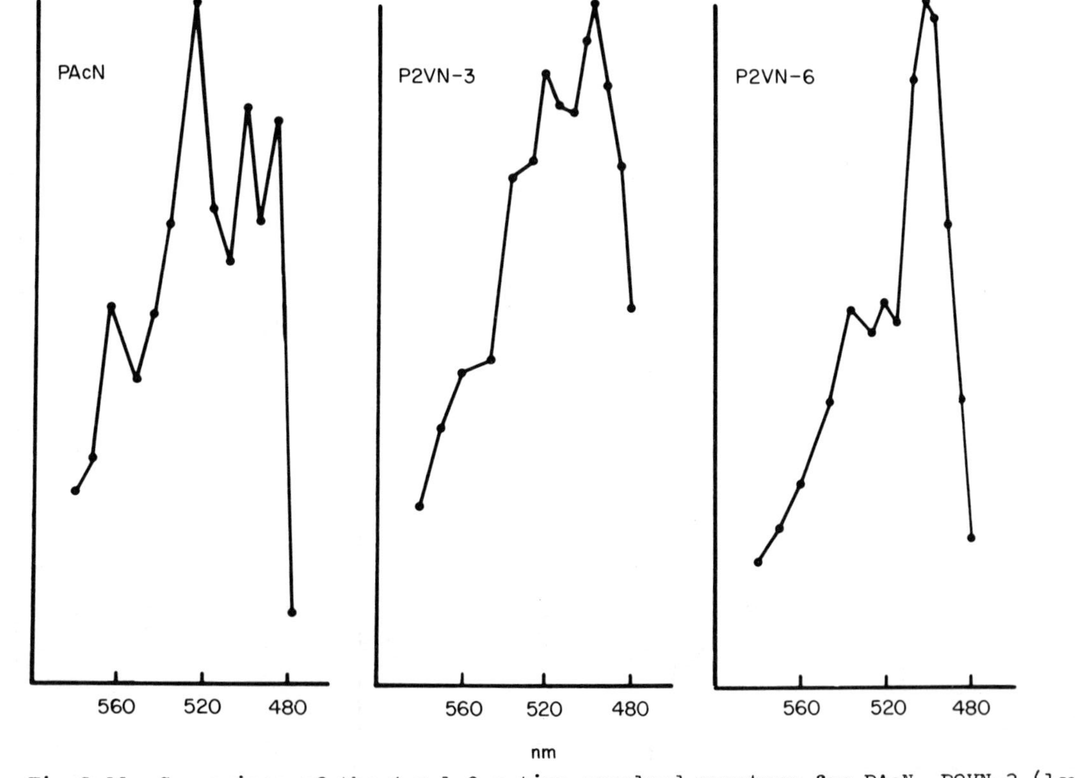

Fig 2.11 Comparison of the t = 1.0 s time-resolved spectrum for PAcN, P2VN-3 (lower MW) and P2VN-6 (higher MW) (all scaled to equal maxima) (Webber and Avots-Avotins [12], reproduced by permission).

There is one ambiguity with the use of an extrinsic triplet trap that cannot be resolved. If at longer times the phosphorescence reverts to that of the pendant chromophore then it is tempting to conclude that the triplet exciton is 'trapped'. However, this may be too narrow an interpretation. If a triplet exciton is confined to a limited region of a polymer coil then there can be a reasonable fraction of triplet excitons that do not have the opportunity to contact a triplet trap. This situation is similar to the effect of molecular weight on quenching or trapping. Confinement of a triplet exciton to a limited region could result from structural defects, e.g. a 'kink' in the polymer chain or a head-to-tail polymerization site.

Another illustrative feature of delayed fluorescence and phosphorescence in P2VN is the time dependence of these two features and the effect of molecular weight on the decay rates. First, the delayed fluorescence decays faster relative to the phosphorescence than one expects from standard solution kinetics. (If annihilation is not important in determining the overall decay then $I_{DF}(t) \propto I_{phos}(t)^2$.) The rationale for this observation first put forward by Pasch and Webber [9] is quite simple: the phosphorescence includes a contribution from singly excited polymer chains (or segments) whereas delayed fluorescence requires multiple excitation. Hence the phosphorescence and delayed-fluorescence sample the triplet exciton population in different ways. Similarly, phosphorescence can originate from trapped triplets but for T-T annihilation to occur at least one triplet must be mobile. Hence trapping serves to reduce the delayed fluorescence lifetime, but not necessarily the phosphorescence lifetime. Perhaps more striking is the effect of molecular weight on the decay kinetics. Simply stated:

1. The rate of decay of delayed fluorescence is much faster for low molecular weight P2VN than for high.
2. The rate of decay of phosphorescence is much slower (and nearly exponential) for low molecular weight P2VN than for high.

These observations are illustrated in Fig. 2.12. Again, the rationale put forward by Pasch and Webber is very straightforward. For low molecular weight polymers the average separation of pairs of triplet excitons (for multiply excited coils) will be smaller than high molecular weight polymers. Hence the average time required for annihilation tends to be diminished. On the other hand for smaller polymers the fraction of multiply excited coils is smaller, such that phosphorescence originates largely from singly excited coils. For higher molecular weights this is not the case such that the annihilation serves to shorten the observed phosphorescence (for a more theoretical treatment of these considerations see Webber and Swenberg [13]). These effects of polymer size on triplet state photophysics appear to be rather general and should be considered by an investigator attempting to interpret polymer phosphorescence or delayed fluorescence.

The case of PAcN is interesting. As can be seen from Fig. 2.4 the structure of the polymer chain permits very little mobility of the naphthalene groups. David *et al* [14, 15] have shown that the efficiency of singlet excimer formation is much lower than is the case for the other vinylnaphthalene polymers. These same workers showed that the phosphorescence quenching in PAcN is approximately as efficient as for polyvinylnaphthalene [15] implying a similar triplet exciton mobility but T-T annihilation leading to delayed fluorescence is not observed.† Based on the results for biacetyl sensitization, one concludes that the triplet exciton in PAcN is trapped or at least confined to relatively short chain segments. If the regions over which the triplet exciton on PAcN can migrate are relatively small the probability of multiple excitations is diminished. (This argument is the same as that proposed for the effect of molecular weight on delayed fluorescence.) Consequently, T-T annihilation may be the most sensitive spectroscopic indicator of the range of triplet exciton migration.

---

† In fact, we have observed a very weak delayed fluorescence from PAcN in a 77K MTHF glass (unpublished results).

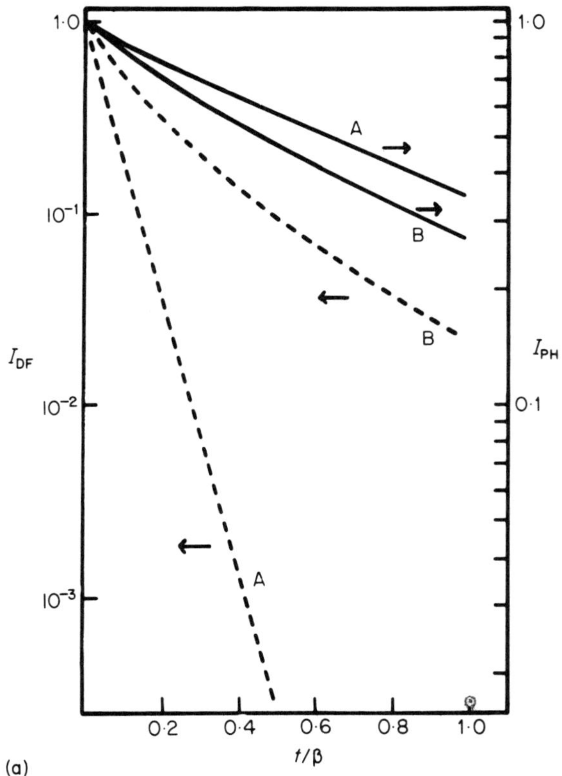

Fig. 2.12

(a) Results from the theoretical treatment of Webber and Swenberg [13]. Semilogarithmic plot of the triplet density ($\langle n\rangle/L$) (solid line) and the annihilation rate (I_A) (dashed line) versus t (in units of β, the unimolecular decay rate of T_1) for L = 100 (A curves) and L = 500 (B curves) (L = number of units in lattice) for a common choice of intrinsic annihilation rate. Note that the logarithmic axis for I_A has twice the logarithmic increment of that for $\langle n\rangle/L$.

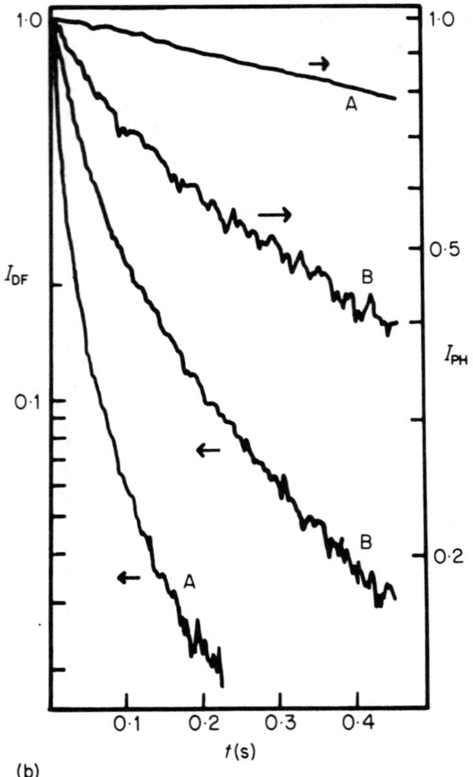

(b) Experimental decay curve for poly(2-vinylnaphthalene) phosphorescence and delayed fluorescence. Note that the logarithmic axes are dissimilar, like part (a). The A-curve corresponds to L = 100 and the B-curve corresponds to L = 3250 (Webber and Swenberg [13], reproduced by permission).

Poly(2-naphthyl methacrylate) (PNMA) was first studied by Somershall and Guillet [16] and later by Pasch and Webber 17 . The observations were analogous to those for P1VN, P2VN:

1. Delayed fluorescence and phosphorescence are both observed.
2. The ratio of delayed fluorescence to phosphorescence increases with molecular weight (although to a smaller extent than for P2VN).
3. The delayed fluorescence decays much more rapidly than the phosphorescence and the rate of decay of both delayed emission features is molecular weight dependent.

The general interpretation of the results for PNMMA was that although triplet migration does occur, it is less facile than for P2VN. However, no complete theoretical model exists to provide a quantitative test of this statement. Recently Nakahira *et al.* [18] have studied triplet energy migration inpoly(1) and poly(2) (see Fig. 2.4) in 77K MTHF glasses. Their observations are in accord with previous work on naphthyl methacrylates, with the interesting observation that poly(2) displays stronger delayed fluorescence than poly(1) by a factor of nearly 1.7. These two polymers have nearly identical phosphorescence decay rates and one expects these polymers to have very similar fluorescence quantum yields in glasses (although no dataare cited on this point). Thus one concludes that rather subtle structural features may strongly influence the rate of triplet migration along the chain (Nakahira and coworkers have also studied these effects on singlet energy migration).

It was found during the course of the phosphorescence studies of PNMMA that this polymer is susceptible to photo-induced changes in the phosphorescence spectrum that are quite striking (see Fig. 2.13). Li and Guillet [19] later showed that this new feature arises from a photo-Fries rearrangement to produce a naphthol ketone.

We conclude this discussion of naphthalenic polymers in low-temperature glasses with brief discussion of two copolymers that suggest interesting possibilities for intrinsic triplet sensitizers and traps.

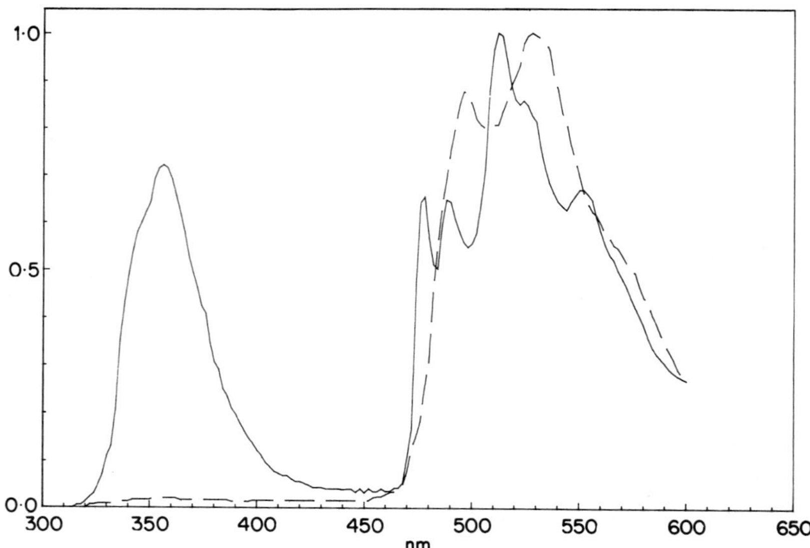

Fig. 2.13 A typical delayed emission spectrum for P2NMA (mol wt = 79 000) for a fresh sample (solid line) and a sample that has developed the impurity emission (dashed line). The latter spectrum is approximately 10x more intense than the former (Pasch and Webber [17], reproduced by permission).

Aikawa *et al.* [20] have studied the phosphorescence of copolymers of phenyl vinyl ketone and 2-vinylnaphthalene, in which the naphthalene content varied from 0.09 to 0.75 mole fraction (note that copolymers of this type will be discussed again in Sections 2.3.3(*c*) and 2.3.3(*d*)). At lower naphthalene mole fraction the 77K phosphorescence originates from phenyl ketone and normal naphthalene (0-0 band at 468 nm). At higher naphthalene contents the phenyl ketone is quenched and the 0-0 band of the naphthalenic component shifts to 482 nm (see Fig. 2.14). The quenching of the phenyl ketone phosphorescence is easily understood as the result of energy transfer to the naphthalene. Aikawa *et al.* interpret the red-shift of naphthalene phosphorescence in terms of the triplet state of naphthalene dimers.

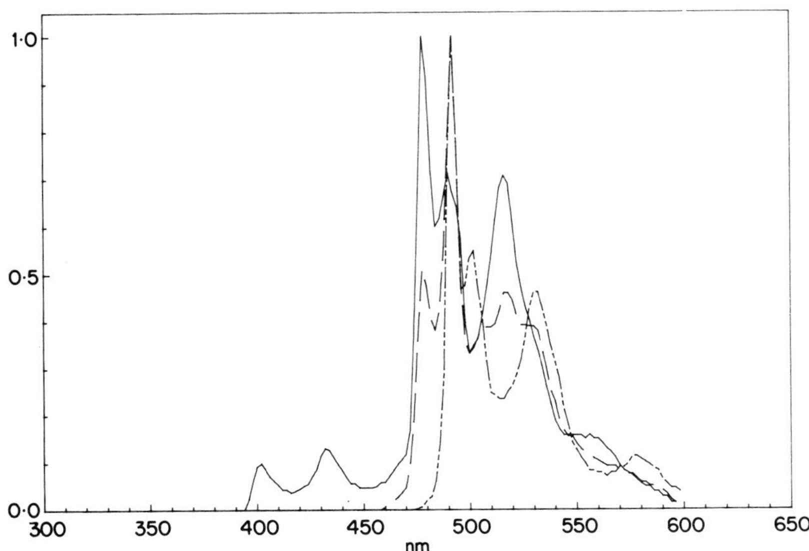

Fig. 2.14 The phosphorescence spectra (uncorrected) of VN-PVK copolymer 9/91 (solid line), 16/84 (dashed line) and of PVN homopolymer (dash-dot line) in MTHF at 77K and exciting at 350 nm. For the copolymer the numbers refer to the mole fraction of naphthalene and ketone respectively (Aikawa *et al.* [20], reproduced by permission).

The decay of the phosphorescence is biexponential, with the slow component remaining very similar to monomeric naphthalene (lifetime in the range 2.0 - 2.7 s). There is a fast component (lifetime ca. 0.5 s) that is ascribed to the dimer triplet. It is possible that in fact this fast component arises from T-T annihilation, but delayed fluorescence would not be observed because the naphthalene singlet would be quenched by the phenyl ketone. Studying the dependence of the decay on excitation intensity would have been very useful to elucidate this point. It is also difficult to ascribe the observed red-shift in the naphthalene phosphorescence solely to the triplet state of naphthalene dimers, since 'chromophore crowding' can easily lead to the same effect (and in fact these two terms are really a matter of semantics). Aikawa *et al.* [20] also demonstrate a very strongly red-shifted and broadened phosphorescence at 187K, which is rather similar to film phosphorescence (see Section 2.3.3(*b*)). This new feature is assigned to triplet excimer phosphorescence.

A copolymer of 2-vinylnaphthalene and 4-vinylpyrene provides an example of an intrinsic triplet trap. The delayed emission was a mixture of naphthalene and pyrene delayed fluorescence and naphthalene and pyrene phosphorescence† (see Fig. 2.15). Preliminary analysis implies that the pyrene delayed fluorescence is a result of intracoil heterogeneous annihilation††:

$$^{3}\text{Naph}^{*}_{(\text{exciton})} + {}^{3}\text{Py}^{*}_{(\text{trap})} \rightarrow \text{Naph} + {}^{1}\text{Py}^{*}_{(\text{trap})} \qquad (2.10)$$

followed by pyrene fluorescence. This polymer system represents an excellent example of the 'antenna effect' in which a photon captured by the main chain chromophore is transferred to a 'reaction centre' (pyrene in this case). We will see another case of this in our discussion of doped films and poly(*N*-vinylcarbazole). (Section 2.3.1(*c*), below).

(c) Poly(*N*-vinylcarbazole)

Because of its commercial application as a photoconducting polymer the excited states of poly(*N*-vinylcarbazole) (PVCz) have been intensively studied, although the preponderance of the studies have focused on the singlet state. The most detailed work has been that of Klopffer and coworkers [21, 22] with special emphasis on the film state (see Section 2.3.3(*c*)). The structure of PVCz (see Fig.2.16) crowds adjacent chromophores and leads to facile singlet excimer formation

---

† Presented at IUPAC Macro 82, Amherst, Massachusetts, July 1982 (preliminary results by J.S. Hargreaves and G. Sowash).

†† The ^{1}Naph* state can result from the annihilation in Reaction (2.10), but sensitization of ^{1}Py* would be fairly efficient.

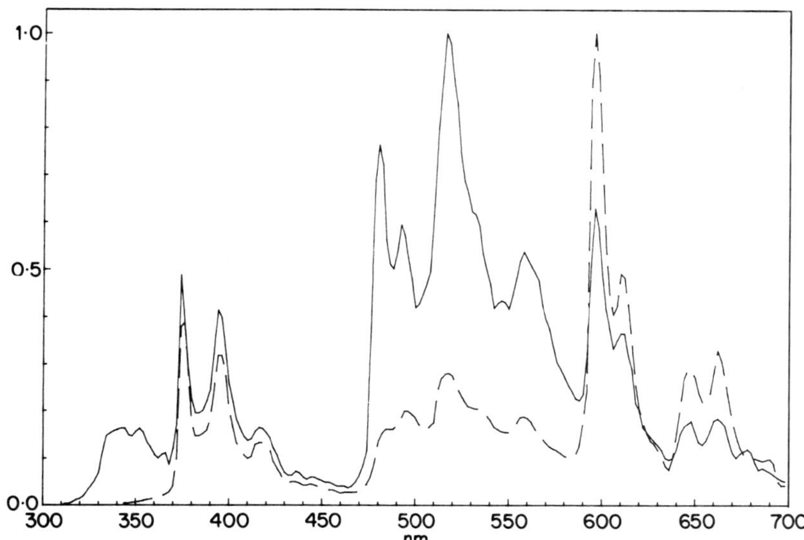

Fig. 2.15 Delayed emission spectra of P2VN-co-Py with X_{py} = 0.001 (solid line) and 0.01 (dashed line) (presented at IUPAC Macro 82, Amherst, Massachusetts, July 1982).

in the fluid or film state.

The delayed emission spectrum of PVCz in 77K toluene glasses shows a very strong component of delayed fluorescence in addition to phosphorescence. The absolute intensity of these two features is very intensity dependent (at low excitation intensities, $I_{DF} \propto I_{ex}^2$, $I_{phos} \propto I_{ex}$) and remarkably dependent on the molecular weight of the polymer [22] (see Figs. 2.17 and 2.18). The general trends of the molecular weight effects are the same as noted for the polyvinyl-naphthalenes, but the dynamic range of the effect is 15-20 times larger. No quantitative explanation of this very strong molecular weight dependence has been presented. Based on the results of sensitized phosphorescence from an intrinsic and extrinsic quencher (see below), it seems likely that the triplet exciton in PVCz is

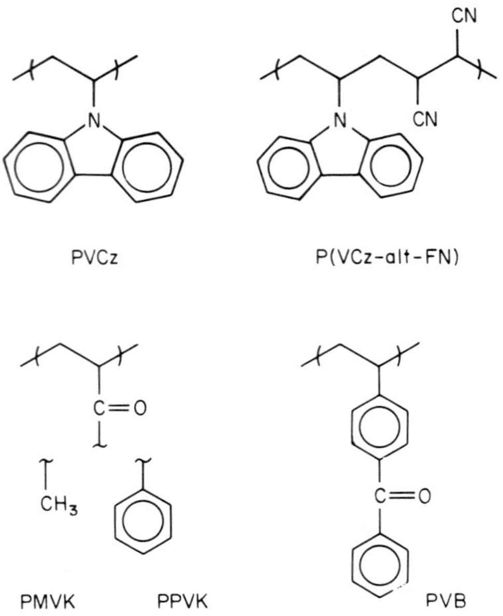

Fig. 2.16 Pendant group structure of carbazole, ketone polymers (see text for abbreviations).

quickly trapped. If one postulates that this trapping is at the end groups, then this 'molecular weight effect' (i.e. mole fraction of ends proportional to M^{-1}, where M is the polymer molecular weight) is in concert with that having to do with the statistics of excitation as was proposed for P2VN. This leads to an unusually strong molecular weight dependence. However, this 'end-group effect' is pure speculation at the present time.

The difference in the rates of decay of delayed fluorescence and phosphorescence is unusually large in PVCz. The lifetime of delayed fluorescence (τ_{DF}) and phosphorescence (τ_{Ph}) are 50 ms and 7.6 s (toluene glass, 77K). This disparity in lifetimes could arise from a very fast rate of triplet migration such that all multiply excited coils decay rapidly via T-T annihilation. Alternatively, as the mobile triplet excitons become trapped, the delayed fluorescence ceases. In this case the decay rate corresponds to the rate of

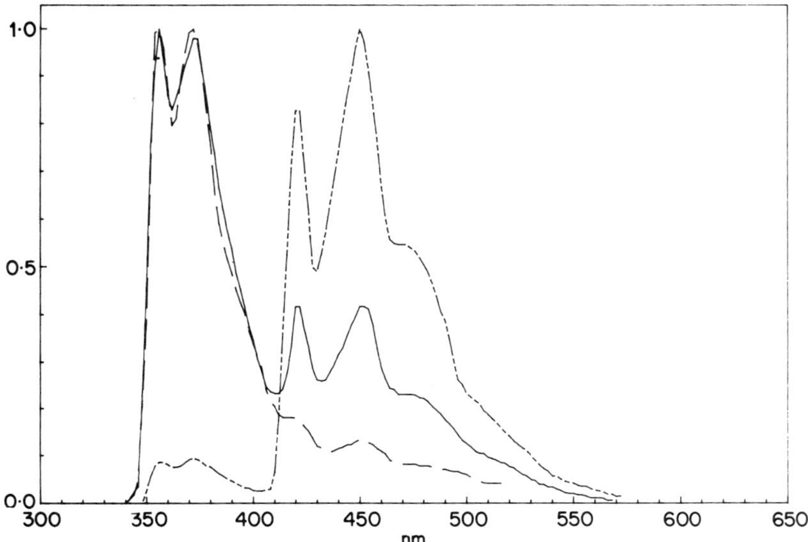

Fig. 2.17 Phosphorescence spectra of three different PVCA fractions, 6×10^{-4} mol basic unit/L in MTHF at 77K, recorded under identical experimental conditions. Excitation wavelength 330 nm (dashed, MW 2×10^{6}; solid, MW 3×10^{5}; dash-dot, MW 9×10^{4}) (Klöpffer *et al.* [22], reproduced by permission).

exciton trapping. Since the decay rate of the PVCz phosphorescence has never been found to be dependent on excitation intensity (which would be the case if T-T annihilation is an important decay mechanism) we favour the triplet exciton trapping mechanism.

Yokoyama and coworkers [23, 24] studied homopolymers and copolymers of PVCz, and were the first to observe a molecular weight effect on the ratio of delayed fluorescence to phosphorescence. However, these reports are perhaps of greater interest because of the observed effects of copolymerization. A series of copolymers of *N*-vinylcarbazole and 1-vinylnaphthalene were prepared. The energy levels of this system have the following relationship:

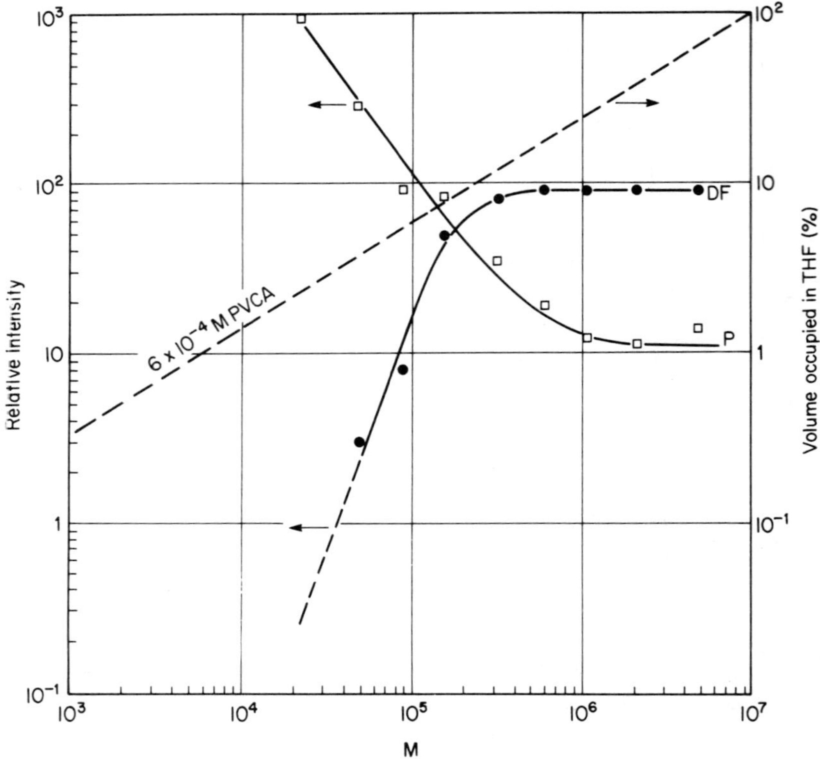

Fig. 2.18 Dependence of delayed fluorescence and phosphorescence intensities on molecular weight of PVCA fractions for identical excitation and recording conditions. The right-hand scale gives the volume occupied by PVCA coils in THF at room temperature (Klöpffer *et al.* [22], reproduced by permission).

E_{S1}(naph)>E_{S1}(Cz)>E_{T1}(Cz)>E_{T1}(naph). Consequently it is possible to excite the carbazole moiety, with subsequent sensitization of the naphthalene triplet state. (This situation is analogous to the P(2VN-co-4VPy) system discussed in the previous section.) The delayed emission spectrum consists of carbazole delayed fluorescence and phosphorescence, and naphthalene delayed fluorescence (very weak) and phosphorescence. The delayed emission spectrum of several

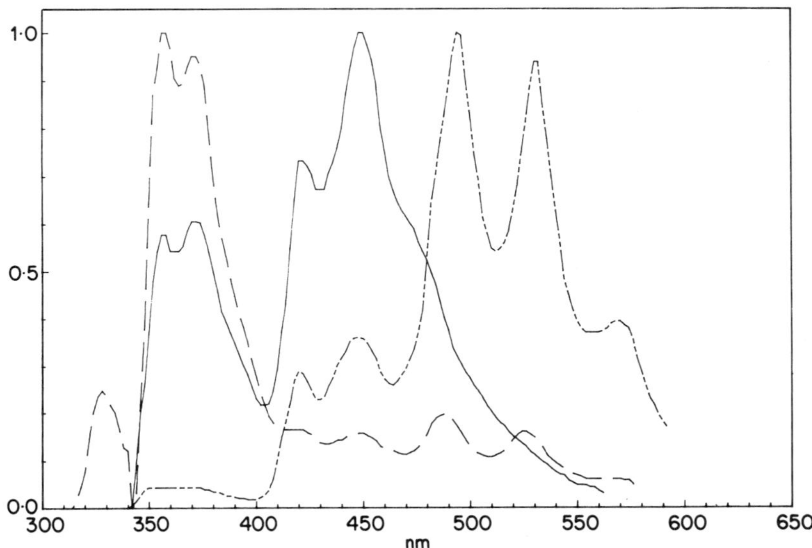

Fig. 2.19 Delayed emission spectra in rigid solution (77K) of PVCz (solid line) and vinyl carbazole (VCz)-vinylnaphthalene (VN) copolymers of different compositions: 100 mole% VCz (solid line), 92 Mole% (dashed line), and 24 mole% (dash-dot line). λ_{ex} = 340 nm. These spectra were measured under almost the same conditions. The naphthalene DF (<340 nm) is enhanced by a factor of 100 (Yokoyama *et al.* [24], reproduced by permission).

different copolymers is presented in Fig. 2.19. Certain features are as expected:

1. Naphthalene phosphorescence becomes more intense at the expense of the carbazole phosphorescence because of $^3Cz^*$ exciton trapping by naphthalene.
2. The naphthalene delayed fluorescence is quite weak because the following process is efficient: $^1N^* \rightarrow {}^1Cz^*$.

However, it is surprising that the most intense carbazole delayed fluorescence occurs not for the homopolymer, but for a polymer which contains 8 mole% naphthalene. Yokoyama *et al.* [24] ascribe this effect to homogeneous T-T annihilation in naphthalene-rich sequences, followed by singlet energy transfer from $^1N^{**}$ (the initially formed singlet state) to $^1Cz^*$, i.e.:

$$^3N^* + {}^3N^* \rightsquigarrow {}^1N^{**} \xrightarrow{Cz} {}^1Cz^* \rightarrow \text{Cz-delayed fluorescence};\quad {}^1N^{**} \rightsquigarrow {}^1N^*;\quad {}^1N^* \xrightarrow{Cz} {}^1Cz^*;\quad {}^1N^* \rightarrow \text{Naphthalene-delayed fluorescence} \qquad (2.11)$$

The $^3N^*$ state in Reaction (2.11) is sensitized via the $^3Cz^*$ exciton. Another plausible mechanism is heterogeneous annihilation:

$$^3N^* + {}^3Cz^* \rightsquigarrow {}^1N^{**} \rightsquigarrow {}^1N^* \rightarrow \text{Naphthalene-delayed fluorescence};\quad {}^3N^* + {}^3Cz^* \rightsquigarrow {}^1Cz^{**} \rightsquigarrow {}^1Cz^* \rightarrow \text{Cz-delayed fluorescence};\quad {}^1N^* \overset{Cz}{\rightsquigarrow} {}^1Cz^* \qquad (2.12)$$

In the mechanism in Reaction (2.12) the naphthalene trap decreases the number of $^3Cz^*$ excitons that are trapped at intrinsic traps normally present in PVCz. Yokoyama *et al.* [23] have suggested that these trapped triplets do not participate in T-T annihilation. However the heterogeneous annihilation process shown in Reaction (2.12) could occur, with a subsequent enhancement of carbazole-delayed fluorescence. The question remains open as to the nature of the intrinsic triplet trap in PVCz and why heterogeneous annihilation involving this trap would be 'forbidden'.

In the earlier work of Yokoyama *et al.* [23] an alternating copolymer of *N*-vinylcarbazole and fumaronitrile (P(VCz-alt-FN)) was prepared. Although there was not a very large range of molecular weights prepared (ca. 4000-8000) this copolymer did display some interesting differences from the PVCz homopolymer:

1. Cz-delayed fluorescence was observed and was more intense than

that of the corresponding homopolymer.†

2. $^3Cz^*$ migration was demonstrated via naphthalene quenching of phosphorescence, and unlike PVCz, the phosphorescence lifetime of $^3Cz^*$ is decreased by the addition of naphthalene.

Perhaps the major conclusions that result from the studies of Yokoyama and coworkers [23, 24] are these:

1. Triplet energy migration can occur in polymers in which the chromophores are not adjacent, although the rate of energy migration may be decreased.
2. PVCz phosphorescence results almost exclusively from a trap (similar to monomeric carbazole) that does not participate in T-T annihilation. Based on the very strong molecular weight effects on the relative intensity of delayed fluorescence and phosphorescence [22], one may speculate that these traps are at chain ends.
3. The rate of triplet energy migration in PVCz prior to the trapping is exceptionally fast.

These general conclusions are consistent with the results of Webber and Avots-Avotins [25] using biacetyl as an extrinsic triplet trap. (This technique was discussed in Section 2.3.1(*b*) and is illustrated in Fig. 2.10.) The biacetyl was found to quench the carbazole phosphorescence and delayed fluorescence. The rate of decay of the carbazole phosphorescence was decreased during the first 500 ms, but then decayed with the same rate as unquenched PVCz at longer times. Perhaps more illustrative of the kinetics of the PVCz-biacetyl system is the time-resolved phosphorescence spectrum (Fig. 2.20). The early time phosphorescence is dominated by the biacetyl, but after 0.8 s the phosphorescence is almost totally carbazole-like. This may be contrasted with Fig. 2.11, where for

---

† In Klöpffer *et al.* [22] no delayed fluorescence was observed in a homopolymer of molecular weight 8000 (the highest molecular weight obtained for P(VCz-alt-FN)). The homopolymer can be prepared with much higher molecular weight, however.

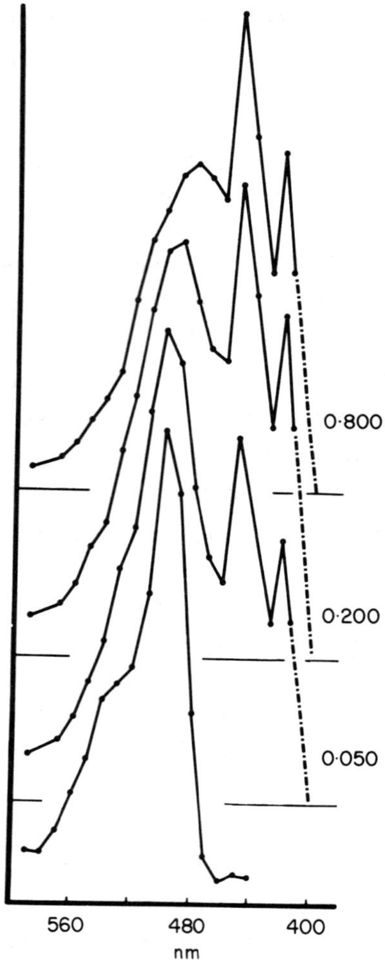

Fig. 2.20 Time-resolved phosphorescence spectra of PVCz + 1.0 x 10^{-2} M biacetyl, with time given in seconds. The lowest trace corresponds to t = 0.0. All spectra have been scaled to the same maximum to facilitate comparison. At 0.800 s the intensity in the 500 nm region is higher than unquenched PVCz, indicating a residual biacetyl contribution (Webber and Avots-Avotins [25], reproduced by permission).

P2VN after 1.0 s the biacetyl still dominates the phosphorescence. Thus our observations on PVCz-biacetyl are consistent with an excitonic model as follows: immediately after excitation there is rapid triplet exciton migration followed by irreversible trapping (at least at 77K) and a cessation of T-T annihilation. The phosphorescence arises primarily from the trapped triplets.

(d) Polyvinylbenzophenone

The benzophenone molecule is one of the most studied moieties in photochemistry and has been used in many studies as a triplet sensitizer. It is not surprising that polymers containing this chromophore were among the early examples of polymer photophysical studies, although as we will see, most studies carried out by David, Geuskens and others emphasized the film state. One slight difficulty with polyvinylbenzophenone (PVB, see Fig. 2.16) is that the polymer is typically prepared by benzoylation of polystyrene. This reaction does not yield complete substitution on the pendant phenyls (although > 90% conversion is claimed) such that 'barriers' to energy migration undoubtedly are present.

The quantum efficiency for intersystem crossing is essentially unity for benzophenone, such that the only spectroscopic observable is phosphorescence. Thus, although it is possible for T-T annihilation to occur, no delayed fluorescence will occur. However, one might expect to observe an excitation dependence of the phosphorescence lifetime. No study of this type for PVB has been reported for 77K glassy matrices, but we will see a related example for the fluid phase.

David and coworkers [26, 27] have characterized triplet energy transfer in copolymers of vinylbenzophenone-styrene using two methods:

1. Quenching of benzophenone phosphorescence by energy transfer to naphthalene.
2. Phosphorescence depolarization.

In the former a Stern-Volmer relationship between the unquenched

(I_o) and quenched (I) phosphorescence intensity and acceptor concentration ([A]) is found:

$$I_o/I = 1 + k_{AT}\tau_D[A] \qquad (2.13)$$

where k_{AT} is the rate constant for energy trapping and τ_D is the energy donor lifetime (i.e. the benzophenone triplet state, τ_D = 5 ms). David *et al.* [26]† then use the theory of Voltz [29] and Voltz and Heisel [30] (that was developed for molecular crystals and other condensed media) to obtain a value for the energy migration rate: Λ_T (cf. Reactions (2.5) and (2.6)):

$$k_{AT} = (4\pi R_o N_A)/1000\ \Lambda_T[1 + R_o/(\Lambda_T\tau_D)^{\frac{1}{2}}] \qquad (2.14)$$

R_o is the distance over which energy transfer is complete (i.e. a 'critical radius'). Λ_T, which has units of cm^2s^{-1}, may be interpreted as an exciton diffusion constant, i.e:

$$\Lambda_T = (1/6)\langle R^2Nk_{mig}\rangle \qquad (2.15)$$

where R is a pair-separation distance, N is the number of nearest neighbours, and k_{mig} is the energy-transfer rate between nearest neighbours. The brackets indicate that Λ_T represents a complex average over various polymer configurations and nearest-neighbour separations. David *et al.* [26] and many other workers frequently assume a reasonable value of R and N and then estimate k_{mig}. Because of the complexity of the physical system, these estimates are of dubious quantitative value, but they do serve as a point of comparison of one polymer with another or of polymer systems with molecular crystals.

The most important point of Reactions (2.14) and (2.15) for the present discussion is the very interesting observation by David *et al.* [26] that Λ_T for P(VB-co-PS) is strongly dependent on the mole

† So far as we know, the first application of the theory of Voltz [29] and Voltz and Heisel [30] to energy transfer in polymers was by David *et al.* [28](to PVB films containing naphthalene).

fraction of vinylbenzophenone. For a vinylbenzophenone (VB) mole fraction up to 0.5 Λ_T increases linearly, but then abruptly increases by a factor of 5 from 0.5 to 0.75 (see Fig. 2.21). This kind of sudden increase in energy migration with mole fraction of chromophore is well known in the molecular crystal field and has been interpreted by Kopelman *et al.*[31]† using percolation theory. The physical idea behind this observation is very simple. As the VB mole fraction is increased the probability of longer sequences of VB units becomes very significant, such that the average distance over which down-chain energy transfer can occur increases sharply. Thus the efficiency of sensitizing an extrinsic energy trap is strongly dependent on the donor mole fraction with a concomitant dependence of the derived Λ_T value. These ideas are similar to those discussed in previous sections on sensitized phosphorescence.

In a related study David *et al.* [27] measured the polarization of benzophenone phosphorescence in P(VB-co-S) polymers as a function of VB mole fraction. The polarization is given by:

$$p = (I_{||}-I_{\perp})/(I_{||}+I_{\perp}) \tag{2.16}$$

where $I_{||}$, $I_{\perp}$ are the intensities of emitted light polarized parallel and perpendicular to the polarization of the excitation light. The numerical value of p depends on the symmetry properties of the absorbing and emitting states, and can be positive, negative or zero. However, if energy transfer occurs between randomly oriented chromophores, then p tends toward zero.††

David *et al.* [27] observe that p^{-1} for the benzophenone phosphorescence increases strongly when the mole fraction of VB exceeds 0.5. This is consistent with the energy-transfer work; at sufficiently high chromophores loading on the polymer chain energy migration is facile. This is an important general principle in

---

† For a recent application to molecular crystal fluorescence, see Parson and Kopelman [32].

†† This technique has been exploited in characterizing singlet energy migration by several workers: see the discussion by Soutar [33].

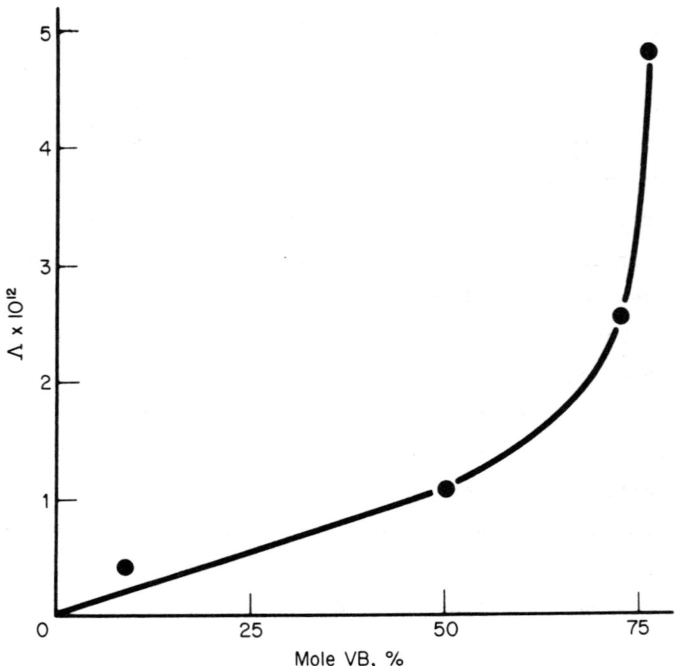

Fig. 2.21 Migration coefficient Λ ($cm^{-2}sec^{-1}$) between benzophenone chromophores as a function of copolymers composition (glassy solution) (David *et al.* [26], reproduced by permission).

polymer photophysics. It is important to note however that 100% chromophore loading of a polymer chain is not required for significant energy transfer to exist.

2.3.2 *Polymer films*

(a) General considerations

For most studies the film is cast from solution. In many cases the film is placed in a good vacuum ($<10^{-5}$ torr) with mild heating (ca. 50°C) for a period in excess of 12 h. It is always hoped that these procedures will serve to remove traces of solvent, but in general the complete absence of solvent cannot be proved. In addition to traces of solvent, small molecule impurities that were present in the solvent may become trapped in the polymer film. It is possible that the configuration of the polymer coils that make up the film may be dependent on the precise casting conditions (temperature, solvent, etc.). Since photooxidation or exciton quenching may occur preferentially at the polymer film surface, the thickness of the polymer film may be relevant to the observed photophysics. Lastly, the molecular weight of the polymer or tacticity can be expected to influence the rate of exciton migration or annihilation. As a consequence of all these factors it would not be surprising if competent investigators obtained slightly different results from precisely the same polymer sample. We will see, however, that agreement is generally satisfactory.

In our film delayed emission work we have found the following to be important:

1. Avoidance of chlorinated solvents (even CH_2Cl_2), using benzene, toluene or tetrahydrofuran.
2. All films are allowed to dry at room temperature (sometimes in a N_2 atmosphere) and then placed on a vacuum line at either room or slightly elevated temperature for at least 12 h, then sealed either under vacuum or He.

(b) Films of naphthalene polymers

In some of the earliest work by Fox *et al.* [34, 35], the prompt and delayed emission of poly(2-vinylnaphthalene) (P2VN) and poly(1-vinylnaphthalene) (P1VN) films was studied. The following general observations were made:

1. Both biphotonic delayed fluorescence and monophotonic phosphorescence are observed. Both emissions are broadened and shifted to longer wavelengths relative to dilute glassy matrices, and may be assigned to singlet and triplet excimer emission.
2. There is a distinct red-shift of the delayed fluorescence

relative to the prompt fluorescence for P2VN. (The same observation was made by Kim and Webber [36] for P2VN - see Fig. 2.22.)

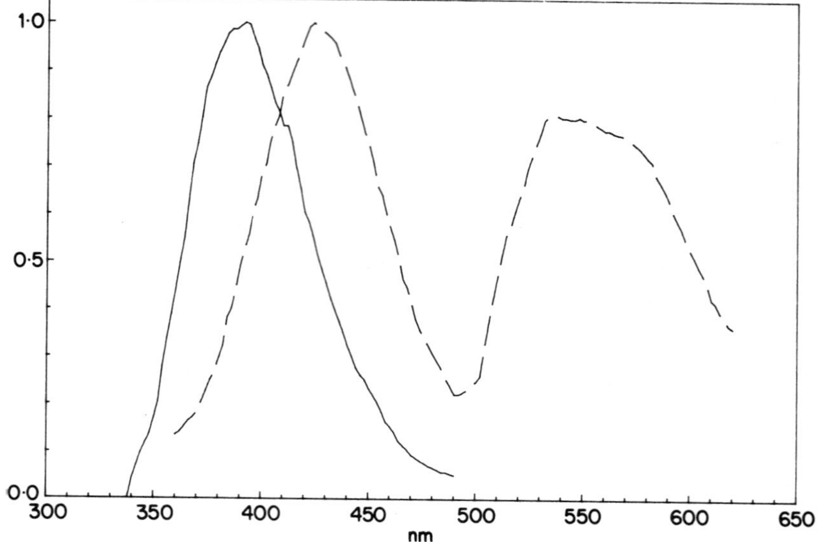

Fig. 2.22 Comparison of prompt fluorescence (solid line) and delayed emission (dashed line) of a film of P2VN with a molecular weight of 505 000. The spectra are corrected for spectral response (Kim and Webber [36], reproduced by permission).

3. Fox *et al.* [35] demonstrated a very strong dependence of the delayed emission spectrum on polymerization or purification conditions and the age and method of storage of the sample.

Kim and Webber [36] studied the delayed emission of P2VN films of various molecular weights. The results were similar to those for P2VN in glasses [17] in that increasing the molecular weight of the polymer increased the ratio of delayed fluorescence to phosphorescence (see Fig. 2.23). This molecular weight dependence was interpreted as implying that the triplet exciton is largely confined to a single polymer coil, such that the excitation statistics favour

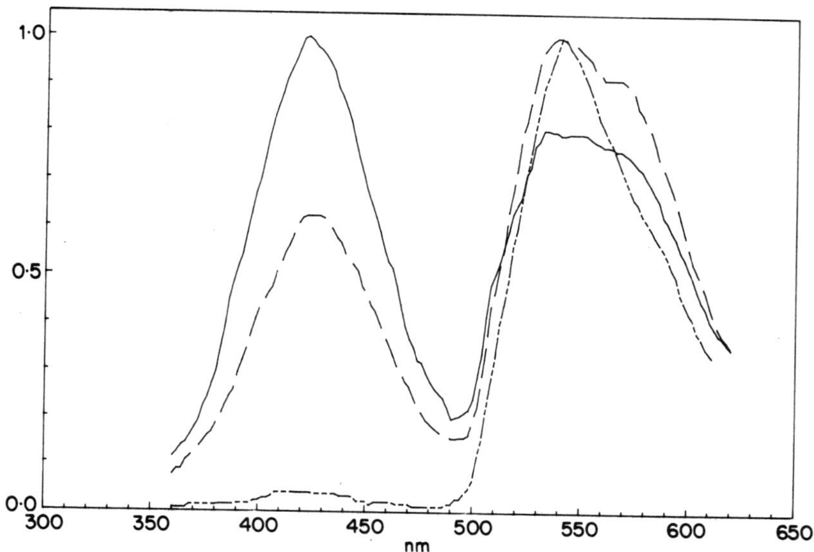

Fig. 2.23 Delayed emission spectra for films of P2VN of different molecular weights: (solid line) 505 000; (dashed line) 106 000; (dashed/dotted line) 49 000 (Kim and Webber [36], reproduced by permission).

T-T annihilation. Other interpretations are possible:

1. The mole fraction of chain ends decreases with increasing molecular weight. Although no evidence for exciton quenching at chain ends currently exists, this possibility cannot be discounted.
2. The morphology of the film will depend on molecular weight and it may be this 'structural' feature that affects T-T annihilation rather than excitation statistics.
3. The effectiveness of polymer purification via multiple precipitations is expected to be somewhat dependent on the polymer molecular weight.

Burkhart *et al.* [37] have studied the temperature dependence of phosphorescence and delayed fluorescence in a sample of P1VN of molecular weight 1.09×10^4. The general appearance of these spectral features similar to films of P2VN. For temperatures above 100K, the phosphorescence shifts to longer wavelengths, whereas the delayed fluorescence is relatively constant in shape (see Fig. 2.24). All delayed emission features are weaker as the temperature is elevated, but interestingly, the ratio of delayed fluorescence to phosphorescence increases. This result implies the presence of a triplet trap that can phosphoresce but which decreases delayed fluorescence by immobilizing triplet excitons. At higher temperatures trap-to-exciton detrapping can occur and the relative importance of triplet-triplet annihilation increases. This is a very plausible mechanism and has been invoked to rationalize the results for poly(*N*-vinylcarbazole) films (see Section 2.3.2(*c*)).

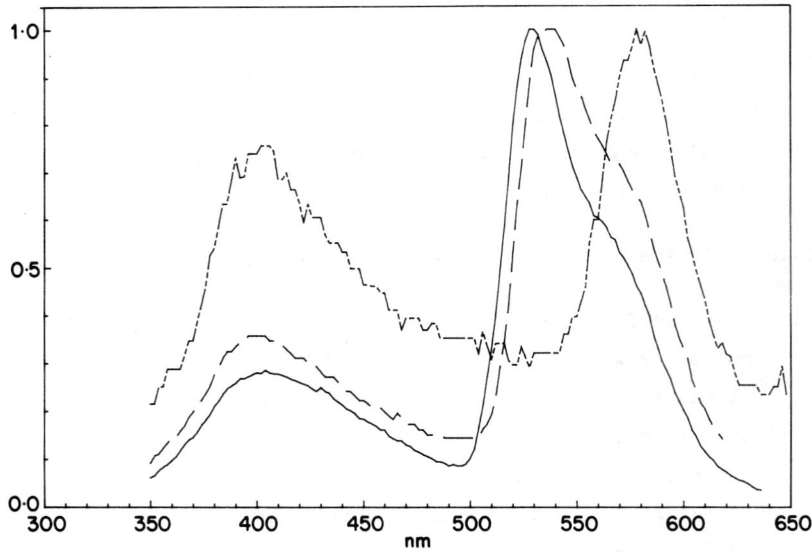

Fig. 2.24 Delayed emission spectra of films of P1VN at 77K (solid line), 100K (dashed line), 180K (dash-dot line). Spectra have been off-set slightly for ease of comparison (Burkhart *et al.* [37], reproduced by permission).

Two observations due to Burkhart *et al.* are somewhat surprising:

1. The phosphorescence and delayed fluorescence decay biexponentially with the same lifetimes, and these lifetimes are not very sensitive to temperature.
2. At temperatures above 230K, the delayed fluorescence is proportional to the first power of the excitation intensity.

The former observation is interpreted as implying that the lifetime of the triplet exciton is much longer than the trapped species. This conclusion seems contrary to the usual expectation, since the exciton can be trapped in addition to the usual unimolecular processes. However, it must be recalled that the phosphorescence is excimer-like and the lifetime of triplet excimers in solid matrices is not known. The second observation of Burkhart *et al.* implies either:

1. E-type delayed fluorescence (thermal excitation of T_1 to S_1).

or 2. T-T annihilation dominates the kinetics.

Because of the large S_1-T_1 energy gap the former is thought to be unlikely. One concludes that triplet exciton mobility may be rather high in these films.

(c) Films of Poly(*N*-vinylcarbazole)

Triplet-state processes in PVCz have been studied primarily by Burkhart and coworkers [38-42], and a very detailed review of experimental results and theoretical interpretation has been presented by Rippen *et al.* [43] and Klopffer [44, 45]. The results may be summarized as follows:

1. At 77K biphotonic delayed fluorescence and monophotonic phosphorescence are observed at low excitation intensities. However at higher excitation intensities, the phosphorescence saturates and the delayed fluorescence becomes proportional to the first power of the excitation (see Fig. 2.25). The prompt fluorescence intensity remains linear in excitation power.

This is an important point in the interpretation of the photophysics of PVCz films.

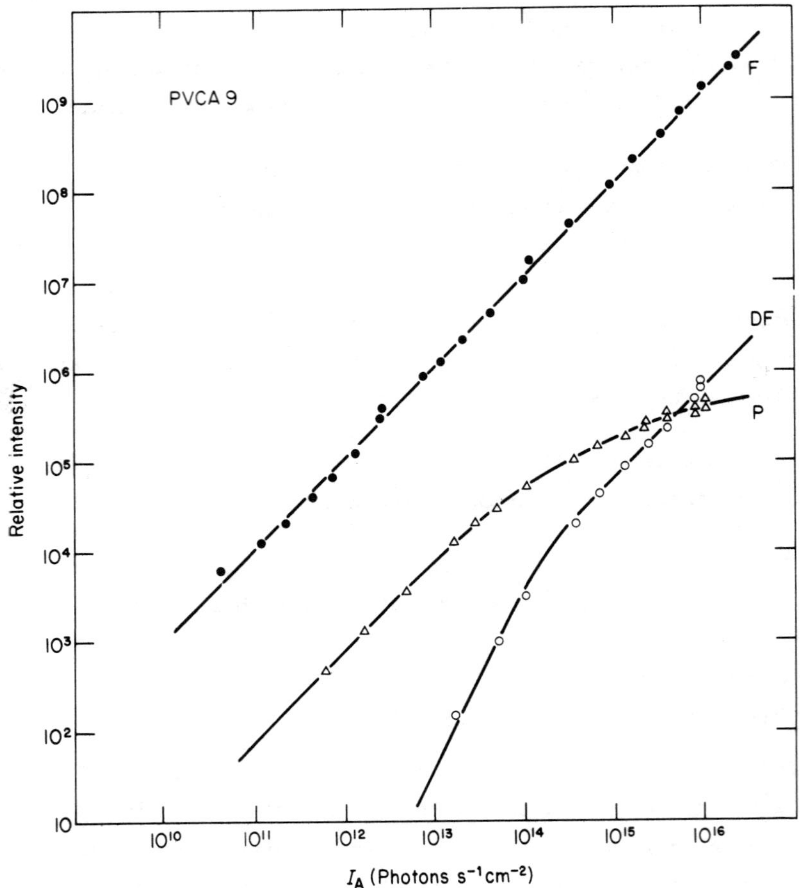

Fig. 2.25 Intensity of prompt fluorescence (F), delayed fluorescence (DF), and phosphorescence (P) of PVCA films at 77K as a function of exciting light intensity (313 nm) (Klopffer [44], reproduced by permission).

2. The delayed fluorescence is broad and red-shifted relative to isolated carbazole, and is similar but not identical† to that of the excimer prompt fluorescence observed from the films. Thus it is implied that the delayed fluorescence does not originate from the usual excimer site.
3. The spectral details of prompt and delayed fluorescence are sensitive to the method of preparing the polymer (i.e. free radical or cationic polymerization), the film casting conditions, and the molecular weight of the polymer. It is suggested (but not demonstrated directly) that the photophysics are strongly affected by the tacticity of the particular polymer sample.
4. The decay of delayed fluorescence and phosphorescence is highly non-exponential, with different rate constants for these two features. In particular it is found that $\tau_{phos} >> \tau_{DF}$ (both lifetimes are 1/e lifetimes).
5. The relative intensity of delayed fluorescence and phosphorescence is highly temperature dependent (see below) whereas the latter monotonically decreases with temperature.

For a complete discussion of the interpretation the reader is referred to the papers of Burkhart and coworkers [38-42] and Klöpffer and coworkers [43-45]. Briefly summarized, a kinetic scheme that is consistent with these observations is:†† (over page)

$$T_m \rightleftarrows T_s \qquad (2.17a)$$

$$T_m \rightleftarrows T_d \qquad (2.17b)$$

$$T_m + T_s \longrightarrow {}^1D_s^* \qquad (2.17c)$$

---

† Just as was the case for films of P2VN (Section 2.3.2(b), the delayed fluorescence is red-shifted relative to the prompt fluorescence see Fig. 2.22).

$$T_m + T_d \longrightarrow {}^1D_d^*. \qquad (2.17d)$$

In Reaction (2.17), T_m represents a mobile triplet (= triplet exciton), T_s and T_d are shallow and deep traps, respectively, and ${}^1D_s^*$, ${}^1D_d^*$ are singlet excimer species formed at T_s, T_d sites, respectively. Neither ${}^1D_d^*$ nor ${}^1D_s^*$ are identical to the 'normal' singlet excimer site. (Simple unimolecular decay of T_m, T_s, T_d is omitted for simplicity.) Hence the delayed fluorescence arises from heterogeneous T-T annihilation of a triplet exciton and a trapped triplet. It is assumed that phosphorescence arises primarily from the trapped triplets, since the delayed fluorescence and phosphorescence decay rates are so different. The trapped triplets are considered to be at excimer-like sites on the basis of the broadened, red-shifted phosphorescence. These triplet excimer sites are not the same as the singlet excimer site because of the shift in delayed fluorescence relative to prompt fluorescence already cited and the absence of singlet-triplet annihilation, i.e:

$$S_m^* + T_d \rightarrow M + M_d \qquad (2.18)$$

(M, M_d are normal and deep trap ground-state sites, respectively). Klöpffer's argument [43-45] that M_d is not the normal excimer site is as follows: suppose singlet excimer fluorescence arose from:

$$S_m^* + M_d \rightarrow M + S_d^*(={}^1D^*). \qquad (2.19)$$

At high excitation intensity the M_d sites are largely converted to T_d, and one expects the intensity of excimer fluorescence to be sublinear. This is contrary to the experimental results. In view of work on

---

†‡ The notation used here is similar to that of Burkhart and Aviles [40], since it is somewhat simpler than that of Klöpffer [44]. The trapping steps (Reactions 2.17a and 2.17b) are pseudo first order, since an exciton must encounter a ground-state trapping site.

dichromorphic model compounds,† it is not surprising that triplet and singlet excimers are different.

The complex temperature dependence of the various processes can be understood in terms of the thermal trapping/detrapping of (reactions 2.17a and 2.17b) [39, 40] (although for the 'deep'trap little thermal detrapping is expected). Burkhart *et al.* noted that this temperature dependence of free radical and cationically prepared polymers is quite different (Fig. 2.26). This may be interpreted in terms of the density of triplet efs and the differences in the average binding energy of the T_s, T_d traps for the different types of polymers. Itaya *et al.* [48] have studied triplet exciton processes in PVCz using extrinsic traps to assess triplet mobility. It was found that exciton diffusion was more rapid for the cationically polymerized species. This is ascribed to the diminished concentration of deep traps. One can conclude that although the scheme presented in Reaction (2.17) may be general, the relative concentrations of efs of each type may be highly dependent on the particular polymer sample and the details of film preparation.

(d) Films of polyvinylketones

Much of the work on this class of polymers has been carried out by David, Geuskens and coworkers. This class of polymers includes poly(vinylbenzophenone) (PVB [28, 49], poly(phenylvinylketone) (PPVK) [50] and poly(methylvinylketone) (PMVK) [51]. The phosphorescence of these films is typical of an $n\pi^*$ triplet state. PVB and PPVK exhibit a clear vibrational progression and in all cases the phosphorescence lifetime is short.††(over page) Although David, Geuskens and coworkers do not compare the film phosphorescence spectra with model compounds, it is obvious that there has not been the gross modification of the phosphorescent state in these films

---

† For a series of constrained naphthalenes, see Schweitzer *et al.* [46]. The effect of the chirality of neighbouring chromophores has been studied by DeSchryver and coworkers. For dicarbazole pentane, see DeSchryver *et al.* [47].

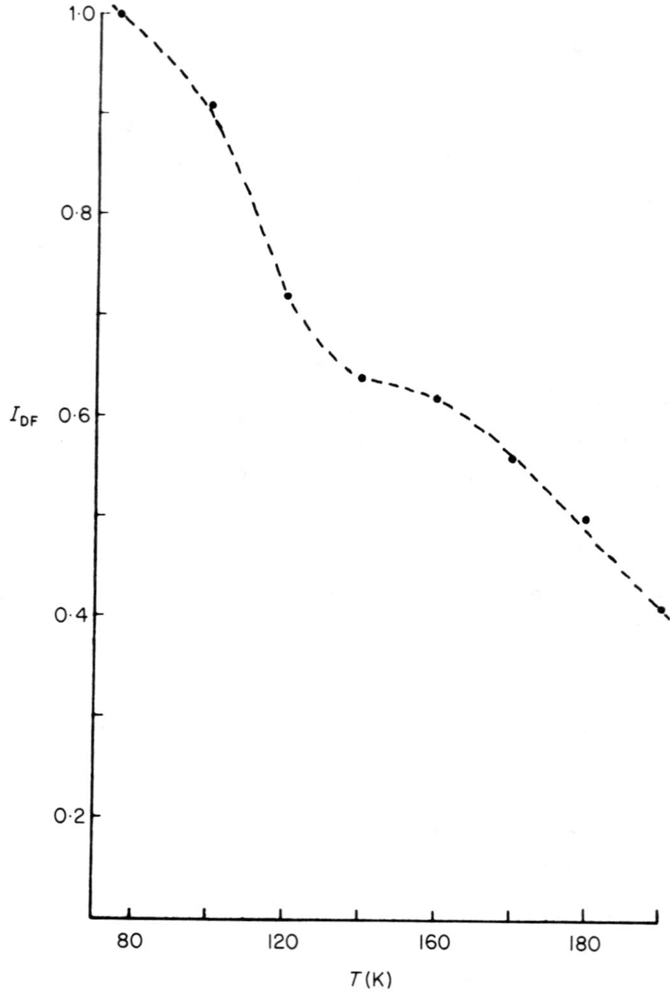

Fig. 2.26(a)

---

†† The short triplet lifetime is implied from some of the studies of glassy matrices, by the interpretation of the results or by analogy to model compounds (e.g. for benzophenone the triplet lifetime is ca. 5 ms). However, this quantity is not determined directly in most cases.

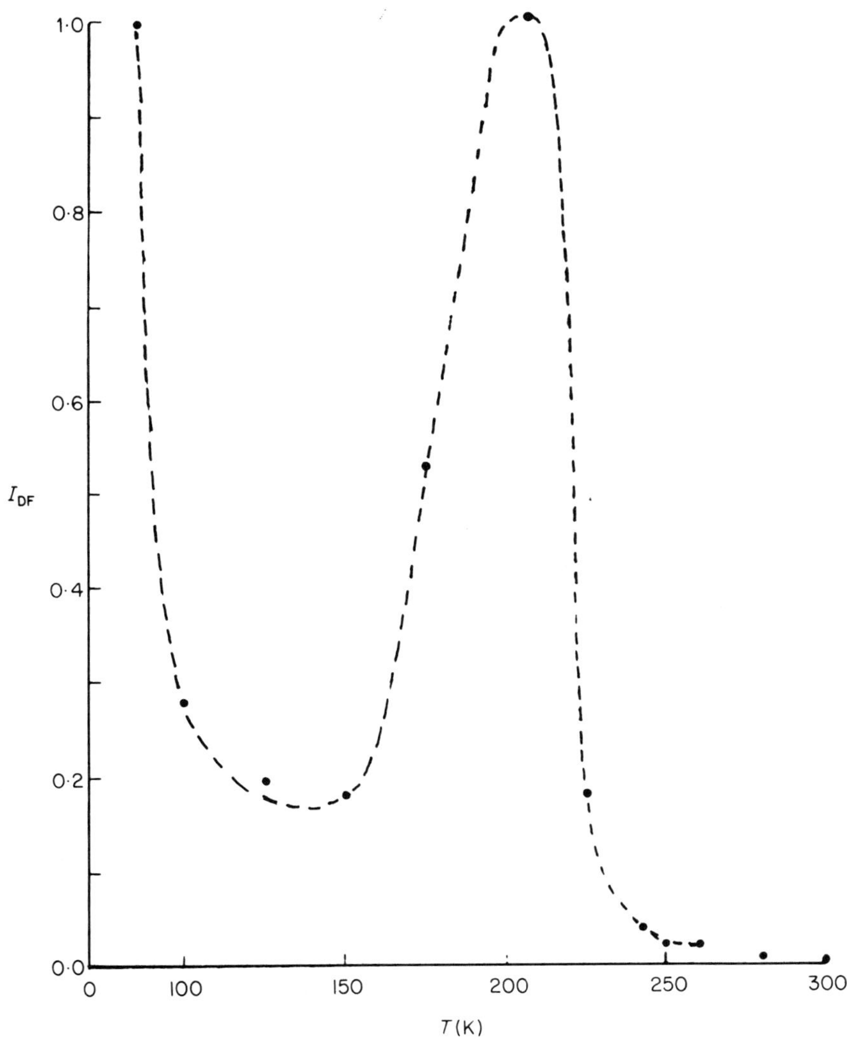

Fig 2.26(b)

Fig. 2.26 Delayed fluorescence intensity of (a) films of PVCz(c) (Burkhart and Aviles [39], reproduced by permission) and (b) PVCz(r) between 77K and 298K (Burkhart and Aviles [40], reproduced by permission).

that is observed in films of polyvinylnaphthalene or PVCz (see Fig. 2.27). The quenching of phosphorescence and the diminution of

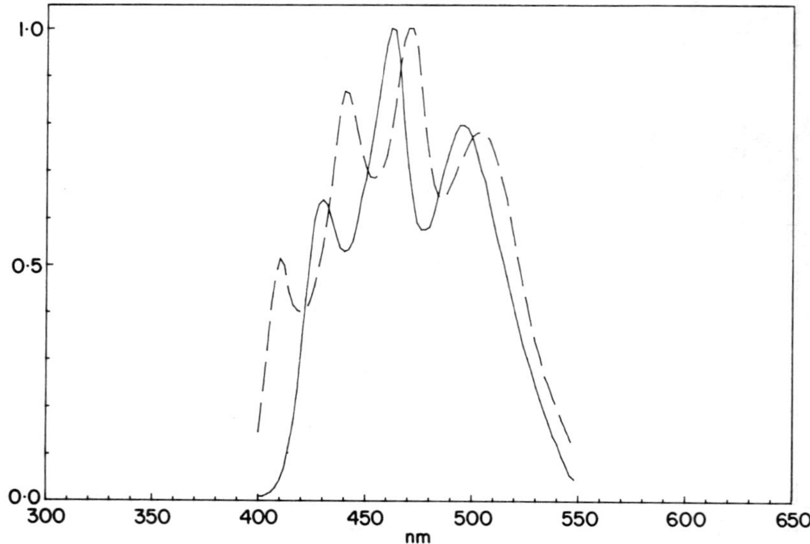

Fig. 2.27 Phosphorescence spectrum of PVB film at 77K (solid line). Excitation wavelength 366 nm (David *et al.* [28], reproduced by permission). Phosphorescence spectrum of polyphenylvinylketone films at 77K (dashed line). Excitation wavelength 366 nm (David [50], reproduced by permission).

the chain scission rate was measured as a function of added naphthalene (a triplet acceptor). The transfer efficiency to the naphthalene can be fitted very well to the theory of Inokuti and Hirayama [4] and a 'critical distance' (R_o) extracted.† (R_o is the separation between energy donor and acceptor for which the transfer rate is equal to all other spontaneous decay processes of the donor.) In Table 2.2 the values for the polymers are compared. We see that $R_o^{PVB} > R_o^{PMVK}$, where the latter is essentially the same as for benzophenone. Thus, it is concluded that triplet energy migration

† The R_o parameter is commonly known as the 'Forster radius', after the original elucidation of singlet-singlet energy transfer [52].

does occur in PVB and PPVK, but not in PMVK (although singlet migration is detected [51]). In the case of PPVK it is also found that addition of naphthalene inhibits chain scission, as one would expect since essentially all photochemistry of these compounds originates from the triplet state.

Table 2.2 Comparison of R_0 values for polyvinylketones and naphthalene

Compound	R_0 (Å)	*Ref.*
Benzophenone†	13	[28, 49]
PVB	36	[28, 49]
PPVK	26	[50]
PMVK	11.3	[51]

† Taken to be a model compound for comparison to all three polymer systems.

In Section 2.3.1(*d*) studies of styrene-vinylbenzophenone copolymers have also been described. The results for glassy matrices and films (or powders) are similar. The phosphorescence decay for copolymers with a low benzophenone content is similar to the model compound (ca. 5 ms), but at higher benzophenone mole fraction a biexponential decay occurs with lifetimes on the order of 2.2 and 20 ms. Likewise, as the benzophenone content increases the quantum efficiency of sensitizing a triplet trap (naphthalene) increases. This has been interpreted in terms of a sharp increase in Λ_T (triplet exciton diffusion constant) for a benzophenone mole fraction greater than ca. 0.70 (see Section 2.3.1(*d*)). This dependency on benzophenone content is qualitatively similar for films and glasses, but the dynamic range of Λ_T is much larger for films. Presumably this is because in films transfer down-chain and cross-chain is enhanced simultaneously.

2.3.3 *Solutions*

(a) General consideration

Observation of phosphorescence and/or delayed fluorescence from fluid-phase triplet-state molecules is much less common than for immobilized molecules because of the ease of quenching. In addition, there is some evidence that for some typical homopolymers, the intersystem crossing from $S_1 \rightsquigarrow T_1$ in the fluid phase is much less efficient than for the appropriate model compounds [53]. Thus there are only two cases of triplet-state luminescence known to us: anthryl- or benzil-tagged polystyrene and delayed fluorescence of triplet-sensitized P2VN and poly(4-vinylbiphenyl). In addition, T-T annihilation in polyvinylketones has been observed. These cases will be discussed in the following subsections.

(b) Anthryl- and benzil-tagged polystyrene

Horie, Mita and coworkers have published a long series of papers in which living polymer techniques have been used to prepare the polymers depicted in Fig. 2.28. The luminescence of these polymers has been used to study some fundamental properties of polymer reaction kinetics. Space does not permit a complete discussion of these experiments, but we note that the kinetics of small molecule-polymer reactions, polymer-polymer reactions and intracoil cyclization have been studied. We will discuss the latter two phenomena.

The benzil molecule is one of the few chromophores for which phosphorescence in fluid solution is easily observed (like glyoxal and biacetyl). The anthracene triplet state is much lower in energy than that of benzil, such that a diffusion-controlled energy-transfer reaction occurs, quenching benzil phosphorescence. Thus, one can study the dependence of the reaction:

$$\text{PSN-}^3\text{B}^* + \text{PSN'-A} \rightarrow \text{PSN-B} + \text{PSN'-}^3\text{A}^* \qquad (2.20)$$

on the degree of polymerization (N and N'); the overall polymer

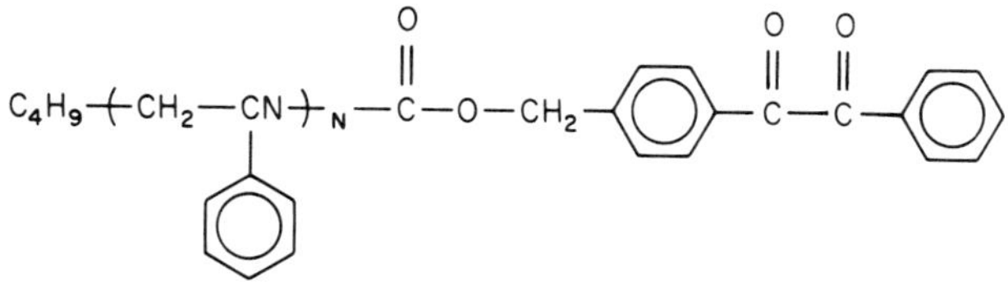

PSN-B

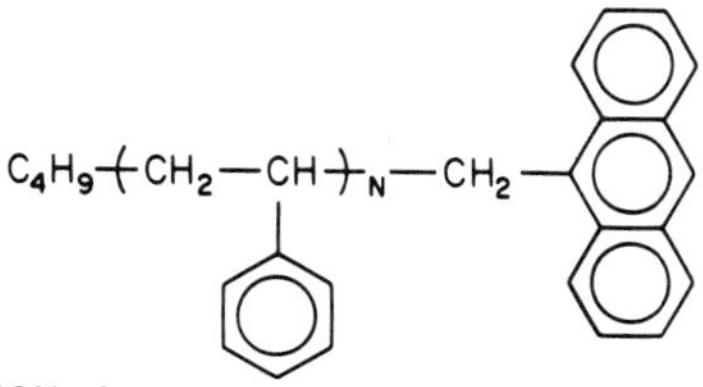

PSN-A

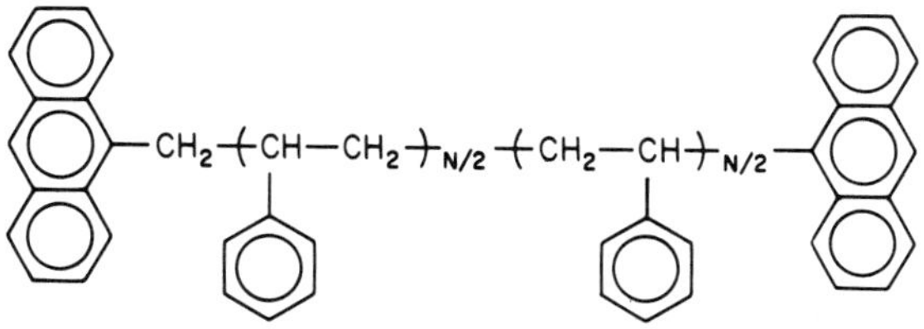

A-PSN-A

Fig. 2.28 Structure of benzyl- and anthryl-tagged polystyrene molecules.

concentration; and solvent and/or temperature. It is found that increasing N and N' decreases rate of the reaction shown in Reaction (2.20) (in the work reported the values of N and N' are essentially equal). However, the precise dependency on the degree of polymerization (P) depends on the solvent and temperature [54] (see Fig. 2.29). Coil-coil interactions can strongly affect the polymer coil diffusion constant. Horie and Mita [55] have discussed different methods for extrapolation of the experimental results to infinite dilution. Conversely these tagged polymers can be used to test theoretical predictions of the concentration effect on polymer-polymer reactions [56]. For example, in the semidilute region

(>0.1 kg dm^{-3}) the reaction rate constant of PSN-$^3B^*$ with PSN-A in the presence of untagged polystyrene decreases with concentration of the latter component like C^{-a}, where a = 1.7-1.9 (in good agreement with the scaling theories of deGennes [57]).

These same workers have investigated the rate of intracoil cyclization using the dianthryl end-tagged polystyrenes A-PSN-A (see Fig. 2.28). There have been two types of approaches, using either anthracene photodimerization (Reaction (2.21a)), which is a singlet state photoreaction, or T-T annihilation (Reaction (2.21b)) (see Fig. 2.30):

$$^1A^* \frown ^1A^* \xrightarrow{k_{SS}^{intra}} A\text{—}A \qquad (2.21a)$$

$$^3A^* \smile ^3A^* \xrightarrow{k_{TT}^{intra}} {}^3A^* \ldots {}^3A^* \rightarrow {}^1A^* \; A \rightarrow DF; \quad \rightarrow {}^3A^* \; A \qquad (2.21b)$$

The photodimerization can be followed by the decrease in anthracene absorption [58]. The T-T annihilation can be followed by either the anthracene delayed fluorescence (DF) [59], or by an analysis of the kinetics of triplet state decay [60] (monitored by T-T absorption). The following dependences of k^{intra} on the degree of polymerization (P) were obtained:

$$k_{SS}^{intra} = P^{-0\cdot 3} \text{ (butanone, 30°C) [58]}$$

$$k_{TT}^{intra}(DF) = P^{-0\cdot 53} \text{ (benzene, 30°C) [59]}$$

$$= P^{-0\cdot 60} \text{ (butanone, 33°C) [59]}$$

$$k_{TT}^{intra}(TT) = P^{-1\cdot 0(\pm 0\cdot 06)} \text{ (benzene) [60]}$$

Since the time scale for the photodimerization is much shorter than the triplet-state processes, it is not surprising that the exponent of P is different for k_{SS}^{intra} and k_{TT}^{intra}. However, Ushiki *et al.* [58] point out that the concentrations of polymer coils in these

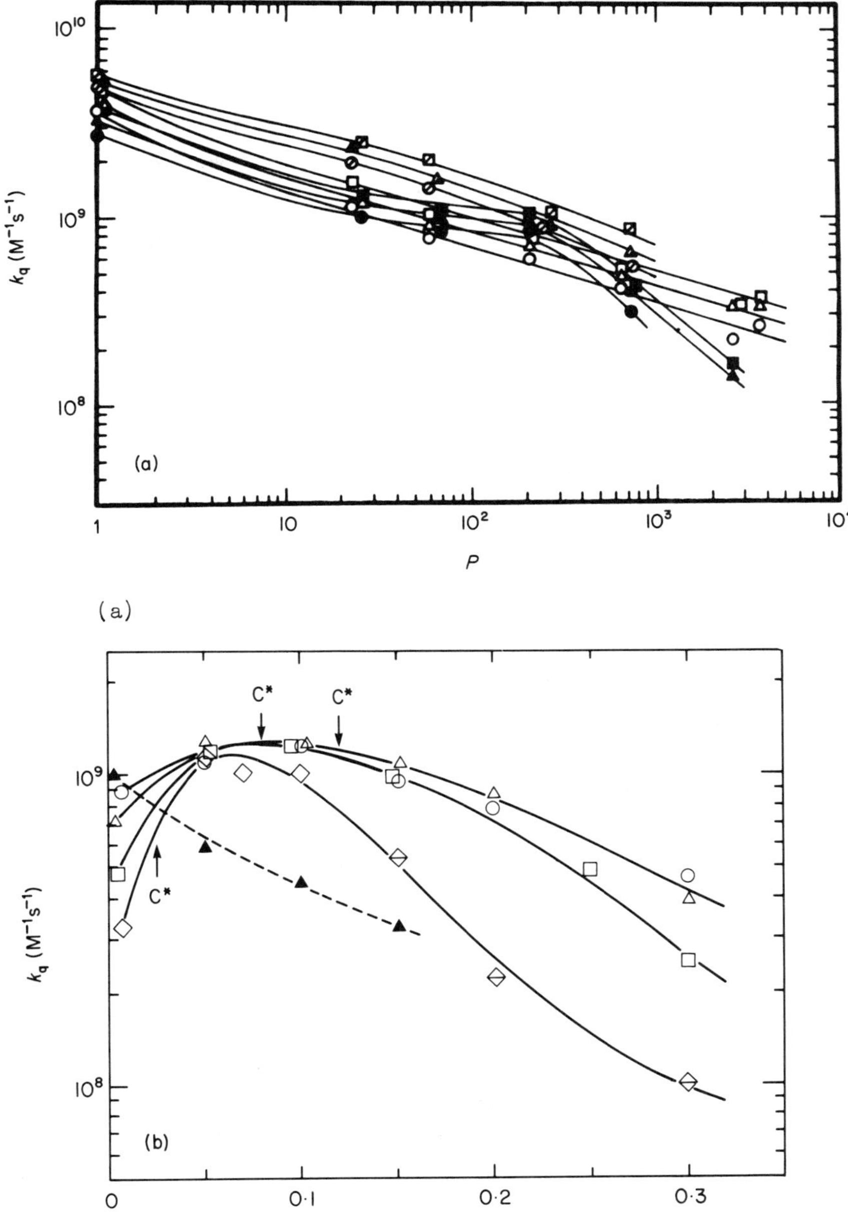

(a)

Fig. 2.29(b)

experiments is not sufficiently low to eliminate all coil-coil interaction. The discrepancy between exponents in the delayed fluorescence method and T-T absorption method has also been ascribed [60] to the effect of coil-coil interaction in the earlier work [61].

In summary, Horie, Mita and coworkers have demonstrated the utility of chromophore-tagged polymers and triplet state spectroscopy in the elucidation of polymer-polymer and intrapolymer kinetics. This work may be considered to be an extension of the use of fluorescence probes attached to polymers. However, there is the important difference that the long lifetime of the triplet state allows the study of processes that involve relatively large molecular motions. It is especially interesting in the case of chromophores like anthracene in which both the singlet state (lifetime on the order of 5 ns) and the triplet state (lifetime on the order of 200 μs) can be used to study the same polymer.

(c) T-T annihilation in polyvinylketones

Schnabel and coworkers have studied the triplet state of poly(phenylvinylketone) (PPVK) and poly(*p*-vinylbenzophenone) (PVBP) in the

---

Fig. 2.29 (a) Dependences of k_q on the degree of polymerization in benzene (open symbols), cyclohexane (closed symbols), and butanone (slashed symbols). Plots for $P = 1$ correspond to the reaction of benzil with 9-methylanthracene. Temperatures: (circles) 20°C; (triangles) 30°C; (squares) 40°C (Horie and Mita [54], reproduced by permission). (b) Dependence of k_q at 30°C on the polymer concentration, C, in benzene (open symbols) and in cyclohexane (closed triangle). The combinations of polymers are PS60B-PS71A-P98 (open circle), PS340B(PS210B)-PS400A-PS410 (open and closed triangles), PS640B-PS800A-PS1000 (square), and PS2900B(PS1200B)-PS3900A(PS3300A)-PS3800 (diamonds). Corresponding overlap concentrations C^* are also shown for the systems (diamond, square, open triangle, from left to right) (Mita *et al.* [56], reproduced by permission).

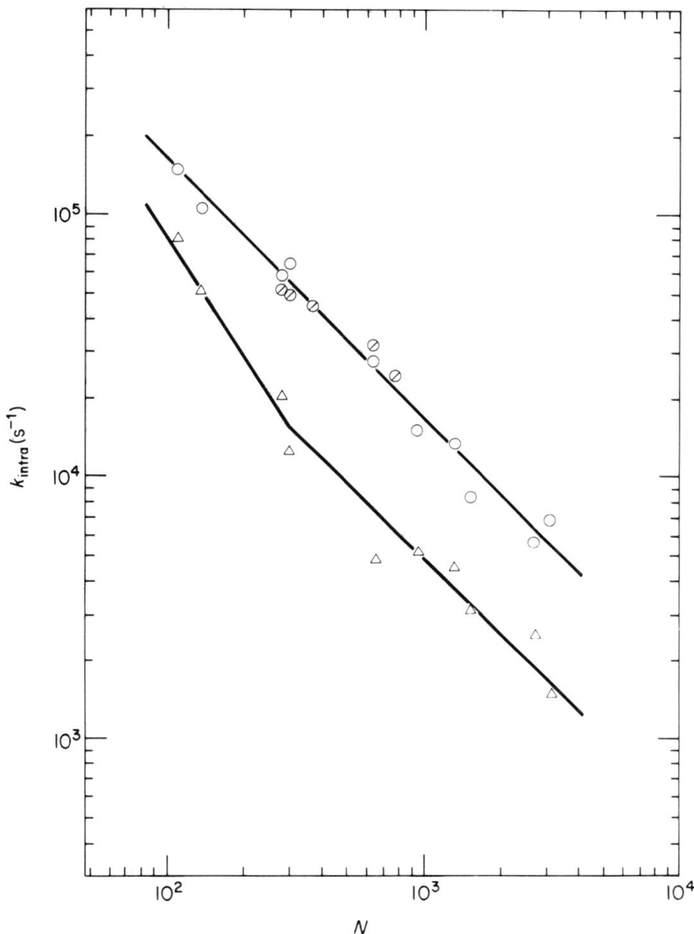

Fig. 2.30 Dependence of k_{intra} for end-to-end collision on the degree of polymerization, N of A-PS-A in benzene (open circle) and in cyclohexane (open triangle). Previous results (slashed circle) obtained for A-PS-A in benzene by delayed fluorescence measurements (N = 280-780, [A-PS-A] = 3×10^{-5} M) are also shown (Horie *et al.* [60], reproduced by permission).

solution phase. In this work a ruby laser excitation source was used (doubled to 347.1 nm, 25 ns half-width) to excite the S_1 state, which undergoes efficient intersystem crossing to T_1. T-T absorption monitored to yield the decay kinetics of the T_1 state.

The results for PPVK are somewhat contradictory. In the earliest work [61, 62] the quenching of the polymer-bound phenylketone triplet state is approximately two and a half times less efficient than a small molecule model. According to Equation (2.9), this implies that Λ_T (the triplet exciton diffusion constant) is approximately zero. However, later work [63, 64] that compared the homopolymer PPVK to vinylphenylketone copolymers with vinylacetate, methylmethacrylate or styrene showed: a steady decrease of $\tau_{\frac{1}{2}}(T_1)$ (half-life of triplet) with the mole fraction of the phenylketone moiety; and, for the homopolymer, $\tau_{\frac{1}{2}}(T_1)$ decreased with excitation intensity. Both observations could be a result of energy migration, which in the former case enhances quenching of the T_1 state and in the latter case leads to the annihilation. On the other hand, quenching of T_1 by 2,5-dimethylhexadiene-2,4 (DMHD) [63] or naphthalene [64] did not change appreciably when the mole fraction of the phenylvinylketone moiety increased. These observations are consistent with $\Lambda_T = 0$. Thus it is reasonable to consider the possibility of intersegmental contacts of phenylketone groups leading to self-quenching or T-T annihilation. This kind of energy migration would be expected to be rather slow and exhibit a rather different dependence on chromophore mole fraction than down-chain energy migration. A recent paper by Das and Scaiano [65] has presented very strong evidence that triplet energy migration **can occur** via cross-coil segmental diffusion. These workers studied the rate of energy transfer from triplet phenyl ketone to pendant naphthalene groups in copolymers of phenyl vinyl ketone, 2-vinylnaphthalene and methylmethacrylate (the majority species). The maximum fraction of phenyl ketone moiety was 0.21, and the naphthalene fraction was <0.039. For this set of copolymers the rate of intracoil energy transfer to the naphthalene was linearly dependent on the mole fraction of naphthalene but independent of the phenyl ketone mole fraction (see Fig. 2.31). Since down-chain excitonic sensitization is expected to increase with the fraction of phenyl ketone, it is concluded that the cross-chain segmental diffusion process is responsible for the naphthalene triplet

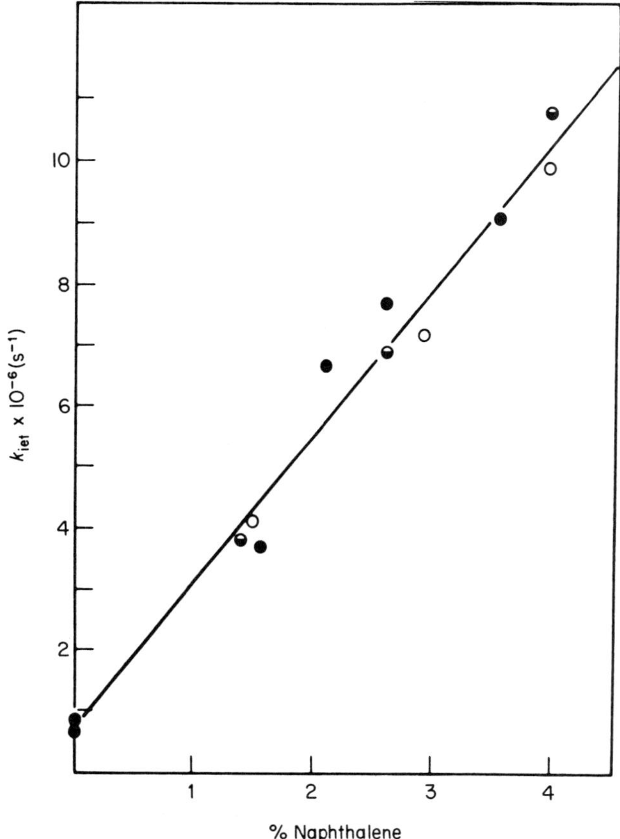

Fig. 2.31 Effect of polymer composition on the values of k_{iet} (intracoil energy transfer) at 22^{0}C in benzene. The points have been identified according to their carbonyl content: (solid circles) over 12%; (half solid circles) between 7 and 9%; (open circles) below 5% (Das and Scaiano, [65], reproduced by permission).

sensitization. Further evidence for this is the enhancement of the intracoil energy transfer as a nonsolvent is added to a good solvent solution of the copolymer (Fig. 2.32), thereby decreasing the size of the polymer coil and presumably enhancing intracoil segmental contacts. Taken together the work of Beck *et al.* [61, 62], Kiwi and

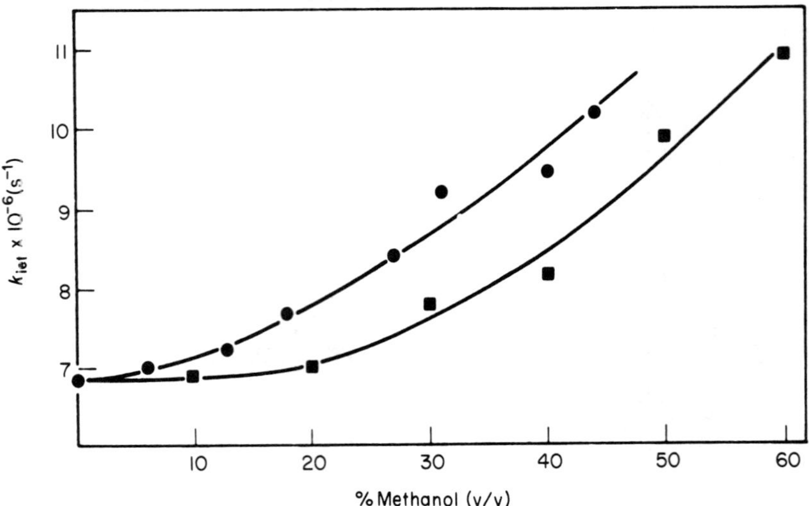

Fig. 2.32 Effect of the addition of a non-solvent (methanol) on the kinetics of intramolecular energy transfer for copolymers of vinylnaphthalene, vinylbenzophenone and methylmethacrylate at room temperature, using as cosolvent acetone (solid circle) and benzene (solid square) (Das and Scaiano, [65], reproduced by permission).

Schnabel [63, 64] and Das and Scaiano [65] demonstrate that segmental diffusion is a mechanism of energy transfer that cannot be excluded from consideration. However the mole fraction of phenyl ketone in the work of Das and Scaiano was below the typical onset of excitonic migration (see Fig. 2.21), such that one cannot unambiguously exclude excitonic triplet migration in PPVK in the solution phase for the homopolymer. It is also possible that some of the apparent contradictions in the PPVK data arise because of 'polymer effects' such as lowering the absolute energy of the T_1 state, thereby invalidating comparisons with model compounds (a similar situation has been shown to exist for polyvinylnaphthalene and polyvinylbiphenyl - see Section 2.3.3(d)).

More recently Schnabel has reported similar studies for PVBP [66].

As in the above it was found that $\tau_{\frac{1}{2}}(T_1)$ decreased with excitation intensity in a fashion consistent with intracoil T-T annihilation. For co-polymers $\tau_{\frac{1}{2}}(T_1)$ of the benzophenone moiety decreased steadily as the benzophenone mole fraction increased. Intracoil energy migration is indicated, but one cannot distinguish the down-chain and cross-chain process. It would be very interesting to have available a solution phase quenching study of PVBP and copolymers of vinylbenzophenone such that Equation (2.9) could be used to estimate Λ_T.

(d) T-T annihilation in polyvinylnaphthalene and polyvinylbiphenyl

Sensitization and quenching studies of P2VN in benzene have demonstrated the following general features [53, 67, 68]:

1. The rate constant of a small molecule sensitizer of the naphthalene triplet (e.g. benzophenone triplet) or a small molecule quencher of the naphthalene triplet (e.g. piperylene) reacting with a polymer is dependent on the molecular weight of the polymer according to the relation:

$$k = P^{(a-2)/3} \tag{2.22}$$

   where a is the Mark-Houwink parameter (i.e. $[\eta] \propto P^a$).
2. The naphthalene triplet state is ca. 1.5 kcal mol^{-1} lower in energy than the small molecule model compound. This obviously can change the value of the product (PR) in Equation (2.7) and this effect has been shown to be important for the case of P2VN-piperylene [69].
3. There are intracoil 'impurity' or 'abnormal groups' that have a lower triplet acceptor state and can quench triplet sensitizers that are quite endothermic with respect to naphthalene [69].
4. The quantum efficiency of intersystem crossing is much lower for P2VN than a model compound (e.g. 2-ethylnaphthalene) and decreases with molecular weight [53]. Hence the T_1 state of P2VN in the solution phase can be prepared quite efficiently by sensitization but not by direct excitation.

Of greater relevance to the previous section is the observation of an excimer delayed fluorescence from P2VN sensitized by benzophenone (or other exothermic sensitizers) [69]. Kinetic modelling has supported the following mechanism:

$$
\begin{aligned}
&{}^3S^* + \mathrm{coil}(0) \rightarrow \mathrm{coil}(1) \\
&{}^3S^* + \mathrm{coil}(1) \rightarrow \mathrm{coil}(2), \text{ etc.} \\
&\mathrm{coil}(n') \qquad \rightarrow \mathrm{DF} \qquad (n'>1).
\end{aligned} \tag{2.23}
$$

In Reaction (2.23), $^3S^*$ is the small molecule triplet sensitizer and coil(n) is a P2VN coil excited n times. In addition to normal unimolecular deactivation of the polymer-bound triplets (not explicitly written in Reaction 2.23), an intracoil T-T annihilation step is possible, leading to delayed fluorescence (DF). The observed DF is essentially identical to the normal excimer fluorescence of P2VN, but experimental considerations (i.e. interference from the laser, absorption by sensitizer, etc.) prevent the observation of any monomer component of the fluorescence. It is tempting to assume that intracoil, liquid phase T-T annihilation would tend to favour the excimer since a bimolecular 'collision' is required; unfortunately the experimental data do not permit a clear conclusion on this point.

In the scheme in Reaction (2.23), heterogeneous annihilation was not included:

$$\mathrm{coil}(n) + {}^3S^* \rightarrow \mathrm{DF} \tag{2.24}$$

While the process in Reaction (2.24) cannot be excluded on energetic grounds, on the basis of kinetic modelling and other physical considerations this process was shown to make little or no contribution to the overall kinetics.

The observation of intracoil T-T annihilation in P2VN is consistent with the observations of Hayashi *et al.* [70]. These workers prepared copolymers of phenyl vinyl ketone and 2-vinylnaphthalene containing a maximum of ca. 0.20 mole fraction of naphthalene chromophores. In all cases the quenching of the ketone triplet by naphthalene was very efficient. For the highest mole fraction of naphthalene clear evidence for naphthalenic T-T annihilation was presented, based on the decay of the naphthalene T-T absorption at 430 nm. No delayed fluorescence was reported however. This is expected because of the efficient quenching of $^1N^*$ by the ketone.

Although different in detail, analogous observations have been made for poly(4-vinylbiphenyl) (P4VBP), including the molecular weight dependence of sensitization and the presence of low energy acceptor states, due almost certainly to intrachain 'abnormal groups' [71]. Thus in addition to the observations of Schnabel and coworkers on polyvinylketones [61-64], there is an additional group of polymers that evidently exhibit facile triplet migration in the liquid phase. These observations obviously extend the possible utilization of polymers as photon antenna to triplet state processes.

ACKNOWLEDGEMENTS

I have greatly benefitted from discussions with my coworkers who have obtained the results from our laboratory presented herein. In particular the experiments of J.F. Pratte, N. Kim and J.S. Hargreaves (aided by G. Sowash and E. Gaffney) have stimulated my consideration of the very interesting and rapidly expanding field of polymer photophysics/photochemistry. I have also enjoyed a lively correspondence with the following: R.D. Burkhart (Reno), C. David (Brussels), C. Frank (Palo Alto), J.E. Guillet (Toronto), A. Gupta (Pasadena), K. Horie (Tokyo), W. Klöpffer (Frankfurt). Finally I would like to acknowledge the financial support of my research by the National Science Foundation (Polymer Program, DMR-8013709) and the Robert A. Welch Foundation (F-356). Finally special acknowledgement must be made of the tireless efforts of Ms Rayna Kolb in the preparation of the manuscript.

APPENDIX

Illustrative kinetic scheme for static and dynamic trapping

We consider a mobile triplet excited state, T_m, and a trapping state, T_t. The activation energy required for trapping and detrapping is $\Delta E_t{}^*$ and $\Delta E_{\bar{t}}{}^*$, respectively. In the absence of annihilative processes the kinetic equations for T_m and T_t are:

$$(d/dt)T_m = -(k_m + K_t)T_m + I_{ex} \quad (A.1a)$$

$$(d/dt)T_t = -(k_t + K_{\bar{t}})T_t + K_t T_m. \quad (A.1b)$$

In Equation (A.1a), K_t is a pseudo first order trapping rate and I_{ex} is the rate of production of T_m by some excitation source. It is assumed that the T_t state is produced only by trapping the T_m state. The steady state solutions are:

$$T_m{}^{SS} = I_{ex}/(k_m + K_t) \quad (A.2a)$$

$$T_t{}^{SS} = (K_t T_m{}^{SS})/(k_t + K_{\bar{t}}) = I_{ex}/(k_t + K_{\bar{t}}) \, [K_t/(k_m + K_t)] \quad (A.2b)$$

The trapping and detrapping rates are given by an Arrhenius expression:

$$K_t = A_t \exp(-\Delta E_t{}^*/kT) \quad (A.3a)$$

$$K_{\bar{t}} = A_{\bar{t}} \exp(-\Delta E_{\bar{t}}{}^*/kT). \quad (A.3b)$$

which can be substituted into Equations (A.2a and A.2b). If we consider the equation for $T_t{}^{SS}$ in the low-temperature limit, defined by the condition that $K_{\bar{t}} \ll k_t$:

$$T_t{}^{SS} = (I_{ex}/k_t)\phi_t(T) \quad (A.4)$$

where $\phi_t(T)$ is the quantum yield of trapping, i.e:

$$\phi_t(T) = K_t/(k_m + K_t) = [(k_m/A_t)\exp(\Delta E_t^*/kT) + 1]^{-1} \qquad (A.5)$$

Thus the steady-state concentration of the T_t state increases with increasing temperature. However, in the high-temperature limit $K_t^- >> k_t$, so:

$$T_t^{SS} = (I_{ex}/K_t^-)\phi_t(T) = (I_{ex}/A_t^-)\phi_t(T)\exp(\Delta E_t^{-*}/kT). \qquad (A.6)$$

Since $\Delta E_t^* > \Delta E_t^{-*}$, T_t^{SS} now decreases with increasing temperature. Thus the steady-state concentration exhibits an 'inverted U' behaviour when plotted against T or (T^{-1}) (see Fig. 2.3). If $\Delta E_t^* = 0$ (ideal static quenching then $\phi_t(T)$ is independent of temperature, and T_t is either independent of T, or decreases, as described by Equation (A.6).

REFERENCES

1. KASHA, M. (1950). *Disc. Faraday Soc.* IX, 14.
2. McGLYNN, S.P., AZUMI, T. and KINOSHITA, M. (1969). *Molecular Spectroscopy of the Triplet State,* Prentice Hall Inc., Englewood Cliffs, New Jersey.
3. BIRKS, J.B. (1970). *Photophysics of Aromatic Molecules,* Wiley-Interscience, New York.
4. INOKUTI, M. and HIRAYAMA, F. (1965). *J. Chem. Phys.* 43, 1978.
5. FRANK, C.W. and HARRAH, L.A. (1974). *J. Chem. Phys.* 61, 1526.
6. COZZENS, R.F. and FOX, R.B. (1969). *J. Chem. Phys.* 50, 1532.
7. COZZENS, R.F. and FOX, R.B. (1969). *Macromol.* 2, 181.
8. FOX, R.B., PRICE, T.R. and COZZENS, R.F. (1971). *J. Chem. Phys.* 54, 79.
9. PASCH, N.F. and WEBBER, S.E. (1976). *Chem. Phys.* 16, 36
10. PASCH, N.F., McKENZIE, R.D. and WEBBER, S.E. (1978). *Macromol.* 11, 733.
11. WEBBER, S.E. (1978). *J. Photochem.* 9, 269.
12. WEBBER, S.E. and AVOS-AVOTINS, P.E. (1979). *Macromol.* 12, 708.
13. WEBBER, S.E. and SWENBERG, C.E. (1980). *Chem. Phys.* 49, 231.

14. DAVID, C., BAEYENS-VOLANT, D. and PIENS, M. (1980). *Eur. Polym. J.* 16, 413.

15. DAVID, C., LEMPEREUR, M. and GEUSKENS, G. (1972). *Eur. Polym. J.* 8, 417.

16. SOMERSALL, A.C. and GUILLET, J.E. (1973). *Macromol.* 6, 218.

17. PASCH, N.F. and WEBBER, S.E. (1978). *Macromol.* 11, 727.

18. NAKAHIRA, T., ISHIZUKA, S., IWABUCHI, S. and KOJIMA, K. (1980). *makromol. Chem. Rapid Commun.* 1, 759.

19. LI, S-K.L. and GUILLET, J.E. (1977). *Macromol.* 10, 840.

20. AIKAWA, M., TAKEMURA, T., BABA, H., IRIE, M. and HAYASHI, K. (1978). *Bull. Chem. Soc. Japan,* 51, 3643.

21. KLÖPFFER, W. and FISCHER, D. (1973). *J. Polym. Sci., Polym. Symp.* 40, 43.

22. KLÖPFFER, W., FISCHER, D. and NAUNDORF, G. (1977). *Macromol.* 10, 450.

23. YOKOYAMA, M., TAMAMURA, T., NAKANO, T. and MIKAWA, H. (1976). *J. Chem. Phys.* 65, 272.

24. YOKOYAMA, M., HANABATA, M., TAMAMURA, T. NAKANO, T. and MIKAWA, H. (1977). *J. Chem. Phys.* 67, 1742.

25. WEBBER, S.E. and AVOTS-AVOTINS, P.E. (1980). *J. Chem. Phys.* 72, 3773.

26. DAVID, C., BAEYENS-VOLANT, D., MACEDO DE ABREU, P. and GEUSKENS, G. (1977). *Eur. Polym. J.* 13, 841.

27. DAVID, C., BAEYENS-VOLANT, D. and GEUSKENS, G. (1976). *Eur. Polym. J.* 12, 71.

28. DAVID, C., DEMARTEAU, W. and GEUSKENS, G. (1970). *Eur. Polym. J.* 6, 537.

29. VOLTZ, R. (1968). *Radiat. Res. Rev.* 1, 301.

30. VOLTZ, R. and HEISEL, F. (1970). *J. Chim. Phys.* 67, 169.

31. KOPELMAN, R. (1976). in *Topics in Applied Physics,* vol. 15 (ed. Fong, F.K.), Springer, Berlin.

32. PARSON, R.P. and KOPELMAN, R. (1982). *Chem. Phys. Lett.* 87, 528.

33. SOUTAR, I. (1981). *Ann. N.Y. Acad. Sci.* 366, 24.

34. FOX, R.B., PRICE, T.R., COZZENS, R.F. and ECHOLS, W.H. (1974). *Macromol.* 7, 937.

35. FOX, R.B., PRICE, T.R., COZZENS, R.F. and McDONALD, J.R. (1972). *J. Chem. Phys.* 57, 2284.
36. KIM, N. and WEBBER, S.E. (1980). *Macromol.* 13, 1233.
37. BURKHART, R.D., AVILES, R.G. and MAGRINI, K. (1981). *Macromol.* 14, 91.
38. TURRO, N.J., CHOW, M-F. and BURKHART, R.D. (1981). *Chem. Phys. Lett.* 80, 146.
39. BURKHART, R.D. and AVILES, R.G. (1979). *Macromol.* 12, 1078.
40. BURKHART, R.D. and AVILES, R.G. (1979). *Macromol.* 12, 1073.
41. BURKHART, R.D. (1976). *Macromol.* 9, 234.
42. BURKHART, R.D. and AVILES, R.G. (1979). *J. Phys. Chem.* 83, 1897.
43. RIPPEN, G., KAUFMANN, G. and KLÖPFFER, W. (1980). *Chem. Phys.* 52, 165.
44. KLÖPFFER, W. (1981). *Chem. Phys.* 57, 75(part II, Kinetic Model).
45. KLÖPFFER, W. (1981). *Ann. N.Y. Acad. Sci.* 366, 373.
46. SCHWEITZER, D., COLPA, J.P., BEHUKE, J., HANSSEN, K.H., HAENEL, M. and STAAB, H.A. (1975). *Chem. Phys.* 11, 373.
47. DESCHRYVER, F.C., VANDERDRIESSCHE, J., TOPPET, S., DEMEYER, K. and BOENS, K. (1982). *Macromol.* 15, 406.
48. ITAYA, A., OKAMOTO, K. and KUSABAYASHI, S. (1976). *Bull. Chem. Soc. Japan* 49, 2037.
49. DAVID, C., DEMARTEAU, W. and GEUSKENS, G. (1969). *Polym.* 10, 21.
50. DAVID, C., DEMARTEAU, W. and GEUSKENS, G. (1970). *Eur. Polym. J.* 6, 1405.
51. DAVID, C., PUTMAN, N., LEMPEREUR, M. and GEUSKENS, G. (1972). *Eur. Polym. J.* 8, 409.
52. FÖRSTER, T. (1946). *Naturwiss.* 33, 166.; (1959) *Disc. Faraday Soc.* 27, 7.
53. BENSASSON, R.V., LAND, E.J., RONFARD-HARET, M. and WEBBER, S.E. (1979). *Chem. Phys. Lett.* 68, 438.
54. HORIE, K. and MITA, I. (1978). *Macromol.* 11, 1175.
55. HORIE, K. and MITA, I. (1980). Institute of Space and Aeronautical Science, University of Tokyo, *Report No.* 580, 87.
56. MITA, I., HORIE, K. and TAKEDA, M. (1981). *Macromol.* 14, 1428.
57. deGENNES, P.G. (1971). *J. Chem. Phys.* 55, 572; (1976),

Macromol. 9, 587 and 594; (1979), *Scaling Concepts in Polymer Physics*, Cornell University Press, Ithaca, New York.

58. USHIKI, H., HORIE, K., OKAMOTO, A. and MITA, I. (1981). *Polym. Photochem.* 1, 303.
59. USHIKI, H., HORIE, K., OKAMOTO, A. and MITA, I. (1981). *Polym. J.* 13, 191.
60. HORIE, K., SCHNABEL, W., MITA, I. and USHIKI, H. (1981). *Macromol.* 14, 1422.
61. BECK, G., KIWI, J., LINDENAU, D. and SCHNABEL, W. (1974). *Eur. Polym. J.* 10, 1069.
62. BECK, G., DOROWOLSKI, G., KIWI, J. and SCHNABEL, W. (1975). *Macromol.* 8, 9.
63. KIWI, J. and SCHNABEL, W. (1975). *Macromol.* 8, 430.
64. KIWI, J. and SCHNABEL, W. (1976). *Macromol.* 9, 468.
65. DAS, P.K. and SCAIANO, J.C. (1981). *Macromol.* 14, 693.
66. SCHNABEL, W. (1979). *Macromol. Chem.* 180, 1487.
67. PRATTE, J.F. and WEBBER, S.E. (1982). *Macromol.* 15, 417.
68. PRATTE, J.F., NOYES, W.A. Jr. and WEBBER, S.E. (1981). *Polym. Photochem.* 1, 3.
69. PRATTE, J.F. and WEBBER, S.E. (1983). *J. Phys. Chem.*, 87, 449.
70. HAYASHI, K., IRIE, M., KIWI, J. and SCHNABEL, W. (1977). *Polym. J.* 9, 41.
71. PRATTE, J.F. and WEBBER, S.E. (1983). *Macromol.* 16, 1193.

CHAPTER THREE

Fluorescence in the solid state

3.1 GENERAL PRINCIPLES

In the field of the fluorescence of polymer solutions, discussed in the previous chapter, a great deal of work has been concerned with determining the value of the individual rate constants for radiative and non-radiative processes, energy transfer and excimer formation and dissociation. Time-resolved methods have been widely used. The existence of different monomer and excimer sites as well as the possibility of excimer dissociation have been investigated and discussed. Energy migration is controversial and seems to be a function of the nature of the chromophores and the distance between them [1]. It is nevertheless generally admitted that it occurs in the usual polymers, such as polystyrene (PS), polyvinyltoluene (PVT), polyvinylnaphthalene (PVN) and polyvinylcarbazole (PVCz). For polymer films, most recent research work has been concerned with the general mechanism of energy migration and transfer using steady-state methods. The purpose of published papers was to answer the following questions:

1. Does energy migration occur and what is the mechanism?
2. What is the role of excimers in the migration process?
3. What is the mechanism of energy transfer: radiative, collisional or non-radiative (exchange or dipole-dipole interaction)?
4. Does energy transfer occur from isolated chromophores, from excimer sites or from both?

This chapter is concerned with the different pathways and mechanisms of energy migration and transfer (Section 3.1). The results given in the literature for singlet transfer in films of vinylaromatic polymers are then reported (Section 3.2 - 3.9). As throughout, the kinetic schemes and symbols in this chapter are those used by Birks unless otherwise specified [2].

Firstly, mechanisms of energy transfer will be discussed more fully than in Chapter 1. The mechanisms referred to in this chapter are:

Mechanism 3.1	Radiative transfer/non-radiative exchange and long-range dipole-dipole interaction.
Mechanism 3.2	Collisional transfer following singlet energy migration.
Mechanism 3.3	Collisional transfer and excimer formation following singlet energy migration.
Mechanism 3.4	Collisional transfer following excimer migration.
Mechanism 3.5	Long-range dipole-dipole transfer from the excimer.
Mechanism 3.6	Long-range dipole-dipole transfer following efficient energy migration.
Mechanism 3.7	Long-range dipole-dipole transfer following intermediate migration.

3.1.1 *Radiative transfer, dipole-dipole and exchange interactions*

The most simple case occurs when neither migration nor excimer formation occurs because the chromophores are too distant from each other to allow these processes. This could occur, for instance, in copolymers with a low content of the emitting chromophore. Energy transfer to an acceptor can nevertheless occur either by radiative, exchange or dipole-dipole interaction. The case of radiative transfer is easy to detect. It is due to the absorption of a fraction a_{YM} of the $^1M^*$ flourescence by the acceptor. Its probability is given approximately by:

$$a_{YM} = \frac{1}{q_{FM}} \int_0^\infty F_M(\bar{\nu}) \; \{1 - 10^{-\varepsilon_Y(\bar{\nu})[^1M]x}\} d\bar{\nu} \qquad (3.1)$$

where $[^1Y]$ is the molar concentration, x is the film thickness and the integral describes the overlap of the fluorescence spectrum $F_M(\bar{\nu})$ of

$^1M^*$ and the absorption spectrum $\varepsilon_Y(\bar{\nu})$ of ^1Y. As a consequence of the absorption of the fluorescence of $^1M^*$ by ^1Y, the fluorescence spectrum of $^1M^*$ is distorted and this can easily be observed by comparing the fluorescence of $^1M^*$ in the absence and in the presence of ^1Y. Quantitative treatment of radiative energy transfer is given in Equation (3.1). This effect is generally considered as a parasitic effect. It can be minimized by measuring the fluorescence of $^1M^*$ by front surface viewing.

In polymer films with a low content of chromophores observed in such a way that radiative energy transfer is negligible, collisional energy transfer is of low probability and only non-radiative exchange and long-range dipole-dipole interactions have to be considered (Mechanism 3.1). The energy transfer probability k'_{YM} depends on the overlap of $F_M(\bar{\nu})$ and $\varepsilon_Y(\bar{\nu})$:

$$k'_{YM} \simeq \int_0^\infty F_M(\bar{\nu})\ \varepsilon_Y(\bar{\nu})\ d\bar{\nu} \tag{3.2}$$

The electron-exchange interaction (Mechanisms 3.2 and 3.3) requires overlap of the electronic wave functions of $^1M^*$ and ^1Y and is therefore short range (< 15 Å). Its magnitude is proportional to the integral but is not related to the optical properties of $^1M^*$ and ^1Y. It is generally dominant on close approach of $^1M^*$ and ^1Y and leads to a relationship of the form:

$$k'_{YM} \sim (2\pi/\hbar)\ Z^2 \int_0^\infty F_M(\bar{\nu})\ \varepsilon_Y(\bar{\nu})\ d\bar{\nu} \tag{3.3}$$

where $Z^2 = K^2 \exp(-2r/L)$, K is a constant with the dimension of energy, L a constant called the 'effective average Bohr radius' and r is the donor-acceptor distance.

Dipole-dipole energy transfer (Mechanism 3.2 and 3.4) on the contrary depends on the optical transition moments according to:

$$k'_{YM} = \frac{(9000 \ln 10)\ K^2}{128\pi^5\ n^4\ N\ r^6\ \tau_M} \int_0^\infty F_M(\bar{\nu})\ \varepsilon_Y(\bar{\nu})\ \frac{d\bar{\nu}}{\bar{\nu}^4} \tag{3.4}$$

where n is the solvent refractive index, K is an orientation coefficient, N the Avogadro Number, z_M the fluorescence lifetime of $^1M^*$ in the absence of ^1Y, and

$$\phi_{FM} = \int_0^{\infty} F_M(\bar{\nu})\, d\bar{\nu} \tag{3.5}$$

Equation (3.4) can be written:

$$k'_{YM} = \frac{1}{\tau_M}\left(\frac{R_o}{r}\right)^6 \tag{3.6}$$

where R_o is the critical transfer distance at which energy transfer by dipole-dipole interaction and $^1M^*$ deactivation by the other processes have equal probabilities. R_o can be calculated knowing the spectroscopic properties of $^1M^*$ and ^1Y using the relation:

$$R_o^{\,6} = \frac{(9000 \ln 10)\, K^2}{128\pi^5\, n^4\, N} \int_0^{\infty} F_M(\bar{\nu})\, \varepsilon_Y(\bar{\nu})\, \frac{d\bar{\nu}}{\bar{\nu}^4} \tag{3.7}$$

R_o can also be determined from the experimental results by plotting the efficiency of transfer $f_{YM} = (\phi_{FM}^{\,o} - \phi_{FM})/\phi_{FM}^{\,o}$ where $\phi_{FM}^{\,o}$ and ϕ_{FM} are the fluorescence quantum yield of M in the absence and in the presence of Y as a function of log Y. Coincidence between the theoretical curve given by:

$$f_{YM} = \pi^{1/2}\, \gamma \exp(\gamma^2)\, (1 - \text{erf}\, \gamma) \tag{3.8}$$

where $\gamma = C/C_o$, and the experimental results indicate that dipole-dipole energy transfer could be operative. R_o is related to the critical concentration C_o by:

$$R_o = \frac{3000}{2\pi^{3/2} N C_o^{\,3}} \tag{3.9}$$

The critical concentration C_o can be shown as the concentration corresponding to $f_{YM} = 0.72$. If the calculated R_o and the value of R_o obtained from the experimental results using C_o and Equation (3.9) are equal, energy transfer can be assigned to Forster long-range interactions. The theoretical curve corresponding to exchange interaction transfer is tabulated in Dexter [3]. It differs from the preceding one at high efficiency of transfer. An experimental value of R_o can be obtained from the results using Equation (3.9) and value of the critical concentration C_o that can be obtained from the results for $f_{YM} = 0.65$.

If energy migration is negligible, the experimental R_o value will be independent of the concentration of the fluorescent donor chromophore in the copolymer. By contrast it can be understood intuitively that the importance of migration, and by consequence the efficiency of transfer, will increase with increasing concentration of donor chromophores.

Energy transfer in this first simple case obeys the same rules it would obey when the chromophore is not linked to a polymer chain (low molecular weight donor and acceptor in glassy solution at low temperature).

3.1.2 *Collisional transfer following singlet energy migration (Mechanism 3.2)*

A second case would be that of a polymer formed from neighbouring chromophores that cannot form excimers. If exchange or coulombic interactions are not possible, then collisional transfer following energy migration can occur. If the migration is sufficiently important that statistical mixing of donor and acceptor occurs during the lifetime of the donor, then a Stern-Volmer type equation is obeyed:

$$\phi_{FM}^{\circ}/\phi_{FM} = 1 + k_{YM}\,\tau_M[^1Y] \tag{3.10}$$

Let us suppose a host film of molar concentration $[^1M]$ contains a molar concentration $[^1Y]$ of guest, so that the guest mole fraction is $C_Y = [^1Y]/[^1M]$. The excitation jump frequency from one host molecule to an adjacent one is [5]:

$$k'_{mig} = k_{mig}\,[^1M] \tag{3.11}$$

Let us also assume that k_{mig} the rate of transfer from $^1M^*$ to an adjacent ^1M is equal to the rate of transfer from $^1M^*$ to ^1Y. In its migration, the exciton† may encounter a guest molecule ^{1}Y. Then, if the probability for the exciton to be captured is p:

$$k_{YM}\,[^1Y] = pk_{mig}\,[^1Y] = pk'_{mig}\,C_Y \tag{3.12}$$

This gives:

$$\phi_{FM}{}^{o}/\phi_{FM} = 1 + pk'_{mig}\,\tau_M\,C_Y \tag{3.13}$$

where $k'_{mig}\,\tau_M$ is the number of jumps of the singlet excitation during the life-time of $^1M^*$. The determination of $\phi_{FM}{}^{o}/\phi_{FM}$ over a great range of guest concentrations is difficult since the fluorescence intensities of different films have to be compared. As an alternative, the ratio I_{FY}/I_{FM} of the fluorescence intensities of guest and host is often used. Indeed [5]:

$$\phi_{FM}{}^{o}/\phi_{FM} - 1 = (I_{FY}/I_{FM})(\phi_{FM}{}^{o}/\phi_{FY}) \tag{3.14}$$

where ϕ_{FY} is the fluorescence quantum yield of Y on direct excitation. This gives for:

$$\frac{I_{FY}}{I_{FM}} = pk'_{mig}\,\tau_M\,\frac{\phi_{FY}}{\phi_{FM}{}^{o}}\,C_Y \tag{3.15}$$

Measurement of $\phi_{FY}/\phi_{FM}{}^{o}$ allows $pk'_{mig}\,\tau_M$ to be obtained from graphs of I_{FY}/I_{FM} as a function of C_Y.

3.1.3 *Collisional transfer and excimer formation following singlet energy migration (Mechanism 3.3)*

A further complication is introduced if excimers can form and compete with ^{1}Y as traps for the excitation‡. We shall consider here that

---

† The exciton refers to mobile singlet energy and not to the delocalized exciton of excited crystals at low temperature.

‡ Energy absorption is always supposed to occur at isolated chromophore sites. Indeed, the number of preformed excimer sites is supposed to be much lower than the number of chromophores, so that direct absorption at excimer sites is neglected.

energy transfer occurs only by collisional transfer from singlet exciton to the acceptor. Energy cannot be transferred from excimers to ^1Y either by long-range transfer or from $^1M^*$, or from $^1D^*$. Thus:

$$\phi_{FD}{}^{o}/\phi_{FD} = 1 + k_{YM}\,\tau'_M\,[^1Y] \qquad (3.16)$$

where $\tau'_M = 1/(k_M + k_{DM}[^1M]$ (3.17)

The physical meaning of k_{YM} can be related to jump frequency as in the preceding case. Note that physical meaning of $k_{DM}[^1M]$ can be different from that of solutions†. If $[^1M]$ in $k_{DM}[^1M]$ is the number of excimer forming sites $[M_D]$ and k_{DM} is supposed to be equal to k_{mig}. Then, assuming

$$k_{DM}[M_D] = pk_{mig}\,[M_D] = pk'_{mig}\,C_D \qquad (3.18)$$

where C_D is the fraction of excimer forming sites.. Equation (3.16) gives:

$$\phi_{FD}{}^{o}/\phi_{FD} = 1 + C_Y/C_D \qquad (3.19)$$

if $k_M << k_{DM}[^1M]$. This transforms into:

$$\frac{I_{FY}}{I_{FD}} = \frac{C_Y}{C_D}\,\frac{\phi_{FY}}{\phi_{FD}{}^{o}} \qquad (3.20)$$

using

$$\frac{\phi_{FD}{}^{o}}{\phi_{FD}} - 1 = \frac{I_{FY}}{I_{FD}}\,\frac{\phi_{FD}{}^{o}}{\phi_{FY}}$$

Equation (3.20) is equivalent to the Equation derived by Kloppfer [6]

---

† The Birks' terminology $k_{DM}[^1M]$ used for excimer formation does not imply any mechanism concerning excimer formation in polymer. It can have any meaning. Excimer dissociation has not been introduced in the simple cases developed. Introduction of this step is straightforward. These simple schemes can also be complicated by the existence of two or more different excimers and of different types of monomer sites.

from the hopping model for energy transfer in polyvinylcarbazole.

3.1.4 *Collisional transfer following excimer migration (Mechanism 3.4)*

A case where an excimer would act as a donor by Stern-Volmer kinetics is of low probability. Indeed, the number of excimer forming sites is generally too low in solid polymers to allow migration of the excimer energy. In any case, the equation corresponding to Equation (3.16) would be:

$$\phi_{FD}{}^{o}/\phi_{FD} = 1 + k_{YD}\ \tau_{D}[{}^{1}Y] \qquad (3.21)$$

In Mechanism 3.1, the long-range resonance transfer occurs in systems that do not form excimers. In Mechanisms 3.2 - 3.4, energy transfer occurs by collisional transfer following migration. More complicated cases include long-range energy transfer from the excimer to the acceptor (Mechanism 3.5) and long-range energy transfer from the isolated first excited singlet state following migration (Mechanism 3.6).

3.1.5 *Long-range dipole-dipole transfer from the excimer (Mechanism 3.5)*

The existence of long-range dipole-dipole transfer from an excimer can be recognized by the following methods:

1. The results show departure from Stern-Volmer linear behaviour when plotted as $\phi_{FD}{}^{o}/\phi_{FD}$ as a function of the acceptor concentration.
2. The results agree with the theoretical equation of Forster when plotted as $\phi_{FD}{}^{o} - \phi_{FD}/\phi_{FD}{}^{o}$ as a function of the logarithm of acceptor concentration. In polymers it is necessarily preceded by a few migration steps. Indeed, the number of excimer sites is always smaller than the total number of chromophores and they cannot be excited directly. The experimental critical distance R_{oexp} obtained from these results using Equation (3.8) is equal to the theoretical critical distance R_{oth} obtained using Equation (3.7).

3.1.6 *Long-range dipole-dipole transfer following efficient energy migration (Mechanism 3.6)*

Different theories have been proposed to interpret some results obtained in low-viscosity media characterised as follows: efficient mixing of donor and acceptor occurs by diffusion during the life-time of the donor and energy transfer occurs by long-range dipole-dipole transfer instead of collisions. The theory proposed by Voltz can be transposed to polymer solids in which energy migration replaces the diffusion of molecules in dilute solutions. The rate parameter for a diffusion controlled collision is [5]

$$k_{diff} = 4\pi N'\ DpR\ [1 + pR\ (\pi Dt)^{-1/2}] \qquad (3.22)$$

where R is the sum of the collision radii, p is the reaction probability per collision and N' is the number of molecules per millimole. In polymer systems, D is replaced by Λ, the migration coefficient. For S-S dipole-dipole transfer, the critical transfer distance $R = R_{oth}$ and $p = 0.5$ since R_o is the distance at which the probability of transfer is 0.5. The energy transfer parameter so obtained is:

$$k_{YM} = 2\pi N'\ \Lambda R_{oth}\ [1 + 0.5\ R_{oth}\ (\pi \Lambda t)^{-1/2}] \qquad (3.23)$$

when:

$$t^{1/2} \gg 0.5\ R_{oth}\ (\pi\Lambda)^{-1/2} \qquad (3.24)$$

it reduces to:

$$k_{YM} = 2\pi N'\ \Lambda R_{oth} \qquad (3.25)$$

This equation allows Λ to be calculated. The rate constant k_{YM} is determined from the slope of the Stern-Volmer plot of the results if τ_M is known. Λ is calculated from Equation (3.25) once the R_{oth} has been calculated.

Energy migration can also result in formation of an excimer from which long-range energy transfer occurs.

3.1.7 *Long-range dipole-dipole transfer following intermediate migration (Mechanism 3.7)*

Mechanism 3.7 implies sufficient energy migration to allow statistical mixing of donor and acceptor during the life-time of the donor. Of course, intermediate cases are very probable: some contribution of migration occurs followed by long-range transfer. These cases are difficult to analyse since the results will show departure from the linear Stern-Volmer plot and from Forster theoretical curve. R_{oexp} determined from a Forster plot will then be larger than the theoretical critical distance.

3.2 POLYSTYRENE (PS) AND POLYVINYLTOLUENE (PVT)

Since the pioneering work of Birks and Kuchela [7] concerning energy transfer from PS to tetraphenylbutadiene (TPBD), this polymer has been investigated by several workers. One of these studies is Basile's report [8]. This author discovered that when the solid polymer contains residual styrene monomer,† the emission spectrum is that of styrene populated by energy transfer (Fig. 3.1). For pure PS, the excimer

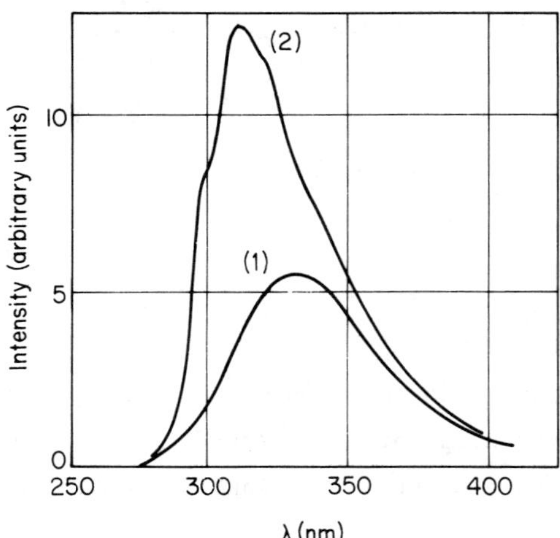

Fig. 3.1 Fluorescence spectra of solid polystyrene (PS): 1, annealed PS; 2, PS containing 1% of styrene (after Heisel and Laustriat [10]).

† See footnote at end of references.

emission is only observed. When TPBD is used as an added acceptor in PS free from styrene, the results for the efficiency of transfer agree with Forster's theoretical curve for long-range transfer. Further, the excimer fluorescence life-time decreases with increasing acceptor concentration. This is in agreement with long-range transfer from excimer to additive (Mechanism 3.5). However, as stated by Basile,the possibility of some contribution of exciton migration cannot be completely exluded.

Jones *et al.* [9] studied energy transfer from pure PS to naphthalene-d_8, *p*-terphenyl and pyrene-d_{10}. The results fit the theoretical Forster curve with R_0 values, respectively, of 1.3, 1.8 and 2.2 nm, whereas the calculated values assuming the excimer as donor are 0.75, 1.30 and 1.83 nm. The calculated values assuming the monomer as donor are very low since the quantum yield of monomer fluorescence is very low. This is evidence against energy transfer exclusively by the Forster mechanism with the monomer as donor (Mechanism 3.1). Mechanism 3.5 is the most probable, although some minor contribution of exciton migration could occur. Decay time measurements of the excimer, if found to be dependent on acceptor concentration, would confirm this mechanism.

Heisel and Laustriat [10] studied energy transfer from PS to *p*-terphenyl (PT), phenyl-2 *p*-biphenyl-5 oxodiazole 1,3,4 (PBD) and di (1-naphthyl)-2,5 oxodiazole 1,3,4 (α-NND). The efficiency of non-radiative energy transfer as a function of acceptor concentration is in agreement with Forster single-step energy transfer (Mechanism 3.5 and Fig. 3.2). In this case also the experimental R_0 values are of the same order of magnitude as those obtained theoretically assuming the excimer as donor. Migration if existent would be of low importance, contrary to what is observed in solution.

Energy migration and transfer has been studied in PVT doped with *p*-terphenyl or diphenylstilbene [11]. In PVT, most of the fluorescence originates from the excimer. Overlap between the absorption spectrum of diphenylstilbene with the excimer emission of PVT is large. The theoretical critical transfer distance from the excimer to diphenylstilbene is equal to the experimental value obtained from the decrease

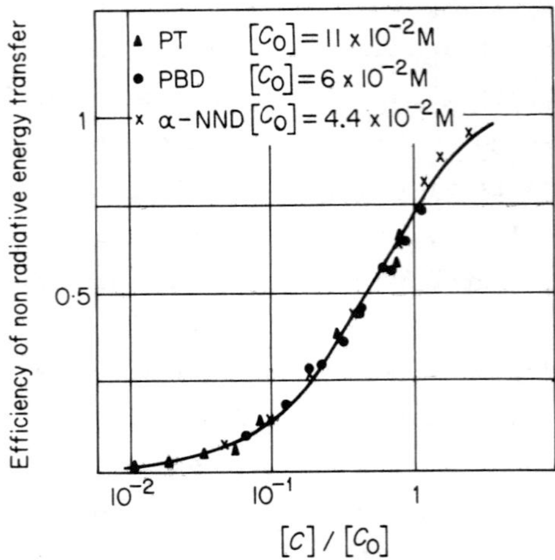

Fig. 3.2 Efficiency of non-radiative transfer as a function of C/C_o. Theoretical curve calculated according to (3.8); points plotted are experimental values (after Heisel and Loustriat [10]).

of decay-time measured in the presence of this additive (Fig. 3.3).

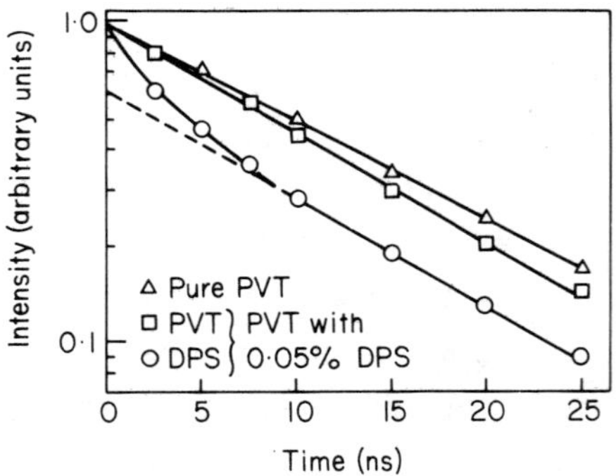

Fig. 3.3 Fluorescence decays of pure PVT and PVT doped with 0.05% DPS (after Powell [11]).

According to these considerations, Mechanism 3.5, with migration of low importance is the most probable, whereas Mechanisms 3.1 and 3.6 are of low probability owing to the poor overlap of the $^1M^*$ fluorescence and additive absorption spectra.

In PVT containing *p*-terphenyl, the overlap of the absorption spectrum of the additive and excimer emission is very poor. Mechanisms 3.5 and 3.6 have thus a low probability. Comparison of experimental and theoretical R_o values (respectively, 2.2 nm and 0.9 nm) for transfer from monomer to *p*-terphenyl indicates that Mechanisms 3.1 and 3.6 are also improbable. Experimental results show that I_{FY}/I_{FD} is linear with C_Y in agreement with Equation (3.20) and Mechanism 3.3. The number of excimer sites can be calculated from the same equation if $\phi_{FY}/\phi_{FD}{}^o$ is known. A value of 2.5×10^{-2} is obtained.

3.3 POLYVINYLNAPHTHALENE AND POLYACENAPHTHYLENE

Poly-1-vinylnaphthalene (P1VN) and poly-2-vinylnaphthalene (P2VN) films have been studied. Although there has not been any systematic study concerning the ratio of excimer to monomer fluorescence, this seems to depend not only on the position (1 or 2) of the chain, but also on the method of synthesis of the polymer. The fluorescence is mainly excimeric between room temperature and 77K although a small contribution of monomeric emission could be present in the high-energy spectral range. The delayed fluorescence observed at 77K in P2VN is red-shifted relative to the prompt fluorescence, in agreement with a mechanism involving triplet-triplet annihilation between a trapped and an excitonic triplet [12].

The emission spectrum of P1VN solution has been much more thoroughly investigated than that of films. On the basis of steady-state and time-resolved emission spectra, the following interpretations have been proposed for solutions. Ghiggino *et al.* [13] assign the short and long wavelength emission respectively to monomer and excimer fluorescence. The delayed emission at 320 nm is due to excimer dissociation. Holden *et al.* [14] assigned the delayed fluorescence to emission from monomer units trapped in such a conformation that excimer formation or energy transfer to neighbouring naphthalene is impossible during their life-time. Recently [15], the existence of a second excimer has been postulated. In a recent analysis of time-resolved spectra of P1VN films excited by electron beam pulse [16], a detailed analysis of the fluorescence of solid poly-1-vinylnaphthalene was proposed. Three emitting species are involved (Fig. 3.4) A short-

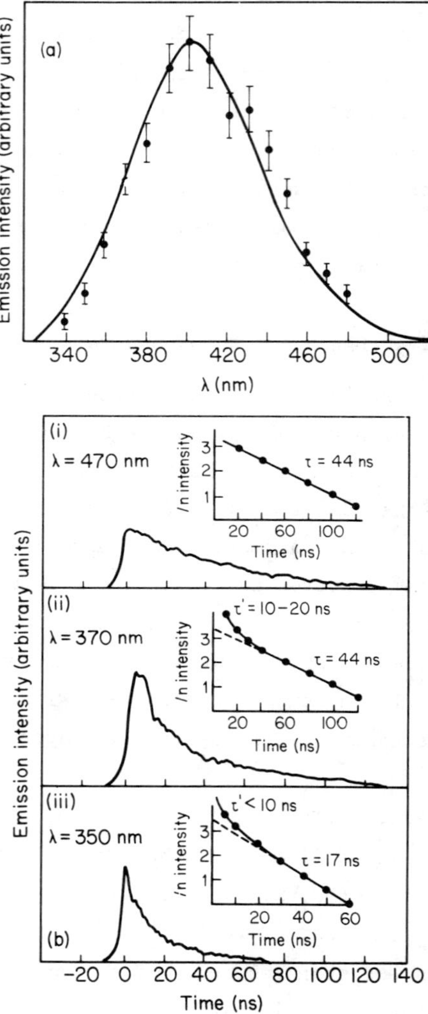

Fig. 3.4 (a) Emission spectrum of solid P1VN film (λ_{ex} = 270 nm). Points are time-averaged total emission intensity following pulse radiolysis. (b) Typical transient emission traces following pulse radiolysis: (i) recorded at 470 nm showing low-energy excimer emission (τ = 44 ns); (ii) recorded at 370 nm showing both low-energy (τ = 44 ns) and high-

lived monomer emission with a life-time of about 5 ns is observed below 360 nm. A slowly decaying emission (44 ns) due to a low-energy singlet excimer is seen in the region 360 - 520 nm. A second excimer with a life-time of 17 ns is detected in the range 330 - 460 nm. The authors suggest that the high-energy edge of this excimer emission could have been considered as delayed emission in solution. Energy transfer to anthracene occurs efficiently [16]. Overlap with monomer emission is important, but R_{oth} is low owing to the low monomer fluorescence yield. Assuming competition between excimer and additive for exciton capture, an excimer trap density of $1/10^3$ is evaluated. This cannot be considered as definitive, since only 3 concentrations of anthracene have been used.

Energy transfer from P1VN to copolymerized vinylpyrene was shown by Webber [17] to be much more efficient in neat films than in solution. According to the author, dynamically formed excimers in solution are more effective energy traps than most preformed excimer sites in solids. This arises from the chromophore-chromophore reorientation that is possible in solution and results in greater stabilization of the excimer state. The author also suggests that the density of excimer-forming sites in films is sufficiently high that energy transfer between these sites is possible and enhances pyrene sensitization.

In polyacenaphthylene (PAcN) films, the contribution of monomer fluorescence is also unimportant at room temperature and 77K [18] although the ratio I_{FM}/I_{FD} is much larger for PAcN solutions at room temperature than it is for PVN solutions. In PAcN, excimer formation is not possible between two neighbouring chromophores as it is in PVN. This is due to the reduced mobility of side groups with respect to the main chain. Examination of molecular models, however, indicates that excimer formation could be possible between chromophores situated as next-to-nearest neighbours. In films, the number of excimer-forming

---

energy excimer (τ = 17 ns) emission; (iii) recorded at 350 nm showing monomer ($\tau \simeq 5$ ns) and high-energy excimer (τ = 17 ns) emissions (after Coulter *et al.* [16]).

sites is sufficiently high and energy migration sufficiently efficient that excimer emission only is observed.

Energy transfer to benzophenone was compared in P1VN [18] and PAcN [19] films at room temperature and 77K. The results are given in Fig, 3.5 and 3.6 . They do not obey the Forster theoretical relation

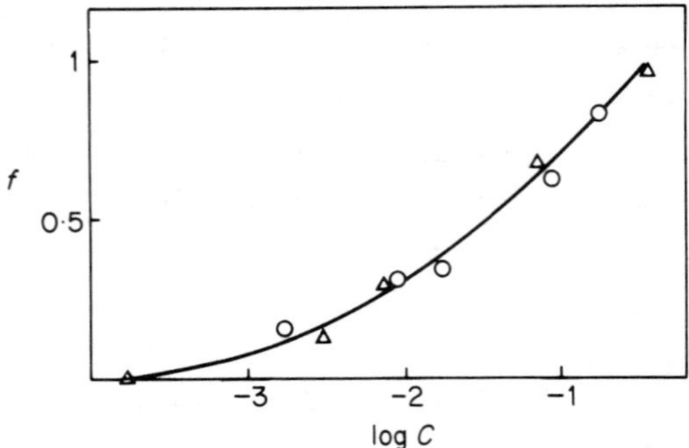

Fig. 3.5 Transfer efficiency as a function of logC (benzophenone concentration) for the system P1VN - benzophenone excited with λ = 313 nm at liquid nitrogen temperature. Δ and O correspond to two independent series of experiments (after David *et al.* [19]).

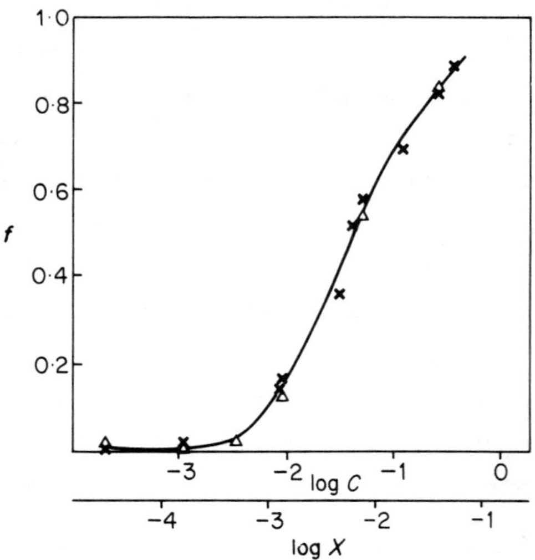

Fig. 3.6 Efficiency of transfer from PAcN and benzophenone as a function of additive concentration in the film λ_{exc} = 300 nm Δ, T = 77K; x T = 298K (after David *et al.*, [18]).

(Equation (3.8)). Approximate experimental R_0 values can however be obtained from the results for comparative purpose. They are 1.5 nm for P1VN and PAcN at room temperature and 77K. The theoretical R_0 values obtained for transfer at room temperature from monomer and excimer respectively, are 6.0† and 8.1‡ nm for P1VN. For PAcN they are 4.5† and 6.7$^{\int}$nm. At 77K, the transfer distances are about 1 nm larger than at room temperature owing to the increase of the quantum yield at low temperature. The theoretical transfer distances are small for transfer from the excimer owing to the low value of the molar absorption coefficient of the $n\pi^*$ lowest transition of benzophenone. They are still lower for transfer from the monomer since its fluorescence yield is very low in films. The experimental results, however, obey a linear relation when plotted as ϕ_{FD}^{0}/ϕ_{FD} as a function of the concentration of benzophenone. Energy migration followed by collisional transfer Mechanism (3.3) is the most probable mechanism, and this could be confirmed by decay-time measurements in the absence and presence of acceptor.

3.4 POLYVINYLCARBAZOLE

The photophysical processes in polyvinylcarbazole (PVCz) have been studied very extensively. There seems now to be a general agreement concerning the main features of the emission spectra. Numerous photophysical processes are involved. Different kinetic models taking into account the experimental observations have been proposed and are being progressively optimized 6, 20, - 25 and references cited therein.

3.4.1 *Room-temperature behaviour* [25]

The emission spectrum at RT is pure excimer fluorescence. Energy transfer to perylene was shown to be much more efficient than predicted by Forster dipole-dipole transfer. Further, energy transfer to hexachloro-*p*-xylene which is not allowed by dipole-dipole interaction and

---

† Assuming that $\phi_{FM} = \phi_{FD}/20$.

‡ Calculated using $\phi_{FD} = 1.5 \times 10^{-2}$ at room temperature.

∫ Calculated using $\phi_{FD} = 0.04$ at room temperature.

can only occur by collisional transfer is also very efficient, demonstrating the importance of energy migration in the host polymer. The following basic model was adopted to explain energy transfer in PVCz [6] Mechanism (3.3):

1. Absorption of UV radiation takes place randomly forming excited singlet states.
2. The lowest excited singlet state forms an exciton that can be trapped either at excimer sites or by guest molecules.
3. The concentration of excimer sites is very small, so that direct excitation is negligible.
4. Quantitatively, the excimer fluorescence intensity in doped films I_{FD} relative to that in the pure polymer I_{FD}^{o} can be expressed in the following equation:

$$Q_D = (I_{FD}^{o} - I_{FD})/I_{FD} = C_Y/C_D \tag{3.26}$$

where C_Y, C_D are the concentrations of guest molecules and excimer forming sites, respectively in mol./mol. basic unit.

This model is valid if direct energy transfer from excimer to guest by dipole-dipole interaction is negligible. Equation (3.26) allows an estimation of excimer forming sites in PVCz of about 2×10^{-3} mol./mol. basic unit. The average random walk distance from the starting point is about 20 nm, compared with 130 nm in the case of *N*-isopropyl-carbazole.

3.4.2 *Low-temperature behaviour*

At low temperature, shallow traps can become effective. A second excimer fluorescence at shorter wavelength (Fig. 3.7) has indeed been discovered by Johnson and Offen [21]. Furthermore, triplet states become observable. Two phosphorescence components with different lifetimes have been clearly identified. The phosphorescence component that is red-shifted with regard to carbazole has been assigned to triplet excimer phosphorescence [23]. The delayed fluorescence is also red-shifted when compared to the undelayed fluorescence of PVCz.

The fluorescence as well as the phosphorescence spectra depend on the mode of polymerization used [26]. In cationic PVCz, both singlet

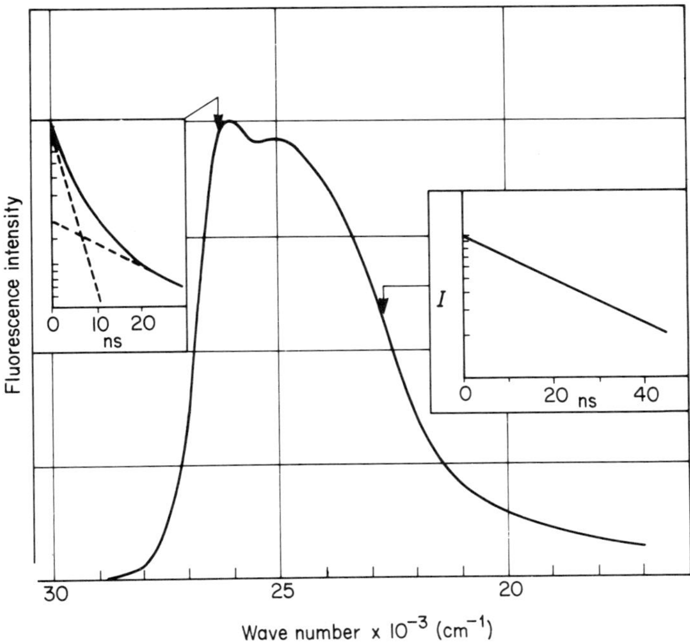

Fig. 3.7 Low-temperature fluorescence spectrum of PVCz and fluorescence decay at two wavelengths (after Klopffer [25]).

excimer sites are observed, whereas the shorter wavelength emission dominates the spectrum of radical polymerization.

A mechanism of photophysical processes in PVCz that takes into account the dependence of fluorescence, delayed fluorescence and phosphorescence as a function of excitation intensity has been proposed. The basic assumptions of this model are [25]:

1. Energy transfer can be described by a simple hopping model, the probability of trapping being given by $k'_{mig}(s^{-1}) \times C_D$, where k'_{mig} is the exciton hopping rate.
2. The difference between radicalic and cationic polymers consists in different concentrations of excimer forming sites, not in their nature. All hopping rates are assumed to be equal in both types of polymers.

3. Annihilation occurs between excitons and filled traps (excimers), the trap depth being much larger than kT at 77K.

An important result of this kinetic analysis is to describe the phosphorescence excimers as unrelated to one of the main singlet excimers. The linearity of the fluorescence intensity versus excitation intensity could not be explained if they were. Indeed, fluorescence should saturate at high intensity due to singlet-triplet annihilation, a Forster type, spin-allowed process, that occurs at high concentration of triplet traps. Direct experimental proof of the existence of such triplet traps is the red-shifted delayed fluorescence, indicating that the singlet created in T-T annihilation and staying at the site of their formation have a lower energy and are therefore different from the states causing the major portion of prompt fluorescence at 77K.

3.5 POLYMERS CONTAINING THE 1,3,5-TRIPHENYL-2-PYRAZOLINE IN THE SIDE CHAIN

A relatively simple system which does not form excimers was studied by Iinuma *et al.* [27]. The polymers studied that contain the 1,3,5-triphenyl-2-pyrazoline as a pendant unit are:

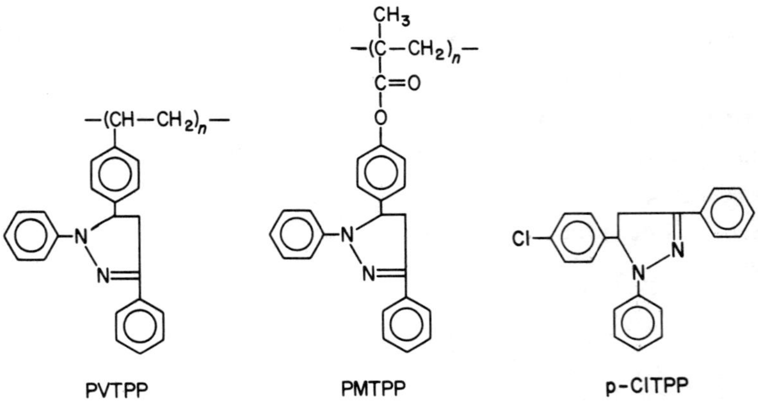

$-(CH-CH_2)_n-$ $-(C(CH_3)-CH_2)_n-$ C=O O Cl N N

PVTPP PMTPP p-ClTPP

Energy transfer to dimethylterephthalate (DMTP) was studied. Either Forster dipole-dipole transfer or exchange mechanism are improbable with this guest-host system, since the lowest excited state of the host lies at much higher energy than that of the guest (Fig. 3.8).

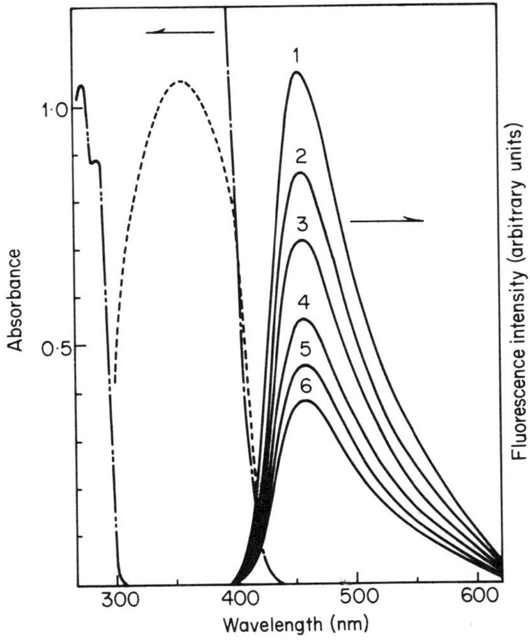

Fig. 3.8 Absorption (transmission and reflection methods) spectra of PMTPP and DMTP, and fluorescence spectra of PMTPP in the absence and presence of DMTP. Absorption spectrum of DMTP (—··—) in a THF solution: absorption spectra of PMTPP film (---- reflection) (—·— transmission); fluorescence spectra of PMTPP film in the absence (1) and presence (2 - 6) of DMTP at $0^{o}C$ (continuous curves). C_M/C_Y (mol./mol. basic unit): 0.006 for 2; 0.01 for 3; 0.02 for 4; 0.03 for 5; 0.04 for 6. C_M and C_Y are the concentrations of host chromophore and guest molecule, respectively (after Iinuma *et al.* [27]).

Quenching of the host fluorescence will thus take place through singlet migration over the host followed by charge transfer from the excited host to the ground-state guest molecule at collisional distance. Assuming that quenching is migration controlled, the authors obtain an equation identical to Equation (3.13), assuming $p = 1$ (Mechanism 3.2):

$$\phi_{FM}{}^{o}/\phi_{FM} = 1 + k'_{mig}\,\tau_M C_Y$$

where $k'_{mig}\,\tau_M$ is the number of hops of singlet excitation migration during the life-time of the host excited state (5 ns). The results are given in Table 3.1. k'_{mig} is shown to be a slightly activated

Table 3.1 Comparison of the number of hops of the singlet excitation energy and the rate parameter (k_{mig}) of its migration between the single crystalline and amorphous glassy state of *p*-CITPP, and polymers, PVTPP and PMTPP.

Compound	$k'_{mig}\,\tau_M$	τ_M (ns)	k'_{mig}(s^{-1})
p-CITPP			
single crystalline state†	3.71×10^5	9.9	3.73×10^{13}
amorphous glassy state ‡	493	9.4	5.24×10^{10}
PVTPP‡	48	ca. 5.0	9.6×10^9
PMTPP‡	45	ca. 5.0	9.0×10^9

† measured at 10°C,

‡ measured at 0°C.

process probably due to the presence of shallow traps in the host chromophores. The value of k'_{mig} and $k'_{mig}\,\tau_M$ obtained for the polymers are compared with those measured for the single crystalline and amorphous glassy state. The k'_{mig} for the polymer is about one fifth of that obtained for the glassy state, and 3.5×10^3 times smaller than those for the single crystal. It is suggested that the molecular order is lower and the chromophore orientation less favourable in the polymer than in the glassy solid.

A separate comparative study was made [28] on singlet energy migration and transfer in 1,3-diphenyl-5-(*p*-chlorophenyl)-2-pyrazoline to dimethyl terephthalate in the glassy and single crystalline phase at room temperature. The higher efficiency of migration (k'_{mig}) in the crystalline state was assigned to a large difference in the preexponential factors of k'_{mig} (respectively 4.7×10^{11} and 6.1×10^{13} (s^{-1}) in the amorphous and crystalline phase). The rate parameter is slightly temperature dependent ($\Delta E \simeq$ 0.052 eV in the glassy state and 0.012 eV in the crystalline state).

3.6 CONJUGATED POLYMERS

The fluorescence emission from polymers containing polyene chromophores has been investigated in solution and in the solid state. The absorption and emission of mixtures of various sequence lengths of conjugated chromophores play an important role in the photophysics of these polymers, which behave as a mixture of linear polyenes, among which exists the possibility of energy transfer from short to longer polyenes. The photophysical properties of these polymers thus strongly depend on the conformation of the chain responsible for the distribution of the lengths of the polyene sequences. These systems have been studied in solution but also as polymer blends in polystyrene and polymethylmethacrylate [29-31]. Polystyrene used as a matrix for styrene-phenylacetylene (20/80) copolymers (SPA) locks the copolymer into a conformation yielding two principal emitting states corresponding to a short and a long sequence of conjugated double bonds, each characteri d by a distinct emission (420 and 490 nm) (Fig. 3.9). Energy transfer was

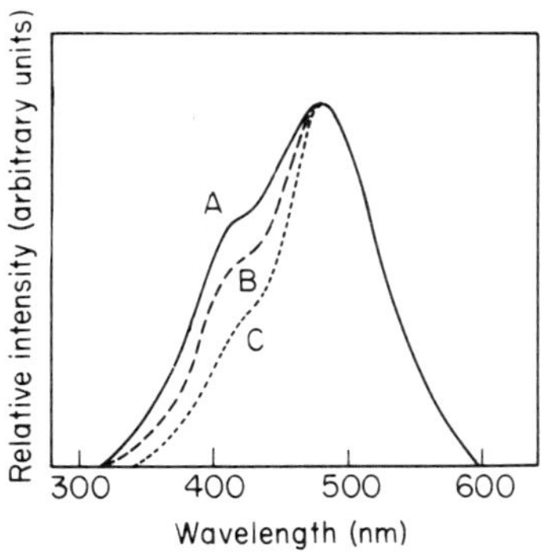

Fig. 3.9 Steady-state emission spectra of SPA in polystyrene film. Excitation wavelengths: A, 290 nm; B, 320 nm; C, 380 nm. Spectra adjusted to fit on same scale (after Guillet *et al.* [31]).

shown to occur from the short to the long sequences (Fig. 3.10).

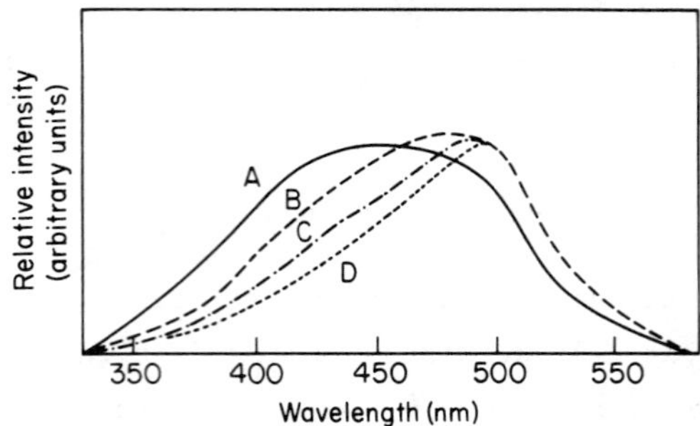

Fig. 3.10 Time-resolved fluorescence spectra of SPA in polystyrene film. Exciting wavelength, 313 nm. The lower and upper limits for the time-resolved spectra are given as the interval from the lamp maximum. A: upper, 0.23 ns; B: lower, 1.15 ns; upper, 4.37 ns; C: lower, 4.6 ns; upper, 7.82 ns; D: lower, 20.2 ns; upper, 25.1 ns. Spectra adjusted to fit on same scale (after Guillet *et al.* [35]).

Indeed, the proportion of the 490 nm peak increases with increasing the excitation wavelength (Fig. 3.9). Furthermore, time-resolved fluorescence spectra and decay-time measurements of the styrene-phenylacetylene copolymer in polystyrene films confirm the existence of a long-lived emission with a maximum at 490 nm and a shorter lived emission with a maximum at 420 nm. According to authors, the difference between the steady-state spectrum and spectrum taken immediately after excitation (Fig. 3.10) results from energy transfer from short to long sequences. Indeed, if no energy transfer would occur between the emitting states, the steady-state spectrum should be comparable to the short time-resolved spectra corrected for the difference in decay-time of the 420 and 490 nm emissions [3].

Use of polymethylmethacrylate as a matrix was shown to favour short sequences of conjugated double bonds, whereas in solution the emission is consistent with the formation of a broader and continuous distribution of double bonds [3].

3.7 COPOLYMERS

Copolymers styrene-acenaphthylene of composition ranging from 7×10^{-2}

to 8 x 10^{-4} mole fraction in acenaphthylene were studied as solid films at room temperature by Schneider and Springer [32]. Fluorescence of the isolated acenaphthylene units is only observed in all cases except for the 8 x 10^{-4} copolymer. In this last case, a contribution of the excimer fluorescence of PS is also apparent. Extrapolation of the results assuming equal quantum yield for the fluorescence of the PS excimer and the acenaphthylene units indicate that equal fluorescence intensity of both components would be observed for copolymers containing a mole fraction of 4 x 10^{-4} in acenaphthylene units.

Singlet energy transfer in films of PS and copolymers styrene-methylacrylate was studied by David *et al.* [33] at RT and 77K. The efficiency of transfer as a function of tetraphenylbutadiene acceptor concentration does not obey the Forster relationship. The values of I_{FM}/I_{FD}, $I_{FM}^{o}/$ and I_{FD}^{o}/I_{FD}, where I_{FM} and I_{FD} are the intensities of monomer and excimer fluorescence in the presence of acceptor and I_{FM}^{o} and I_{FD}^{o} are the corresponding fluorescence in the absence of acceptor, were measured as a function of acceptor concentration. The equations relating these parameters to the acceptor concentration were derived using the complete kinetic treatment [33] and assuming transfer to additive, either from monomer or from excimer or from both of them. The results agree with a model where both monomer and excimer transfer to the additive. The efficiency of transfer is about the same in PS and in the copolymers at RT.

The values of the monomer and excimer fluorescence intensities were measured by David *et al.* [34] at RT and 77K in a broad range of styrene-methylacrylate and styrene-methylmethacrylate copolymers in the absence of additives in order to investigate energy migration in the PS sequences and transfer to the excimer sites. Kinetic equations corresponding to different models were solved. The results were found to agree with a model that assumes that energy absorbed by any styrene unit can migrate to a pair of neighbouring styrene units suitably orientated to form an excimer. The number of these sites is proportional to the number of styrene pairs in the polymer.

To investigate the dependence of the fluorescence properties of polymers on the number of carbons separating the aromatic rings, the fluorescence spectra of copolymers of general formula:

$$-\underset{\mathrm{C_6H_5}}{\overset{\mathrm{CH_3}}{\mathrm{C}}}-CH_2-CH_2-\underset{\mathrm{C_6H_5}}{\overset{\mathrm{CH_3}}{\mathrm{C}}}-(CH_2)_n- \quad \text{with } n = 3\text{-}6$$

recorded at RT and 77K, were analysed by David *et al.* [35] and compared with those of poly-α-methylstyrene (Figs. 3.11 and 3.12). For PS,

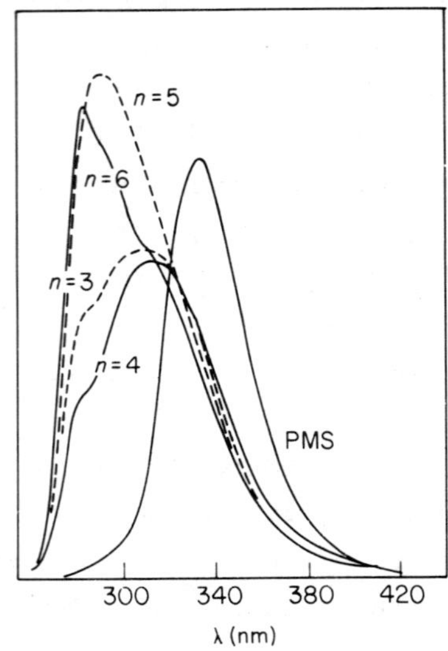

Fig. 3.11 Fluorescence spectra of films at room temperature. λ_{exc} = 253.7 nm. Incident energy totally absorbed (after David *et al.* [35]).

excimer fluorescence with λ_{max} at 330 nm is recorded at room temperature. At 77K, two emissions, assigned as in polystyrene (PS) to monomer and excimer fluorescence, are poorly resolved. Some of the condensation copolymers (n = 3, n = 4) show at room temperature a maximum at about 310 nm and a poorly resolved component at 280-290 nm. The others have a λ_{max} at 280-290 nm but a non-resolved component at 310 nm is probably contained in the broad fluorescence band. At 77K, all the copolymers have their maximum of fluorescence at 280-290 mm. A very small component at 310 nm could be involved in the tail although unresolved. For

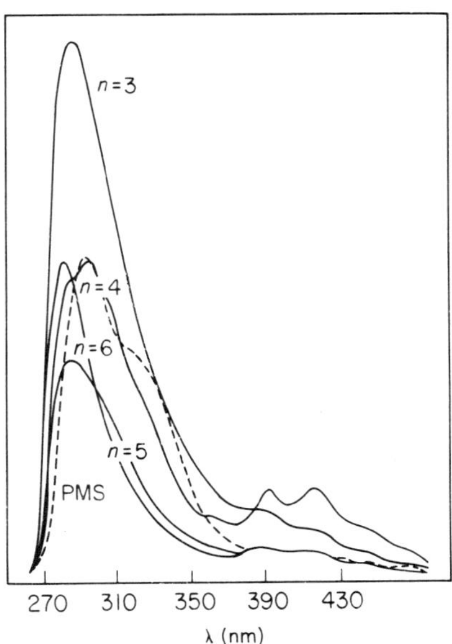

Fig. 3.12 Emission spectra of films at 77K. λ_{exc} = 253.7 nm. Incident energy totally absorbed (After David *et al.* [35]).

a given copolymer, the contribution at 310 nm is lower at 77K than at room temperature. The peaks around 390 nm and 425 nm are due to phosphorescence.

Those results show that the excimer fluorescence observed at 330 nm in PMS and PS is not emitted by the copolymers. This excimer corresponds to intrachain interactions between neighbouring chromophores. Intrachain excimer could not form in the copolymer. The broad and structureless emission observed at 310 nm at room temperature can be assigned to another energy trap resulting from either intermolecular or intramolecular interaction between chromophores separated by more than 3 carbon atoms. The number of these sites is, nevertheless, insufficient to trap quantitatively the energy absorbed in the film since a component at 280-290 nm is observed in all samples at room temperature. The contributions of these traps to the emission spectrum at 77K is either low or non-existent.

Energy transfer from PVCz to perylene and dimethylterephthalate in films occurs by energy migration followed by collisional transfer

(Mechanism 3.3). Energy transfer to these additives in films of copolymers vinylcarbazole-styrene and vinylcarbazole-vinylacetate rich in vinylcarbazole (from 100% to 79% in vinylcarbazole) was studied by Okamoto *et al.* [36]. Using the hopping model (Mechanism 3.3) and Equation (3.19), the number of excimer sites was found to be larger when dimethylterephthalate is used as additive (3×10^{-3} mol./mol.$^{-1}$) than when perylene is used (1.5×10^{-3} mol./mol.$^{-1}$).

3.8 DOPED POLYMER FILMS

Doped polymer films constitute a very convenient medium for detailed investigation of the migration or transfer of energy between like chromophores. Indeed, a high concentration of low molecular weight donor or acceptor molecules can often be incorporated in polymer films without altering their optical properties. This approach has been used recently by Johnson [37] to investigate S-S energy migration between donors. The conclusions obtained can be extrapolated to the neat donor and compared with those obtained for the same donor in the crystalline state or in neat films of homo- or copolymers containing the same chromophore. This author has investigated energy migration between *N*-iso propylcarbazole (NIPC) molecules as energy donor in polystyrene films. Two acceptors were successively studied: dimethylterephthalate and perylene. Energy transfer from NIPC to the first and second acceptor occurs respectively by the collisional charge transfer interaction and by the Forster dipole-dipole interaction. Fluorescence decay times were measured as a function of acceptor concentration for different NIPC concentrations.

The value of τ_M/τ_{YM} (where τ_M and τ_{YM} are decay times of NIPC respectively in the absence and presence of acceptor) at a given acceptor concentration depends on NIPC concentration according to:

$$\frac{\tau_M}{\tau_{YM}} = 1 + K_q\,[\mathrm{DMT}] \tag{3.27}$$

where K_q depends on NIPC concentration (Fig. 3.13). Also classically:

$$\frac{\tau_M}{\tau_{YM}} = 1 + k_{YM}\;\tau\;[\mathrm{DMT}] \tag{3.28}$$

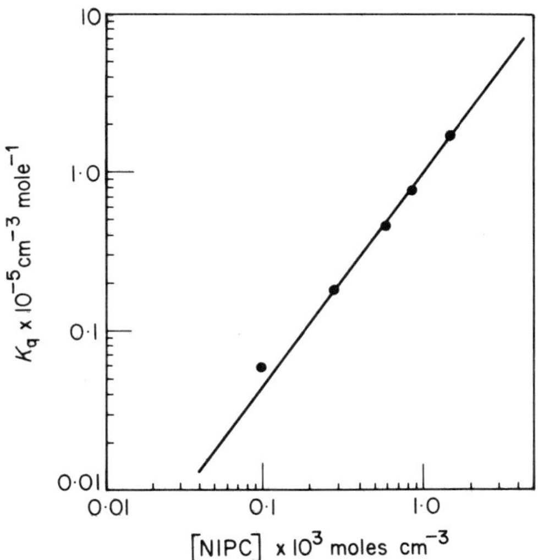

Fig. 3.13 Donor concentration dependence of the fluorescence quenching constant for the NIPC-DMT-polystyrene system (after Johnson [37]).

with $k_{YM} = 4\ \pi\ DRN/1000 = K_q/\tau_M$ (3.29)

where D which is called the diffusion coefficient by the authors is in fact the migration coefficient for energy displacement between NIPC molecules and R is the critical interaction radius for quenching (either collisional radii or Forster critical transfer distance).

Using these equations, it was found that migration occurs in a series of random walk steps involving Forster dipole-dipole resonance transfer. K_q and D were measured for the system NIPC - perylene as a function of NIPC concentration and extrapolated to the neat donor. A value of the diffusion length l of 25 nm was obtained using:

$$l = (2\ D\tau_M)^{1/2} \qquad (3.30)$$

Values of 20 and 130 nm were obtained by Kloppfer for polyvinyl carbazole and crystalline NIPC, respectively.

The migration of electronic excitation in a system consisting of molecularly dispersed NIPC in a polystyrene matrix thus follows the behaviour in the amorphous polymer more closely than in the crystalline polymer as reasonably expected.

Tetramethyl-1,2-dioxetane was used by Turro *et al.* [38] to generate excited singlet acetone. Analysis of the quenching data in terms of the Forster model indicates that the critical transfer distances from singlet acetone to 9,10-diphenylanthracene and biphenyl in PS films are 2.3 and 2.6 nm respectively. These distances are much larger than the collisional radii. Participation of the pendant phenyl group in singlet-singlet energy transfer is ruled out by energetic considerations (acetone singlet: 82 kcal. $mol.^{-1}$; benzene singlet: 105 kcal. $mol.^{-1}$). Long-range Forster mechanism is thus dominant.

3.9 POLYMER BLENDS

In recent years considerable attention has been paid to the extent of compatability of components of polymer blends. This notion is, however, inextricably bound up with the choice of the experimental method used. One technique may lead to the conclusion that a blend is compatible, whereas another one may allow detection of domains of smaller size than the preceeding one. Therefore, there is a large interest in finding new methods to investigate such systems. Fluorescence methods were recently developed for that purpose. Three main approaches were used. In the first one, developed by Morawetz and co-workers [39-41]; compatibility is related to the extent of donor-acceptor energy transfer between the two components of the blend which were selectively 'marked' either with the donor or with the acceptor. In the second approach, the ratio of monomer to excimer fluorescence of one of the components which has to be a vinylaromatic polymer has been demonstrated by Franck and co-workers [42-44] to be a measure of the compatibility. In the third one, proposed by Ledwith [45], the ratio of exciplex to monomer carbazole has been measured in mixtures of two copolymers, one of them containing copolymerized carbazole and the other containing copolymerized terephthalate.

3.9.1 *Measurement of energy transfer: polymers marked with donors or acceptors* [*39-41*].

Each polymer of the blend is labelled either with the donor or with the acceptor. The final composition of donor and acceptor in the film is 0.01M. This is obtained by dilution of the labelled polymer with the corresponding unlabelled one. The fluorescence intensity of the donor (I_D) and of the acceptor (I_A) is measured as a function of the composition of the blend. The donor-acceptor pairs used are: carbazole-anthracene or naphthalene-anthracene. The ratio I_D/I_A was found to increase with the incompatibility of the blend. Many polymer blends were studied. The technique demonstrates a slight departure from random mixing even in the polyphenylene-oxide-polystyrene system which has been described as perfectly compatible.

3.9.2 *Measurement of excimer fluorescence* [*42-44*]

The ratios of excimer to monomer fluorescence for blends of P2VN (0.2% by weight, solvent cast at 22°C) with different polyalkylmethacrylates, polyvinylacetates and polystyrenes (Table 3.2) are given in Fig. 3.14.

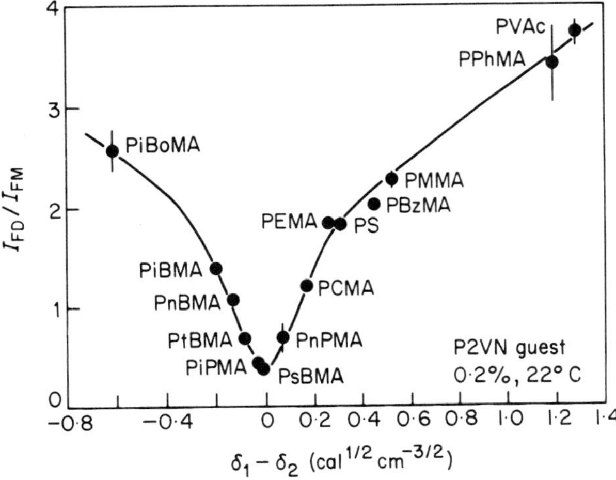

Fig. 3.14 Dependence of observed excimer fluorescence in the guest polymer on enthalpic interactions with the host polymer (After Frank and Gashgari [42]).

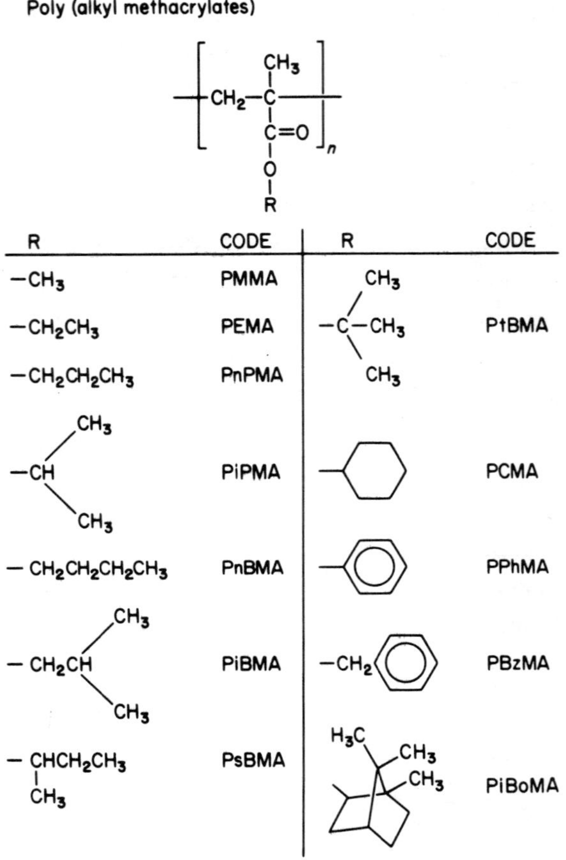

Table 3.2

A correlation is observed between I_{FD}/I_{FM} and the difference $(\delta_1 - \delta_2)$ of the solubility parameters of host and guest polymers. These solubility parameters are related to X_{12}, the binary interaction parameter of the components, by the relation:

$$X_{12} = (V_r/RT)\ (\delta_1 - \delta_2)^2 \tag{3.31}$$

X_{12} is related to the free energy of mixing of a blend which is given by

$$\Delta G_{mix} = \frac{RVT}{V_r} \left[\frac{\phi_1}{X_1} \ln \phi_1 + \frac{\phi_2}{X_2} \ln \phi_2 + X_{12}\, \phi_1\, \phi_2\right] \qquad (3.32)$$

where ϕ_i is the volume fraction of component i, X_i is the degree of polymerization of component i and V_r is a reference volume.

Fig. 3.14 shows that enthalpic interactions dominate the fluorescence behaviour of guest P2VN. Similar results have been reported for poly-acenaphthylene and poly 4-vinylbiphenyl in the same host series.

How can I_{FD}/I_{FM} correlate with the compatibility of the mixture? There are three types of excimer forming sites. The first one arises from association between rings on adjacent units in the same chain. Such an association is possible for racemic g^-t and tg^- dyads in syndio-tactic sequences and meso tt and g^-g^+ dyads in isotactic sequences, (Fig. 3.15). However, only tt meso dyads are populated appreciably in

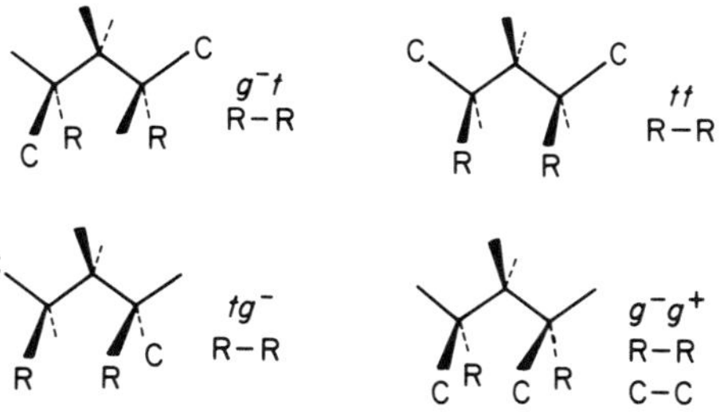

Fig. 3.15 (After Frank and Harrah [46]).

aromatic vinyl polymers. The number of intramolecular adjacent excim-er sites is independent of the guest concentration and of the nature of the host matrix. The second type of excimer forming site is also intra-molecular but results from interaction between remote segments due to bending of the chain. It is independent of concentration but could depend on the host. The intramolecular contribution of excimer forma-tion in films of P2VN (0.2%) and PS was investigated a few years ago [46]. It was shown that the concentration of excimer forming sites in the solid state is fixed by the temperature at which the film is cast. The fraction of excimer forming sites is related to the excimer site conformation energy which is found to be 2500 cal. for P2VN. Excimer sampling is shown to be consistent with thermally activated exciton migration to preformed excimer sites. The mechanism of this migration is attributed to a modulation of interchromophore separation resulting from longitudinal acoustical vibrations of the polymer chain. The third type of site is intermolecular between segments of different chains.

A simple model was developed, the main features of which are given below. If the guest and host are molecularly compatible and energy migration is not the limiting factor, the intermolecular contribution of I_{FD}/I_{FM} is:

$$I_{FD}/I_{FM\,(\text{miscible blend})} = AC_{\text{bulk}} \tag{3.33}$$

where C is the bulk concentration of the guest and A a constant. If the blend is thermodynamically incompatible, the intermolecular contribution of I_{FD}/I_{FM} in the cluster is:

$$I_{FD}/I_{FM\,(\text{cluster})} = BM_{\text{host}}\,X_{12}\,C_{\text{bulk}} \tag{3.34}$$

where M is the molecular weight of the host, and B another constant. The combined expression gives:

$$I_{FD}/I_{FM\,(\text{intermolecular})} = (A + BM_{\text{host}}\,X_{12})\,C_{\text{bulk}} \tag{3.35}$$

This allows us to determine A and B, which are independent of the host, and hence to calculate the bulk concentration of guest at which the local guest concentration in the cluster is equal to that of the pure guest polymer.

$$C_{\text{local}} = (1 + \frac{B}{A}\,M_{\text{host}}\,X_{12})\,C_{\text{bulk}} \tag{3.36}$$

This corresponds to a calculated value of I_{FD}/I_{FM} in the blend that is in agreement with that of pure P2VN giving support to the proposed model. Above this concentration the films are cloudy.

An example of the variation of I_{FD}/I_{FM} with concentration is given in Fig. 3.16. At low guest concentration of 0.003% or less, the P2VN guest polymer chains are molecularly dispersed in the host matrix and are not interacting. I_{FD}/I_{FM} results from intramolecular interactions

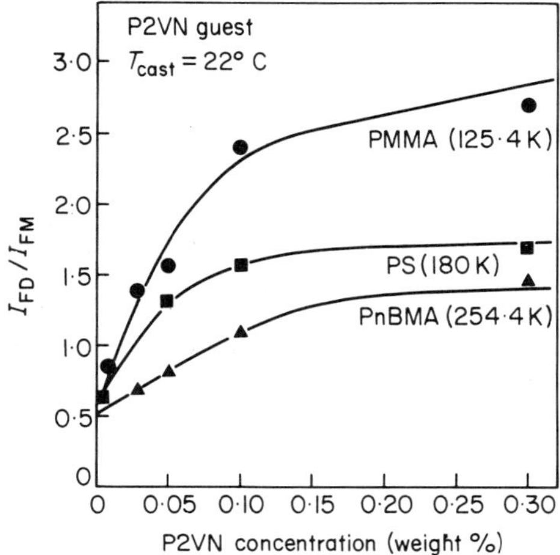

Fig. 3.16 Dependence of observed excimer fluorescence on the host polymer at very low values of concentration (after Frank and Gashgari [42]).

and is independent of the host polymer in that case. Energy migration is unidimensional along the chain. In the range of 0.003 to 0.1%, there is a rapid rise in I_{FD}/I_{FM} consistent with incipient phase separation. The major contribution to the increase of I_{FD}/I_{FM} is probably a change from uni- to tridimensional energy migration. The number of excimer-forming sites becomes the limiting factor of I_{FD}/I_{FM} at higher concentration. I_{FD}/I_{FM} varies linearly with the guest concentration (Fig. 3.17). The abrupt change of I_{FD}/I_{FM} was shown to occur before any sign of bulk phase separation is visible. Such an abrupt change of I_{FD}/I_{FM} also occurs as a function of host molecular weight at given guest concentration, and also precedes the appearance of cloudiness. The demarcation between clear and cloudy films is indicated by the dashed line in Fig. 3.17. The intermolecular contribution to I_{FD}/I_{FM} is obtained by substracting from the experimental data, the intercept of Fig. 3.17.

3.9.3 *Measurement of exciplex fluorescence* [45]

The monomers used in the synthesis of the copolymers are shown over:

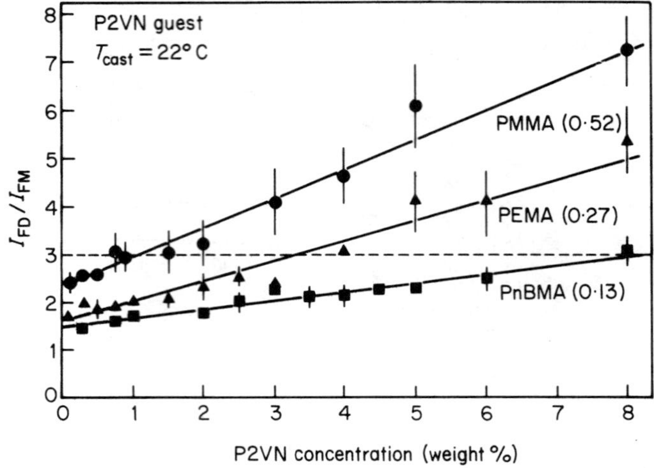

Fig. 3.17 Dependence of observed excimer fluorescence on guest concentration (after Frank and Gashgari [42]).

CH_2–CH_2–O–$COC(CH_3)$=CH_2 (on carbazole N)

CARB

CH_2=$C(CH_3)COOCH_3$

MMA

CH_3OOC–C_6H_4–$COOCH_2CH_2OCOC(CH_3)$=CH_2

TEREPH

CH_2=$CHCOOBu^n$

BA

Compatibility is measured between a copolymer containing the copolymerized energy donor (ex. MMA-co-CARB) and a second copolymer containing the copolymerized energy acceptor (ex. MMA-co-TEREPH). The intensity of the intermolecular exciplex formed between CARB and TEREPH is weak, and as a drawback of the method the amount of copolymerized donor and acceptor has to be rather high (see Fig. 3.18). The CARB emission is measured at 365 nm and the exciplex emission at 420-450 nm. Typical spectra are shown in Fig. 3.18. Examples of the obtained results of I_{FE}/I_{FM} as a function of the composition of the blends are given in Fig. 3.19. For the compositions indicated, BA-co-TEREPH is completely miscible with MMA-co-CARB.

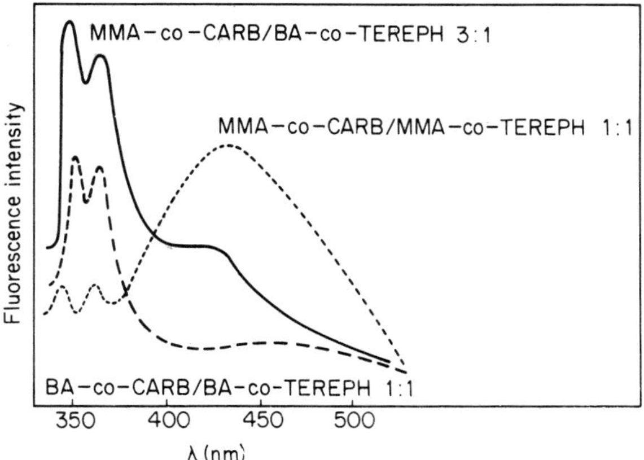

Fig. 3.18 Emission spectra of cast polymer films (after Ledwith [45]).

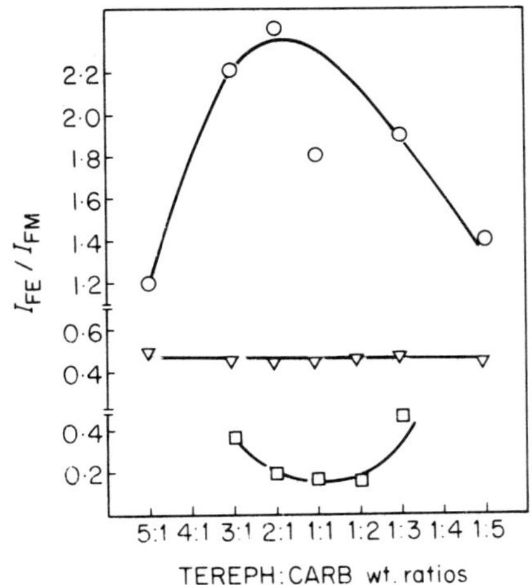

Fig. 3.19 Exciplex/carbazole emission ratios as a function of blend composition: O, MMA-co-TEREPH (18%)/MMA-co-TEREPH (23%), I_{FE} measured at 420 nm; ∇, BA-co-TEREPH (8%)/MMA-co-CARB (23%), I_{FE} measured at 425 nm; □ BA-co TEREPH (8%)/BA-co-CARB (10%), I_{FE} measured at 450 nm (after Ledwith [45]).

As a conclusion, fluorescence methods afford a convenient way to have an insight of the inter-polymer compatibility at levels much below those where phase separation would occur. Each of the proposed methods has advantages and drawbacks. The first and third one imply synthesis: incorporation of fluorescent groups or copolymerization. The first one requiring only a small amount of chromophore, the results can be related to the polymers to which the groups are attached whereas this is not true for the third method. The second method is only applicable to some special case of blends in which one of the components is a vinylaromatic polymer.

3.10 ENERGY MIGRATION AND FLUORESCENCE POLARIZATION

Fluorescence polarization, whose kinetics are described fully in Chapter 4, has been applied to the study of energy migration in films by David *et al.* [47]. By measuring the degree of polarization P of the fluorescence as a function of molar composition, energy migration was demonstrated to occur in three series of random copolymers films: styrene-methylmethacrylate, styrene-methylacrylate and vinylnaphthalene-methylmethacrylate.

In a fluorescent polymer or copolymer chain, the luminescent chromophores cannot be considered as uniformly distributed owing to the spatial requirements imposed by the main chain. The orientation of neighbouring chromophores is determined by the conformation of the chain and by steric factors. If energy migration, which is the consequence of a given number of energy transfer steps between monomer units, during the lifetime of the excited state takes place and depends on the relative orientation of the initially and finally excited chromophores, depolarization will occur and will depend on the copolymer composition. Table 3.3 shows that the depolarization and hence energy migration is not important for copolymers containing 5% or less of the fluorescent comonomer. In copolymers of low aromatic contents, long sequences of luminescent chromophores do not exist as can be calculated according to Harwood [48]. Transfer between non-neighbouring units is thus an unlikely event. When the mole fraction of the fluorescent comonomers attains or exceeds 15%, the number of long aromatic sequences rises abruptly, consequently, the number of migration steps along the chain during the lifetime of the excited state may become very high. Transfer

Table 3.3 Polarization* of the fluorescence of copolymer films.

(A) Styrene copolymers

λ_{exc} (nm)	St-MA films† St: 5%	St-MA films† St ⩾ 15%	St-MMA films† St ⩾ 25%	St-MA glassy sol.‡ St: 5%	St-MA glassy sol.‡ St: 15%
260	0.10	⩽ 0.02	⩽ 0.02	0.076	0.065
253.7	0.10	⩽ 0.02	⩽ 0.02	0.075	0.036

(B) 1-Vinylnaphthalene-methylmethacrylate copolymers

λ_{exc} (nm)	VN: 3% films§	VN ⩾ 35% films§‖	EtN¶ in PMMA films‡	VN: 3% glassy sol.‡	VN: 35% glassy sol.‡	EtN¶ glassy sol.‡
305	0.21	⩽ 0.02	0.19	0.09	0.065	0.11
290	0.19	⩽ 0.02	0.20	0.08	0.047	0.12
265	0.11	⩽ 0.02	0.16	0.05	0.038	0.09

* $p = (I_{VV} - GI_{VH}/I_{VV} + GI_{VH})$ where $G = (I_{HV}/I_{HH})$. The maximum theoretical depolarizing effect is 0.5 when the transition moments for absorption and emission are parallel.

† Measured at 290 nm - (monomer fluorescence).

‡ easured at λ_{max} - monomer fluorescence only is emitted by homo- and copolymers in glassy solution.

§ Measured at 330 nm - (monomer fluorescence).

‖ Measured at 430 nm - (excimer fluorescence).

¶ EtN = 1-ethylnaphthalene.

between neighbouring units situated sufficiently close to each other to allow exchange or multipole-multipole interaction can occur. This transfer contributes to depolarization since the chain conformation prevents parallelism of the transition moments to occur. It is to be noted that the efficiency of migration is much more important for films than for the same systems in dilute glassy solutions [49]. In the films, the important depolarization has been attributed to intrachains interactions in the statistical coil as in glassy solutions but also to an additional pathway due to interchain interactions. Neutron diffraction experiments on solid amorphous polymers [50] has indeed demonstrated the interpretation of different statistical coils in the bulk polymers.

3.11 CONCLUSION

Our knowledge concerning singlet energy migration and transfer in films is presently very approximate and incomplete. Indeed in most cases, only steady-state fluorescence measurements have been performed and the results interpreted according to rather rough models as described in part 1 of this chapter. The interpretation of the results would have to be confirmed by decay-time measurements in all cases. Time-resolved fluorescence spectra and study of the growth and decay of the emission, especially, in the short time scale (< 1 ns) have only been very recently applied to films and could bring extremely valuable information. The study of the photophysical properties of films of known tacticity as determined for instance by C^{13} NMR would also be very interesting. Indeed, it has been shown that in solid solution [46] the conformations of the chains which are favourable to excimer formation and hence are responsible for the ratio of excimer to monomer fluorescence are a function of the tacticity of the polymer. One of the main difficulties when studying the fluorescence of polymer films is however the presence of anomalous structures either situated at chain ends or resulting from accidents during the chain growth process. Owing to the importance of energy migration, these anomalous units could act as efficient energy traps even if their number is very low. Although our knowledge of the fundamental processes involved in energy migration and transfer is very superficial, it is nevertheless sufficient to be applied in different research fields. The contribution of the fluorescence to the study of the miscibility of polymer blends and to the orientation of polymer films is of considerable importance.

As a conclusion, much remains to be done concerning singlet excited state processes in polymer films.

3.12 REFERENCES

1. MacCALLUM, J.R. (1981) *Eur. Pol. J.* 17, 209.
2. BIRKS, J.B. (1970) *Photophysics of Aromatic Molecules*, Wiley/ Interscience, New York.
3. DEXTER, D.L. (1953) *J. Chem. Phys.* 21, 836.
4. FORSTER, T.H. (1948) *Ann. Physik* 2, 55.
5. KLOPFFER, W. (1969) *J. Chem. Phys.* 50, 1689.
6. KLOPFFER, W. (1969) *J. Chem. Phys.* 50, 2337.
7. BIRKS, J.B. and KUCHELA, K.N. (1959) *Disc. Faraday Soc.* 27, 57.
8. BASILE, L.J. (1964) *Trans. Faraday Soc.* 60, 1702.
9. JONES, P.F., SIEGEL, S. and NESBITT, R.S. (1972) Paper presented at the IUPAC Symposium on Photochemical Processes in Polymer Chemistry, Leuven.
10. HEISEL, F. and LAUSTRIAT, G. (1969) *J. Chim. Phys.* 66, 1895.
11. POWELL, R.C. (1971) *J. Chem. Phys.* 7, 1871.
12. WEBBER, S.E. and AVOTS-AVOTINS, P.E. (1981) *Macromolecules* 14, 105.
13. GHIGGINO, K.P., WRIGHT, R.D. and PHILLIPS, D. (1979) *J. Pol. Sci. Polym. Phys.* 16, 1499.
14. HOLDEN, D.A., WANG, P.Y.K. and GUILLET, J.E. (1980) *Macromolecules* 13, 295.
15. GUPTA, A., LIANG, R., MOACANIN, J., KLIGER, D., GOLDBECK, R., HORWITZ, J. and MISKOWSKI, V. (1981) *Eur. Pol. J.* 17, 485.
16. COULTER, D.R., LIANG, R.H., Di STEFANO, S., MOACANIN, J. and GUPTA, A. (1982) *Chem. Phys. Letters* 87, 594.
17. HARGREAVES, J.S. and WEBBER, S.E., Private Communication.
18. DAVID, C., LEMPEREUR, M. and GEUSKENS, G. (1972) *Eur. Pol. J.* 8, 417.
19. DAVID, C., DEMARTEAU, W. and GEUSKENS, G. (1970) *Eur. Pol. J.* 6, 1397.
20. BURKHART, R.D. and AVILES, R.G. (1979) *Macromolecules* 12(6), 1073.
21. JOHNSON, P.C. and OFFEN, H.W. (1971) *J. Chem. Phys.* 55, 2945.
22. JOHNSON, G.E. (1975) *J. Chem. Phys.* 62, 4697.

23. BURKHART, R.D. and AVILES, R.G. (1979) *J. Phys. Chem.* 83(14), 1897.

24. VENIKOUAS, G.E. and POWELL, R.C. (1975) *Chem. Phys. Letters* 34, 601.

25. KLOPFFER, W. (1981) *Annals N.Y. Acad. Sci.* 366, 373.

26. BURKHART, R.D. and AVILES, R.G. (1979) *Macromolecules* 12, 1078.

27. IINUMA, F., MIKAWA, H. and SHIROTA, Y. (1981) *Mol. Cryst. Liq. Cryst.* 73, 309.

28. SANO, Y., YOKOYAMA, M., SHIROTA, Y. and MIKAWA, H. (1979) *Mol. Cryst. Liq. Cryst.* 53, 291.

29. NORTH, A.M., ROSS, D.A. and TREADAWAY, M.F. (1974) *Eur. Pol. J.* 10, 411.

30. SAMEDOVA, T.G., KARPOCHEVA, G.P. and DAVYDOV, B.E. (1972) *Eur. Pol. J.* 8, 599.

31. GUILLET, J.E., HOYLE, C.E. and MacCALLUM, J.R. (1978) *Eur. Pol. J.* 54, 337.

32. SCHNEIDER, F. and SPRINGER, J. (1971) *Makromol. Chem.* 146, 181.

33. DAVID, C., PIENS, M. and GEUSKENS, G. (1973) *Eur. Pol. J.* 9, 533.

34. DAVID, C., LEMPEREUR, M. and GEUSKENS, G. (1973) *Eur. Pol. J.* 9, 1315.

35. DAVID, C., LEMPEREUR, M. and GEUSKENS, G. (1974) *Eur. Pol. J.* 10, 1181.

36. OKAMOTO, K., YANO, A., KUSABAYASHI, S. and MIKAWA, H. (1974) *Bull. Chem. Soc. Japan* 47, 749.

37. JOHNSON, G.E. (1980) *Macromolecules* 13, 145.

38. TURRO, N.J., KOCHEVAR, I.E., NOGUSHI, Y. and CHOW, M.F. (1978) *J.A.C.S.* 100(10), 3170.

39. AMRANI, F., HUNG, J.M. and MORAWETZ, H. (1980) *Macromolecules* 13, 649.

40. MIKES, F., MORAWETZ, H. and DENNIS, K.S. (1980) *Macromolecules* 13, 969.

41. MORAWETZ, H. (1981) *Annals. N.Y. Acad. Sci.* 366, 387.

42. FRANK, C.W. and GASHGARI, M.A. (1981) *Annals. N.Y. Acad. Sci.* 366, 387.

43. SEMERAK, S.N. and FRANK, C.W. (1981) *Macromolecules* 14, 443.

44. FRANK, C.W. and GASHGARI, M.A. (1978) *Macromolecules* 12, 165.

45. LEDWITH, A., (1982) *Pure Appl. Chem.* 54, 549.

46. FRANK, C.W. and HARRAH, L.A. (1974) *J. Chem. Phys.* 61, 1526.

47. DAVID, C., PUTMAN-de LAVAREILLE, N. and GEUSKENS, G. (1977) *Eur. Pol. J.* 13, 15.

48. HARWOOD, H.J. (1965) *Angew. Chem. Int. Ed.* 4, 394.

49. DAVID, C, BAEYENS-VOLANT, D. and GEUSKENS, G. (1976) *Eur. Pol. J.* 12, 71.

50. BALLARD, D.G., WIGNALL, G.D. and SCHALTEN, J. (1973) *Eur. Pol. J.* 9, 965.

51. VOLTZ, R., KLEIN, J., HEISEL, F., LAMI, H., LAUSTRIAT, G. and COCHE, A. (1966) *J. Chim. Phys.* 63, 1259.

Note to Section 3.2

In all this chapter except here, monomer emission refers to the fluorescence of the $^1M^*$ chromophore and has nothing to do with residual unsaturated molecules.

CHAPTER FOUR

Polarized fluorescence in oriented polymers

4.1 INTRODUCTION

There are two major aspects to the importance of polarized fluorescence studies of oriented polymers. Firstly, there is the possibility that this technique will provide a different measure of orientation from X-ray diffraction,optical birefringence, broad line NMR, etc, because the fluorescent probe molecules are associated with a specific part of the structure. A most useful situation would be that the probe molecules characterized the orientation of the non-crystalline regions in a semi-crystalline polymer, and it will be shown that this expectation is probably realized in the case of oriented polyethylene terephthalate. Secondly, because the polarized fluorescence method can be used to obtain fourth-order orientation averages as well as second-order averages, it should provide greater insight into the distribution of molecular orientations of the polymer or part of the polymer structure that is labelled by the probe molecules. This is important in obtaining greater understanding of the deformation mechanisms, for example the relationship of the measured molecular orientation distribution function as a function of the draw ratio to that predicted on the basis of theoretical models such as that for a rubber-like network.

A central issue in the application of the fluorescence technique is the precise situation of the fluorescent probe and its direction with respect to the molecular chain axis. A related issue is the fact that the absorption and emission axes of the probe will not in general be parallel. The fluorescent probe can be introduced into the polymer system in three ways:

1. By homogeneous dispersion in the polymer melt or in a solution containing both polymer and probe.
2. By diffusion into the oriented structure.
3. By direct chemical incorporation into the polymer chain.

In the first two methods it is assumed that the presence of small quantities of probe molecules does not perturb the structure to be monitored. In all cases it will still be required to establish how the probe orientation typifies the orientation of the polymer or part of the polymer structure. Polarized fluorescence measurements will therefore be likely to be most useful when they are used in combination with other structural techniques, and this will be illustrated later.

In a typical experimental arrangement a sample containing the fluorescent molecules is illuminated with plane polarized light of a suitable wavelength. The absorbed light is re-emitted at a longer wavelength and its intensity and degree of polarization measured by passing it through an analyser to a light-detection system. A general theory that relates the fluorescence intensities to the distribution of orientations, motion and local environment of the fluorescent molecules for any possible experimental arrangement of illumination and observation directions is very complex. Suitable choice of experimental conditions greatly simplifies the theory and for this reason two types of experiment evolved. Molecular motion was commonly studied in isotropic systems where the polarization of the fluorescence depends on the ratio of the relaxation times of the various molecular motions to the excited state lifetime of the fluorescent molecule [1-3]. Molecular orientation was studied in anisotropic solid polymer systems in which molecular motion could be neglected so that the fluorescence intensity depends on the distribution of orientations of the fluorescent molecules [4,5]. Although in this latter case some consideration was given to the possible influence of molecular motion [6-8], it is only recently that formal treatments combining the two aspects have been developed in order to allow the study of liquid crystals and oriented polymer systems at elevated temperatures [1,5,9,10]. In parallel to the study of the properties of the host material, spectroscopists have been using the technique of incorporating fluorescent molecules into oriented systems to obtain information about the optical properties such as transition moment directions [11,12].

The development of the theory of the polarized fluorescence technique will be presented first, followed by a review of its application to the study of oriented polymers. For simplicity, the theory for the specific application to determine molecular orientation in anistropic solid polymer materials will be presented. We therefore assume that there is negligible motion of the fluorescent molecule during its excited-state lifetime.

4.2 THE FLUORESCENCE PHENOMENON [13,14]

As explained in Chapter 1, when a fluorescent molecule is illuminated with radiation in a suitable region of the spectrum the ground singlet electronic state S_0 is raised to a higher energy level S_n where n will depend on the wavelength used. Each $S_0 \rightarrow S_n$ absorption will have a different transition moment vector associated with it, so that the observed absorption spectrum will depend on the polarization of the incident light. If the incident radiation consists of a fixed band of wavelengths, then the absorption can be described by a single oscillating electric dipole with a dipole moment parallel to a unit vector **A** fixed within the molecule. The probability of excitation of the molecule when the incident light is plane polarized along a direction **T** is given by:

$$(\mathbf{A}.\mathbf{T})^2.$$

The transition $S_n \rightarrow S_1$ occurs rapidly by vibrational decay, and it is the $S_1 \rightarrow S_0$ transition that results in the emission of radiation known as fluorescence. The molecules fall to higher vibrational levels in the S_0 levels and the emission is red-shifted. In a similar way to the absorption process, the emission behaviour for a fixed band of wavelengths can be described by an emission oscillator with dipole moment parallel to unit vector **B** fixed within the molecule. **A** and **B** may or may not be parallel. The component of emitted radiation transmitted by a perfect analyser that transmits light polarized along direction **T'** is given by:

$$(\mathbf{B}.\mathbf{T}')^2.$$

The observed intensity will be proportional to:

$$\cos^2 A \times \cos^2 B \qquad (4.1)$$

When A and B are angles between A and T and B and T', respectively. It is apparent from this expression that if a relationship exists between A and B then the measured fluorescence intensities from an assembly of fluorescent molecules should be capable of providing average values of both $\cos^2 A$ and $\cos^4 A$. Absorption dichroism measurements would only provide an average value of $\cos^2 A$.

Competing with the emission process are radiationless transitions to the ground state, and the relative magnitude of the rate constants for the decay processes determines the quantum yield. Implicit in the development of the theory is the assumption that the quantum yield is independent of probe-molecule orientation. For several of the decay processes the rate constant, and hence quantum yield, is sensitive to the local environment. Oriented polymers are very heterogeneous systems at the molecular level, and so it is feasible that the quantum yield from a probe molecule in a highly ordered oriented local environment will be significantly different from that in a disordered one [15].

The complete problem is to calculate the total fluorescent intensity from a sample consisting of an array of fluorescent molecules dispersed in a solid polymer. This will depend on the orientations of the fluorescent molecules, the birefringence of the sample and the rate of absorption of the exciting light by the sample. The problem involves therefore, transformations between systems of coordinates on each fluorescent molecule and a system of coordinates fixed within the sample. Since the emission is incoherent, each fluorescent molecule is treated independently, and the resulting fluorescent light intensities are added to find the total light intensities emitted by the array. The approach adopted by most workers is to ignore initially the effects of an anisotropic refractive index on the measured intensities and treat these, together with other possible corrections, separately.

4.3 THE DESCRIPTION OF THE DISTRIBUTION OF MOLECULAR ORIENTATIONS

It is sensible to present the information obtained from different experimental methods in a way that allows the easy comparison of results and permits the testing of theoretically predicted distribution functions. Roe [16] and Krigbaum [17] developed a method for obtaining information about the orientation distribution function for crystallites from X-ray diffraction measurements based on its expansion in a series of generalized spherical harmonics [18]. The expansion is well suited to handling problems relating to the orientation distribution of structural units in anisotropic samples, and Roe and other workers extended the application to the analysis of polarized fluorescence and wide-line NMR [19-21]. The spherical harmonic series expansion has since become the most common way of representing the distribution function.

A coordinate system $Ox_1x_2x_3$ is imagined fixed within each of the molecular units under consideration. The orientation of a particular unit M with respect to a coordinate system $OX_1X_2X_3$ fixed in the sample is described by the Euler angles θ,ϕ and ψ as shown in Fig. 4.1. θ and ϕ denote the polar and azimuthal of the Ox_3 axis with respect to $OX_1X_2X_3$ and ψ denotes the extent of rotation of the structural unit around Ox_3. The two coordinate systems are usually chosen so that their axes are coincident with the respective symmetry axes of the polymer sample and structural unit (if there are any).

The distribution of structural unit orientations is described by the function $N(\theta,\phi,\psi)$ normalized so that $N(\theta,\phi,\psi)\sin\theta\,d\theta\,d\phi\,d\psi$ represents the fraction of units whose axes lie within the solid angle $\sin\theta\,d\theta\,d\phi\,d\psi$. $N(\theta,\phi,\psi)$ may be expressed by an expansion in spherical harmonics:

$$N(\theta,\phi,\psi) = \sum_{\ell=0}^{\infty} \sum_{m=-\ell}^{\ell} \sum_{n=-\ell}^{\ell} a_{\ell mn} P_{\ell mn}(\cos\theta) \exp(-im\phi)\exp(-in\psi). \qquad (4.2)$$

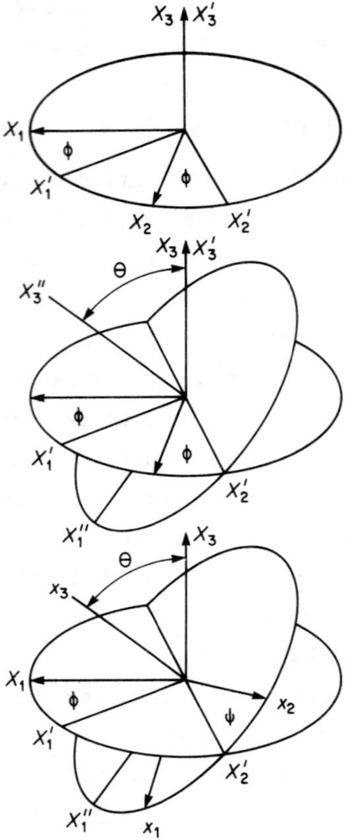

Fig. 4.1 The specification of the orientation of the axes $x_1x_2x_3$ of a structural unit with respect to the reference frame $X_1X_2X_3$.

Here $P_{\ell mn}(\cos\theta)$ is a generalization of Legendre functions [17], and:

$$P_{\ell m0}(\cos\theta) = P_{\ell}^{m}(\cos\theta).$$

$P_{\ell}^{m}(\cos\theta)$ being the normalized associated Legendre function. The functions $P_{\ell mn}(\cos\theta)\exp(-im\phi)\exp(-in\psi)$ are mutually orthogonal and the coefficients $a_{\ell mn}$ are determined from:

$$a_{\ell mn} = \frac{1}{4\pi^2} \int_0^{2\pi} \int_0^{2\pi} \int_0^{\pi} N(\theta,\phi,\psi)\, P_{\ell mn}(\cos\theta)\exp(-im\phi)\exp(-in\psi)\sin\theta\, d\theta d\phi d\psi \tag{4.3}$$

When all the $a_{\ell mn}$ are known, the distribution function is known completely. In principle the values of $a_{\ell mn}$ up to $\ell = 4$ may be determined from fluorescence meaurements. The number of independent coefficients depends on the symmetry of the structural unit and of the orientation distribution. When the unit has orthorhombic symmetry, that is the distribution remains unchanged when any unit is rotated through 180° about any one of the axes Ox_3 Ox_2 and Ox_1, only even positive values of n are allowed [21]. If the orientation distribution also has orthorhombic symmetry then only $a_{\ell mn}$ for even values of ℓ, m and n are non-zero [21,22].

4.4 UNIAXIAL OR FIBRE SYMMETRY

For distribution of uniaxial symmetry about the OX_3 axis with no preferred orientation of the structural unit around OX_3, the distribution function is independent of ϕ and ψ and all coefficients with ℓ odd, $m \neq 0$ and $n \neq 0$ vanish. The redundant subscripts are usually ommitted from the expression:

$$N(\theta) = \sum_{\ell=0}^{\infty} a_\ell P_\ell\,(\cos\theta). \tag{4.4}$$

As before the function is normalized so that:

$$\int_0^{\pi} N(\theta)\,\sin\theta\; d\theta = 1 \tag{4.5}$$

and P_ℓ $(\cos\theta)$ are the associated Legendre polynomials P_ℓ^m $(\cos\theta)$ with $m = 0$. Information about the distribution can be presented in any of three ways:

Firstly by the coefficients a_ℓ; secondly by the average values of the Legendre polynomials $\langle P_\ell\,(\cos\theta)\rangle$; and thirdly by $\langle\cos^n\theta\rangle$, where the angled brackets are used to denote the average values. Each form can be generated from any one of the others, for example:

$$\langle P_\ell(\cos\theta)\rangle = \frac{2}{(2\ell+1)} a_\ell \qquad (4.6)$$

and from the definitions of Legendre polynomials:

$$\langle\cos^2\theta\rangle = 1/3\ (2\langle P_2(\cos\theta)\rangle + 1) \qquad (4.7)$$

$$\langle\cos^4\theta\rangle = 1/35\ (8\langle P_4(\cos\theta)\rangle + 20\langle P_2(\cos\theta)\rangle + 7)$$

and conversely:

$$\langle P_2(\cos\theta)\rangle = 1/2\ (3\langle\cos^2\theta\rangle - 1) \qquad (4.8)$$

$$\langle P_4(\cos\theta)\rangle = 1/8\ (35\langle\cos^4\theta\rangle - 30\langle\cos^2\theta\rangle + 3).$$

Of the three forms, $\langle P_\ell(\cos\theta)\rangle$ is to be preferred [23], since the orthogonality of the $P_\ell(\cos\theta)$ makes the inter-relationship of orientation distributions or the rotation of reference axes simpler to express. In addition, the values of $\langle P_\ell(\cos\theta)\rangle$ all have the simple limits $0 \leqslant \langle P_\ell(\cos\theta)\rangle \leqslant 1$, whereas the limits on a_ℓ and $\langle\cos^\ell\rangle$ depend on ℓ.

X-ray scattering [24] can in principle measure the value of the averages $\langle P_\ell(\cos\theta)\rangle$ for all ℓ, but in practice it is limited, especially in oriented amorphous and partially crystalline polymers. Optical birefringence, infra-red, visible and ultra-violet dichroic measurements are limited to the determination of $\ell = 2$ averages. Both the second- ($\ell = 2$) and forth-order ($\ell = 4$) averages are obtainable by suitable analysis of polarized fluorescence, polarized Raman spectroscopy and NMR, from which higher moments can also be obtained in favourable circumstances. Although the series expansion with only the first three coefficients known cannot accurately describe even a moderately oriented sample, the measured orientation averages can be directly related to the anisotropy in the mechanical properties of the specimen [25].

The spherical harmonic expansion is one of many possible methods of representing the distribution function. However, Anderson and Norden [26] have shown that the values of a_ℓ corresponding to the average values are those that minimize the square of the deviation of the truncated series from the true function. The truncated spherical harmonic expansion, limited though it is, is the best available approximation to the distribution function. Several authors have considered the constraints on the distribution function imposed by its physical interpretation. $N(\theta,\phi,\psi)$ must be everwhere positive (or zero), continuous and of even symmetry. In addition there are limits on the possible value of each coefficient for given values of the lower-order coefficients, as discussed by Nomura *et al.* [20], Nobbs *et al.* [27] and Bower [23]. Bower [23,28] has also developed the concept of the most probable form of $N(\theta)$ as the one for which the entropy of orientation is a maximum consistent with the observed $\langle P_\ell(\cos\theta)\rangle$. This leads to:

$$N_{mp}(\theta) = \exp\left(\sum_\ell \alpha_\ell \, P_\ell(\cos\theta) \right) \qquad (4.9)$$

where the α_ℓ are Lagrange multipliers chosen so that the value of $\langle P_\ell(\cos\theta)\rangle$ for the distribution of orientations should be equal to the observed values.

4.5 THE MODEL OF THE FLUORESCENCE MOLECULE

The complexity of the equations relating the fluorescence intensities to the orientation averages depends on the symmetry of the orientation distribution of the fluorescent molecules and on the model used to relate the absorption and emission moment directions. The first treatments of Nishijima and coworkers [29-37] assume that the emission and absorption moments are coincident. The more detailed theories of Desper and Kimura [19] and Roe [21] retain this assumption. However, experimental studies provided evidence that this assumption was not valid in most cases,and at around the same time Rupprecht *et al.* [38] and Kimura *et al.* [6] published theories with non-coincident absorption and emission moment directions. Non-coincidence can arise from fixed but non-parallel absorption and emission axes, parallel axes with resonance energy transfer between non-parallel probes [1,39] or motion of the probe molecule during the fluorescence lifetime. Some treatments of this model define the direction of the absorption and emission moments by their respective polar and azimuthal angles in the $OX_1X_2X_3$ frame and then consider either a

Fig. 4.2 (a) Orientation of a unique axis **M** of a molecule with respect to the symmetry axes $OX_1X_2X_3$ of the sample. (b) Orientation of absorption and admission axes, AB respectively, with respect to axes $Ox_1x_2x_2$ fixed in the fluorescence molecule.

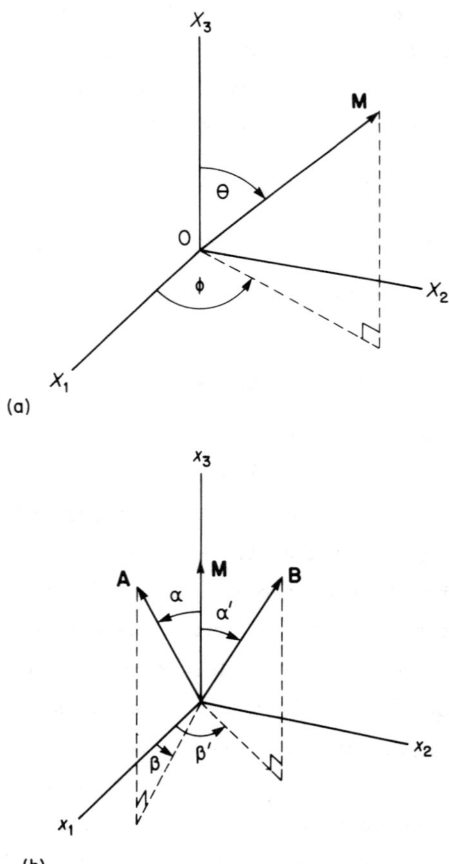

fixed or a distribution of values for the angle α'' between them. The relationship between absorption and emission moment directions is more easily treated by defining their directions relative to a system of coordinates fixed within the fluorescent molecule itself. For example, the molecular frame $Ox_1x_2x_3$ shown in Fig. 2(b) with Ox_3 aligned parallel to a unique axis **M**

fixed within the fluorescent molecule (e.g. the long axis of a cylindrically shaped molecule). The absorption and emission moment directions are then defined by the polar and azimuthal angles α, β and α', β', respectively, as shown in Fig. 2(b). An angle ε is defined as the difference between the azimuthal angles $\varepsilon = \beta - \beta'$. A specific model of this type was considered by Kimura *et al.* [6] in which the absorption and emission moments each have independent, cylindrical symmetric distributions around **M**. A more general treatment was given by Nobbs *et al.* [27]. The unique axes **M** are assumed to have a distribution of orientations that is related to that of the polymer chains and so has the same symmetry. (The actual nature of the relationship is uncertain and is considered later.) This latter model incorporates the previous two as special cases, and versions differing only in the assumptions concerning the distributions of the transition moment directions around **M** have been used by Margulies and Yogev [11], Chapoy and Dupré [5] and by Jarry and Monnerie [9].

4.6 DEVELOPMENT OF THEORY

To avoid the complication from refraction of the light a the sample surfaces we shall assume that the incident plane-polarized monochromatic light enters the sample, and the observed fluorescent light leaves the sample, normal to its surface.

It follows from Equation (4.1) that the observed intensity from the array of **N** fluorescent molecules is given by:

$$I = I_0 \ \Sigma \cos^2 A \cos^2 B$$

where I_0 is a constant depending on instrumental factors, the exciting light intensity and the quantum yield of the fluorescent molecules. The summation is over all the fluorescent molecules contributing to the observed intensity. Let the unit vector **T** describing the direction of polarization of the exciting light have polar coordinates γ, η in the $OX_1X_2X_3$ frame. Its represention in matrix form (T_i) is:

$$(T_i) = \begin{pmatrix} \sin\gamma\ \cos\eta \\ \sin\gamma\ \sin\eta \\ \cos\gamma \end{pmatrix}. \qquad (4.10)$$

In a similar manner the unit vector **A** describing absorption transition moment direction in the $Ox_1x_2x_3$ frame is given by (A_i) where:

$$(A_i) = \begin{pmatrix} \sin\alpha\ \cos\beta \\ \sin\alpha\ \sin\beta \\ \cos\alpha \end{pmatrix}. \qquad (4.11)$$

By making use of a transformation matrix (O_{ij}) that relates quantities expressed with respect to $OX_1X_2X_3$ to those expressed with respect to $Ox_1x_2x_3$, $\cos^2 A$ can be written as:

$$\cos^2 A = \left(A_i\ O_{ij}\ T_j \right)^2 \qquad (4.12)$$

where the Einstein repeated subscript convention is used to denote summation. In a similar manner $\cos^2 B$ can be written as:

$$\cos^2 B = \left(T_j'\ O_{ij}'\ B_j \right)^2 \qquad (4.13)$$

where O_{ij}' is the transpose of O_{ij}. (T_i') is the matrix form of the unit vector **T** parallel to the transmission direction of the analyser. **T'** has polar and azimuthal angles γ' and η' in the $OX_1X_2X_3$ frame:

$$(T_i') = \begin{pmatrix} \sin\gamma'\ \cos\eta' \\ \sin\gamma'\ \sin\eta' \\ \cos\gamma' \end{pmatrix} \qquad (4.14)$$

and

$$(B_i) = \begin{pmatrix} \sin\alpha'\ \cos\beta' \\ \sin\alpha'\ \sin\beta' \\ \cos\alpha' \end{pmatrix} \qquad (4.15)$$

Since $Ox_1x_2x_3$ can be imagined to be brought from initial coincidence with $OX_1X_2X_3$ into actual orientation by successive rotations of ϕ about OX_3, θ about Ox_2 and ψ about Ox_3, (O_{ij}) is given explicitly by:

$$(O_{ij}) = \begin{pmatrix} \cos\theta\cos\phi\cos\psi - \sin\phi\sin\psi & \cos\theta\sin\phi\cos\psi + \cos\phi\sin\psi & -\sin\theta\cos\psi \\ -\cos\theta\cos\phi\sin\psi - \sin\phi\cos\psi & -\cos\theta\sin\phi\sin\psi + \cos\phi\cos\psi & \sin\theta\sin\psi \\ \sin\theta\cos\psi & \sin\theta\sin\psi & \cos\theta \end{pmatrix} \tag{4.16}$$

Expressing the fluorescence intensity in terms of these matrices the following equation is obtained:

$$I = I_o \,\Sigma\, (A_i\, O_{ij}\, T_j)^2\, (T_p{}'\, O_{pq}{}'\, B_q)^2. \tag{4.17}$$

The summation may be replaced by the average values of the independent terms. $(O_{ij}{}')$ is the transpose of (O_{ij}) and thus Equation (4.17) becomes:

$$I = G \langle A_i A_\ell B_q B_t \rangle \langle O_{ij} O_{\ell m} O_{pq}{}' O_{st}{}' \rangle\, T_j T_m T_p{}' T_s{}' \tag{4.18}$$

where G is I_o N.

Performing the sum over all values of i, ℓ, q and t leads to:

$$I = G\, F_{jmps}\, T_j\, T_m\, T_p{}'\, T_s{}' \tag{4.19}$$

where:

$$F_{jmps} = \langle A_i A_\ell B_q B_t \rangle \langle O_{ij} O_{\ell m} O_{qp} O_{ts} \rangle. \tag{4.20}$$

The intensity is now expressed as a sum of the products of pairs of terms. One term of each pair, F_{jmps} contains the information about the distribution of orientations and is in fact a component of a fourth-rank tensor [F_{jmps}] . From the definition of F_{jmps} it is seen that interchange of j and m or p and s in Equation (4.20) does not change its value and so there are at most only 36 independent components [4,8,9]. For a sample with orthotropic symmetry, only 12 of these terms are non-zero, nine of the type F_{jjpp} and three of the type F_{jpjp} [4,27]. The nine components of the form F_{jjpp} correspond to the nine intensities I_{jp} defined by Desper and Kimura [19]. Each of these components can be measured directly. When the polarization vectors of the polarizer and analyser are parallel to axes of the $OX_1X_2X_3$ frame, one of the quantities $T_jT_jT'_pT'_p$ is unity and all the other quantities $T_jT_mT'_pT'_s$ are zero. The intensity measured is therefore proportional to the value of the corresponding component F_{jjpp}. The determination of the components F_{jpjp} requires measurements in which both polarization vectors are inclined to the axes of the $OX_1X_2X_3$ frame.

For such measurements the influence of birefringence of the polymer sample must be considered. All the information available about the distribution of orientations from measurements of fluorescence intensities is contained in these twelve quantities.

Each component F_{jmps} is the product of two terms. The first of these $\langle A_iA_\ell B_qB_t\rangle$ depends only on the properties of the fluorescent molecules and the second $\langle O_{ij}O_{\ell m}O_{qp}O_{ts}\rangle$, depends only on functions of θ, ϕ and ψ averaged over the distribution of orientations; it is these averages that we wish to determine.

Each of the twelve components F_{jmps} depends, in the most general case, on many unknown quantities. In order to reduce the number of unknown quantities so that it is equal to or less than the number of independent intensities that can be measured, assumptions about the optical behaviour of the fluorescent molecule must be made [27]. A great simplification is made by assuming that the fluorescent molecules have no preferred orientation around their Ox_3 axes and that if every fluorescent molecule is rotated through 180° around any axis normal to Ox_3 the distribution of orientations remains unchanged. These assumptions appear to be reasonable for a long thin fluorescent molecule that does not bind chemically to the polymer chains. The Ox_3 axis would then be expected to be the geometric long axis of the molecule.

Making these assumptions is equivalent to setting $\psi = 0$ in Equation (4.16) and regarding β and β' as random while retaining the relationship $\varepsilon = \beta - \beta'$ between them. These assumptions lead to a reduction of the number of unknown quantities to eleven, which are five averages of functions of θ and ϕ, the unknown constant G and the quantities $\langle\cos^2\alpha\rangle$, $\langle\cos^2\alpha'\rangle$ $\langle\cos^2\alpha\cos^2\alpha'\rangle$, $\langle\cos^2\varepsilon\sin^2\alpha\sin^2\alpha'\rangle$ and $\langle\cos\varepsilon\sin^2\alpha\sin^2\alpha'\rangle$. The number of independent components F_{jmps} is correspondingly reduced to eleven, since the following relationship now holds:

$$(F_{1122} - F_{2211}) + (F_{2233} - F_{3322}) + (F_{3311} - F_{1133}) = 0. \qquad (4.21)$$

In principle the eleven independent F_{jmps} could be used to evaluate the eleven unknown quantities unless $\langle\cos^2\alpha\rangle = \langle\cos^2\alpha'\rangle$, when each of the differences in Equation (4.21) is individually zero.

It may be noted that if $\alpha = \alpha' = 0$, the results of the present treatment are equivalent to those of Bower [22]. If α and α' are non-zero and ε is assumed to be random, then the treatment is equivalent to one given by Kimura *et al.* [6], although they disregard the components F_{jpjp}.

When the absorption and emission axes coincide ($\alpha = \alpha'$) then $F_{jpjp} = F_{jjpp} = F_{ppjj}$ and, as first pointed out by Desper and Kimura [19], only six independent components determine the observed intensity:

$$\begin{aligned}
F_{1111} &= G\langle\sin^4\theta\ \cos^4\phi\rangle \\
F_{2222} &= G\langle\sin^4\theta\ \sin^4\phi\rangle \\
F_{3333} &= G\langle\cos^4\theta\rangle \\
F_{1122} &= G\langle\sin^4\theta\ \sin^2\phi\ \cos^2\phi\rangle \\
F_{2233} &= G\langle\sin^2\theta\ \cos^2\theta\ \sin^2\phi\rangle \\
F_{3311} &= G\langle\sin^2\theta\ \cos^2\theta\ \cos^2\phi\rangle.
\end{aligned} \qquad (4.22)$$

As pointed out by Bower [4] these equations contain five independent orientation averages.

For a sample of uniaxial symmetry each component F_{jmps} must be invariant under rotation through any arbitrary angle around the Ox_3 axis, which is chosen to be parallel to the unique axis of the specimen. This reduces the number of independent components to six and the corresponding quantities F_{jmps} may be shown to be given by:

$$\begin{aligned}
F_{1111} = F_{2222} &= G\ (A_0\langle\cos^4\theta\rangle + A_1\langle\cos^2\theta\rangle + A_2) \\
F_{3333} &= 8G\ (A_3\langle\cos^4\theta\rangle + A_4\langle\cos^2\theta\rangle + A_5) \\
F_{2233} = F_{1133} &= 4G\ (A_6\langle\cos^4\theta\rangle + A_7\langle\cos^2\theta\rangle + A_8) \\
F_{3322} = F_{3311} &= 4G\ (A_9\langle\cos^4\theta\rangle + A_{10}\langle\cos^2\theta\rangle + A_{11}) \\
F_{2323} = F_{1313} &= 4G\ (A_{12}\langle\cos^4\theta\rangle + A_{13}\langle\cos^2\theta\rangle + A_{14}) \\
F_{1122} = F_{2211} &= G\ (A_{15}\langle\cos^4\theta\rangle + A_{16}\langle\cos^2\theta\rangle + A_{17}).
\end{aligned} \quad (4.23)$$

In addition, the following relationship holds:

$$2F_{1212} = F_{1111} - F_{1122}. \quad (4.24)$$

The eighteen coefficients A_0 to A_{17} depend only on the following averages:

$\langle\cos^2\alpha\rangle$, $\langle\cos^2\alpha'\rangle$, $\langle\cos^2\alpha\cos^2\alpha'\rangle$, $\langle\cos^2\varepsilon\sin^2\alpha\sin^2\alpha'\rangle$ and $\langle\cos\varepsilon\sin 2\alpha\sin 2\alpha'\rangle$.

Unfortunately the number of unknown quantities again exceeds the number of independent intensities that can be measured. It is necessary therefore, to make some further assumptions. One possibility is to assume that if α and α' are not the same for every fluorescent molecule the distribution of values is sufficiently narrow that some effective mean values can be used [27], so that the unknown quantities that depend on the properties of the fluorescent molecules but not on their orientations become α, α', $\langle\cos^2\varepsilon\rangle$, $\langle\cos\varepsilon\rangle$.

The forms of the coefficients A_0 to A_{17} that result when this assumption is made are given by Table 4.1. If A represents one of the coefficients then:

$$A = D + K(\cos^2\alpha + \cos^2\alpha') + L(\cos^2\alpha - \cos^2\alpha') + M\cos^2\alpha\cos^2\alpha' + N\langle\cos\varepsilon\rangle\sin^2\alpha\sin^2\alpha'$$

If $\langle\cos^2\varepsilon\rangle$ and $\langle\cos\varepsilon\rangle$ are fixed, the six Equations (4.23) contain six unknowns and can be solved.

Several simpler relationships can be produced from these equations. The intensity constant of fluorescence, G, is given by:

$$G = \frac{1}{64}\left[F_{3333} + 2(F_{1111} + F_{1122} + F_{3322} + F_{2233})\right]. \tag{4.25}$$

Comparison of the equations show that when $\alpha \neq \alpha'$ then $F_{3322} \neq F_{2233}$:

$$F_{3322} - F_{2233} = 16G\,(3\langle\cos^2\theta\rangle - 1)\,(\cos^2\alpha - \cos^2\alpha'). \tag{4.26}$$

The measured fluorescence intensities often show such an anisotropy; however it may also be caused by anisotropic absorption [27] or by scattering [40] of the exciting and fluorescent light as they pass through the sample. The intensities must be corrected for these effects before any conclusions concerning $(\cos^2\alpha - \cos^2\alpha')$ can be drawn.

A simple relation exists for $\langle\cos^2\theta_A\rangle$, where θ_A is the angle between the absorption direction and the OX_3 axis:

$$F_{3333} + 2F_{3322} = 32G\,(\langle\cos^2\theta\rangle\,(3\cos^2\alpha - 1) + 1 - \cos^2\alpha) = 64G\langle\cos^2\theta_A\rangle. \tag{4.27}$$

If $\cos^2\alpha$ is known then $\langle\cos^2\theta\rangle$ can be immediately calculated.

For the long, cylindrical molecules used as fluorescent probes, experimental results indicate that often the distribution of absorption transition moment directions is identical to that of the emission transition moments, but the distributions are independent of each other ($\cos^2\alpha = \cos^2\alpha'$, ε = random). In this case orientation averages can be determined from three fluorescence components and measurements of an isotropic sample as follows.

For a specimen in which the molecules are randomly oriented there are only two independent components F_{jjjj} and F_{jjpp}:

$$I_{||} = F_{1111} = F_{2222} = F_{3333} \tag{4.28}$$

$$I_{\perp} = F_{jjpp} \text{ for } j \neq p \tag{4.29}$$

where $I_{||}$ and $I_{\perp}$ denote the intensity observed when the polarizer and analyzer transmission directions are parallel and crossed, respectively.

From Equations (4.28) and (4.29) we obtain:

$$R = I_{||}/I_{\perp} = (1+2\langle\cos^2\alpha''\rangle)/(2-\langle\cos^2\alpha''\rangle) \tag{4.30}$$

where:

$$\langle\cos^2\alpha''\rangle = \chi(1-\cos^2\alpha - \cos^2\alpha') + (1+\chi)\cos^2\alpha\cos^2\alpha' + \xi/2 \tag{4.31}$$

where $\chi = \langle\cos^2\varepsilon\rangle$ and $\xi = \langle\cos\varepsilon\rangle\sin 2\alpha\sin 2\alpha'$. $\langle\cos^2\alpha''\rangle$ is simply the average of the square of the cosine of the angle between the absorption and emission axes of the fluorescent molecule.

When $\cos^2\alpha = \cos^2\alpha'$ and ε is random:

$$\cos^2\varepsilon = \tfrac{1}{3}[(10r_o)^{\frac{1}{2}}+1]. \tag{4.32}$$

r_o is the emission anisotropy:

$$r_o = \frac{I_{||} - I_{\perp}}{I_{||} + 2I_{\perp}} = \frac{R-1}{R+2} \tag{4.33}$$

$\cos^2\alpha$ may then be used to calculate the intensity constant G:

$$G = \frac{1}{64}\left[\frac{F_{3333}(1 + 2\cos^2\alpha) + 8F_{2222} + 4F_{3322}(2 + \cos^2\alpha)}{2 + (\cos^2\alpha)^2}\right] \quad (4.34)$$

and the orientation averages:

$$\langle\cos^2\theta\rangle = \frac{2F_{3333} + 4F_{3322} - (1 - \cos^2\alpha)K}{(3\cos^2\alpha - 1)K} \quad (4.35)$$

$$\langle\cos^4\theta\rangle = \frac{4\cos^2\alpha\, F_{3333} - 8(1 - \cos^2\alpha)F_{3322} + (1 - \cos^2\alpha)^2 K}{(3\cos^2\alpha - 1)^2 K}. \quad (4.36)$$

Where $K = 64G$. Equations (4.34) - (4.36) are similar in form to those developed by Hennecke and Fuhrmann [15], the choice of a different method of expressing them has led to simpler equations.

Chapoy and Dupré [5] have developed a set of relations for the absorption axis being parallel to **M** ($\alpha = 0$, $\alpha' \neq 0$) and ε is random.

The simplest possible case is when the absorption and emission axes coincide so that $\alpha = \alpha' = 0$ and there remain only three components:

$$F_{1111} = F_{2222} = 3F_{1122} = 24G(1 - 2\langle\cos^2\theta\rangle + \langle\cos^4\theta\rangle) \quad (4.37)$$

$$F_{3333} = 64G\,\langle\cos^4\theta\rangle \quad (4.38)$$

$$F_{2233} = F_{3311} = 32G\,(\langle\cos^2\theta\rangle - \langle\cos^4\theta\rangle) \quad (4.39)$$

and:

$$G = \frac{1}{192}(8F_{1111} + 12F_{2233} + 3F_{3333}). \quad (4.40)$$

$\langle\cos^4\theta\rangle$ and $\langle\cos^2\theta\rangle$ can thus be determined from three suitable intensity measurements.

Table 4.1

	D	K	L	M	N
A_o	$3+6\chi$	$-15-6\chi$		$51+6\chi$	-12
A_1	$2+4\chi$	$6-4\chi$		$-62+4\chi$	8
A_2	$3+6\chi$	$1-6\chi$		$19+6\chi$	4
A_3	$1+2\chi$	$-5-2\chi$		$17+2\chi$	-4
A_4	$-2-4\chi$	$6+4\chi$		$-10-4\chi$	4
A_5	$1+2\chi$	$-1-2\chi$		$1+2\chi$	
A_6	$-1-2\chi$	$5+2\chi$		$-17-2\chi$	4
A_7	$-2+4\chi$	-4χ	-6	$10+4\chi$	-4
A_8	$3-2\chi$	$-1+2\chi$	2	$1-2\chi$	
A_9	$-1-2\chi$	$5+2\chi$		$-17-2\chi$	4
A_{10}	$-2+4\chi$	-4χ	6	$10+4\chi$	-4
A_{11}	$3-2\chi$	$-1+2\chi$	-2	$-1-2\chi$	
A_{12}	$-1-2\chi$	$5+2\chi$		$-17-2\chi$	4
A_{13}	2	-6		18	-3
A_{14}	$-1+2\chi$	$1-2\chi$		$-1+2\chi$	1
A_{15}	$1+2\chi$	$-5-2\chi$		$17+2\chi$	-4
A_{16}	$22-20\chi$	$-30+20\chi$		$22-20\chi$	8
A_{17}	$1+2\chi$	$11-2\chi$		$-15+2\chi$	-4

$\chi = \langle \cos^2 \varepsilon \rangle$

4.7 INTENSITY CORRECTIONS

The observed fluorescence intensities need to be corrected for several effects before they can be equated to the components F_{jmps}. The complex relationship between the error in the measured intensities and $\cos^2\alpha$ on the error in $\langle\cos^4\theta\rangle$ and $\langle\cos^2\theta\rangle$ has been investigated by Hennecke and Fuhrmann [41]. In general $\langle\cos^4\theta\rangle$ is more sensitive to measurement error than $\langle\cos^2\theta\rangle$ and the sensitivity increases as the value of $\cos^2\alpha$ decreases. When $\cos^2\alpha = 0.8$ ($r_o \sim 0.2$) the intensities must be accurate to within 5% for a corresponding error of about 8% in $\langle\cos^4\theta\rangle$. Corrections for instrumental factors such as the polarization dependence of the optical components used to generate the exciting light and observe the emitted light are easily determined. The half angle of the aperture of the detector only weakly influences the relative intensities [27,42,43] and can be neglected if less than 35°.

Corrections for the anisotropic nature of the sample itself can be divided into those that depend on the sample thickness and those that do not. The first category includes absorption, birefringence and light scattering. The effect of the anisotropy of absorption and birefringence on the propagation of light through a transparent sample is in general quite complex. It is relatively simple only for light propagating along one of the symmetry axes OX_j, when only the influence of absorption need be considered [11,27]. In such cases the measured intensity is the product of the appropriate component F_{jjpp} and a correction term R_{jjpp}:

$$I_{jjpp} = R_{jjpp}\ F_{jjpp}. \tag{4.41}$$

In most cases the specimen is completely transparent to the fluorescent light and:

$$R_{jjpp} = [\ 1 - \exp(-\alpha_j d)\ ]\ /\ \alpha_j d. \tag{4.42}$$

$\alpha_j d$ is the optical density, D_j, of the sample for incident light polarized parallel to OX_j and can easily be determined. Anisotropic absorption often accounts for the observed inequality $I_{jjpp} \neq I_{ppjj}$ even when it is known that $F_{jjpp} = F_{ppjj}$. In the limit of low optical density the corrections can be expressed by the following approximations:

$$R_{jjpp} = 1 - \tfrac{1}{2}D_j \qquad (4.43)$$

thus:

$$R_{jjpp} - R_{ppjj} = R_{jjjj} - R_{pppp} = \tfrac{1}{2}\,(D_p - D_j) \qquad (4.44)$$

and:

$$I_{jjpp} - I_{ppjj} \doteq \tfrac{1}{2}\,(D_p - D_j)\;F_{jjpp}. \qquad (4.45)$$

The optical densities are related to the distribution of orientations and for a uniaxial sample:

$$R_{33pp} - R_{pp33} = \tfrac{1}{2}\,D_3 . \left(\frac{1 - 3\langle\cos^2\theta_A\rangle}{2\langle\cos^2\theta_A\rangle} \right). \qquad (4.46)$$

For a highly oriented sample with $\langle\cos^2\theta_A\rangle = 0.8$ the optical density must be less than 0.09 to reduce the error in the uncorrected relative intensities to less than 5%. A very low concentration of probes is necessary before the absorption correction can be neglected, for example less than 50 ppm of a probe in a sample 1mm thick.

The components F_{jpjp} can only be determined by setting both the polarizer and analyser transmission directions away from the symmetry axes of the specimen. For these measurements the effects of birefringence must be considered in addition to absorption anisotropy [9,22,27,44,45]. The factor R_{jpjp} involves many terms and can only be calculated if the refractive indices at the wavelengths of absorption and emission are accurately known. R_{jpjp} must usually be regarded as unknown. An insight into the behaviour of R_{jpjp} can be obtained by assuming that the absorption is negligible, in which case [9,46]:

$$R_{jpjp} = \frac{1}{d} \int_0^d \cos(\theta \ x) \ \cos\left[\theta' \ (d-x)\ \right] \, dx \tag{4.47}$$

where d is the sample thickness and:

$$\theta = (n_j - n_p)/\lambda \quad \text{and} \quad \theta' = (n_j' - n_p')\lambda'.$$

n_j is the refractive index for exciting light of wavelength polarized along OX_j, the dashed terms refer to light of the fluorescent wavelength. Integration gives:

$$R_{jpjp} = \frac{1}{\theta d(1-p^2)} \left[\sin(\theta d) - \sin(P\theta d)\ \right] \tag{4.48}$$

where $P = \theta'/\theta$.

In the limit of low birefringence R_{jpjp} tends to 1, as θd increases the value oscillates about zero with diminishing amplitude until in the limit of high birefringence $R_{jpjp} = 0$. The effects of birefringence and absorption can drastically change the observed angular dependence of the fluorescence intensity, especially in samples of low birefringence [27]. In transparent uniaxially oriented, samples the orientation averages can be determined from four measurements with polarizations parallel to the axes OX_j and the value of R for an isotropic sample. These four intensities are independent of the birefringence, but not of the effects of anisotropic absorption.

The effect of the depolarization of the incident and emitted light by scattering as they pass through a uniaxially oriented sample in which absorption is negligible has been considered by Pinaud *et al.* [40]. They concluded that 4 correction factors were required when both incident and observed light is polarized along one of the OX_j directions. The corrections are no longer single terms but sums of terms:

$$I_{jjjj} = A_{jj}\, F_{jjjj} + (B_j + B_j' - A_{jj} - A_{jp})\, F_{jjpp} + (1 - B_j - B_j' + A_{jp})\, F_{pppp} \quad (4.49)$$

$$I_{jjpp} = (B_j - A_{jj})\, F_{jjjj} + (1 - B_j - B_j' + A_{jj} + A_{jp})\, F_{jjpp} + (B_j' - A_{jp})\, F_{pppp} \quad (4.50)$$

where:

$$A_{jp} = \frac{\exp(-\tau_j d) - \exp(-\tau_p' d)}{\tau_p' d - \tau_j d} \quad (4.51)$$

$$B_j = \frac{1 - \exp(-\tau_j d)}{\tau_j d}. \quad (4.52)$$

τ_j is the turbidity of the sample to the exciting light polarized along the OX_j direction and the dashed terms refer to light of the fluorescence wavelength. One effect of anisotropy in the turbidity is to make $I_{jjpp} \neq I_{ppjj}$ when $\cos^2\alpha = \cos^2\alpha'$. The turbidities can be measured and the correction can be regarded as known.

The following effects are independent of the sample thickness:

1. Partial reflection of the incident light at the surface of the sample.
2. The effect of the medium on the absorption cross-section of the fluorescent molecules.
3. The effect of the medium on the emission probability of the fluorescent molecules.
4. The effect of the refractive index of the medium on the solid angle of the cone of observed light within the medium.

Since these effects all depend on the refractive index of the medium they will depend on the polarization of the incident and observed radiations.

Several of these effects influence the intensities in opposite ways so that the net effect is small. Nobbs *et al.*[27] estimated that their influence on the orientation averages of even a highly birefringent material such as PET is less than the experimental error and they can be ignored.

4.8 REMAINING PROBLEMS

It can be concluded from the above that the theory of the polarized fluorescence experiment has developed to the stage where the orientation averages $\langle P_2(\xi)\rangle$ and $\langle P_4(\xi)\rangle$, ($\xi = \cos\theta$), can be reliably obtained from the intensity measurements. The questions remain as to the location of the probe molecules and the nature of the relationship between the distribution of orientations of the probe molecules and that of the polymer segments. If the fluorescent chromophore is a chemical part of the chain then fairly precise answers can be given; however, when the fluorescent probe is simply dispersed within the polymer matrix several possible locations and relationships exist. The probe may preferentially align itself with a particular segment or group of segments in a particular conformation and so directly reflect their orientation. Alternatively, the probe may adopt some average orientation of the neighbouring polymer segments. In these cases the nature of the probe and of the polymer will influence the relationship, and the form of the relationship must be determined for each system.

The problem is stated mathematically by means of the addition theorem for Legendre polynomials. Letting the subscripts M and P denote that the orientation function refers to the fluorescent probe axis and some direction fixed within the polymer segment, respectively, then:

$$P_\ell(\xi_P) = P_\ell(\xi_M)\, P_\ell(\cos\gamma) + 2\sum_{m=1}^{\ell} \frac{(\ell-m)!}{(\ell+m)!} P_\ell^m(\xi_M)\, P_\ell^m(\cos\gamma)\cos m\phi' \qquad (4.53)$$

where:

$$\sin\phi' = \frac{\sin\theta_P \sin(\phi_P - \phi_M)}{\sin\gamma} \quad \text{and} \quad \xi_i = \cos\theta_i.$$

θ_i and ϕ_i are the polar and azimuthal angles of the relevant axis with respect to the sample frame $OX_1X_2X_3$ and γ is the angle between the segment and probe axes. Difficulties occur when the equation is averaged over all the fluorescent probes. For example, if it is assumed that the segment and probe axes are uniaxially distributed around OX_3 then the equation for $\langle P_2(\xi_P)\rangle$ becomes:

$$\langle P_2(\xi_P)\rangle = \langle P_2(\xi_M)\ P_2(\cos\gamma)\rangle + \frac{1}{12}\langle P_2^2\ (\xi_M)\ P_2^2(\cos\gamma)\ \cos 2\phi'\rangle \tag{4.54}$$

where:

$$\cos 2\phi' = 1 - \frac{2\ P_2^2\ (\xi_P)\ \sin^2(\phi_P - \phi_M)}{P_2^2\ (\cos\gamma)}. \tag{4.55}$$

For a chemically linked fluorescent chromophore it can be assumed that the values of $\langle P_\ell^m(\cos\gamma)\rangle$ and $\langle\cos^2\phi'\rangle$ are fixed and independent of the degree of orientation of the sample. The form of the relationship can then be derived by the measurement of samples with known $\langle P_\ell(\xi_P)\rangle$. For a dispersed probe assuming some local average segment orientation, the averages in Equation (4.54) can only be evaluated if the distribution of orientations of the segment and probe molecules are completely known. It can be said however that in partially oriented samples the orientation averages of the fluorescent probe will be greater than those for the polymer segments [46,48]. For a given sample, the greater the number of polymer segment orientations over which the fluorescent probe averages then the greater the value of the probe orientation averages. A relationship between the length of the fluorescent probe and its degree of orientation is thus expected, and has been observed in a study by Hennecke and Fuhrman [15].

An associated problem is uniqueness. It is clear that for a chemically linked fluorescent chromophore there is only one set of values of $\langle P_2(\xi_M)\rangle$ and $\langle P_4(\xi_M)\rangle$ for given values of $\langle P_2(\xi_P)\rangle$ and $\langle P_4(\xi_P)\rangle$. In the local average case Equation (4.54) implies that this is no longer true. Results of a theoretical study by Nobbs *et al.* [48] show that the values of $\langle P_2(\xi_M)\rangle$ and $\langle P_4(\xi_M)\rangle$ for fixed $\langle P_2(\xi_M)\rangle$ and $\langle P_4(\xi_P)\rangle$ depend strongly on $\langle P_6(\xi_P)\rangle$ and by implication higher orientation averages. As far as we are aware, no experimental work has yet been reported examining the uniqueness of the measured fluorescence orientation averages.

4.9 EXPERIMENTAL STUDIES OF POLARIZED FLUORESCENCE IN ORIENTED POLYMERS.

In many of the experimental studies to be reported, a primary aim of the research was to gain greater understanding of the physical properties of oriented polymers, including such features as stiffness and thermal shrinkage. In several instances a quantitative theoretical analysis of the data has been attempted, but often the approach has been more empirical, although useful correlations have been established.

Because the application of the polarized fluorescence technique has developed from a somewhat empirical base to increasingly sophisticated treatments in more recent years, it is appropriate to present a review in which the experimental work is considered chronologically.

In the very early studies of polarized fluorescence in oriented polymers by Morey and his collaborators, it was considered that the fluorescence phenomenon was analogous to absorption dichroism. A pad of fibre specimens was illuminated by unpolarized radiation and I_{max} and I_{min}, the maximum and minimum values of the fluorescent intensity determined when the analyser was parallel and perpendicular to the fibre axis respectively. Morey defined a quantity:

$$P_m = \frac{I_{\text{max}} - I_{\text{min}}}{I_{\text{max}} + 2I_{\text{min}}} \quad \text{x}100$$

which he termed the percentage orientation. A good correlation was obtained between p_m and the crystalline orientation, from which it was concluded that the fluorescent dye molecules may become attached to the surfaces of the crystallites, it seeming unlikely that they could actually be present in the crystalline regions.

The first systematic attempts to explore the application of the polarized fluorescence technique to oriented synthetic polymers are due to the pioneering work of Nishijima and his colleagues. Nishijima recognised that the technique provided, in principle, more detailed information regarding the distribution of chain orientation than other available methods at the time (notably infra-red dichroism and birefringence). His approach was to calculate the theoretical angular distributions of the polarized fluorescence intensity for specific experimental arrangements on the basis of idealized distributions of molecular orientations [31-37]. In a typical arrangement [34] the polarization vector of the exciting light is

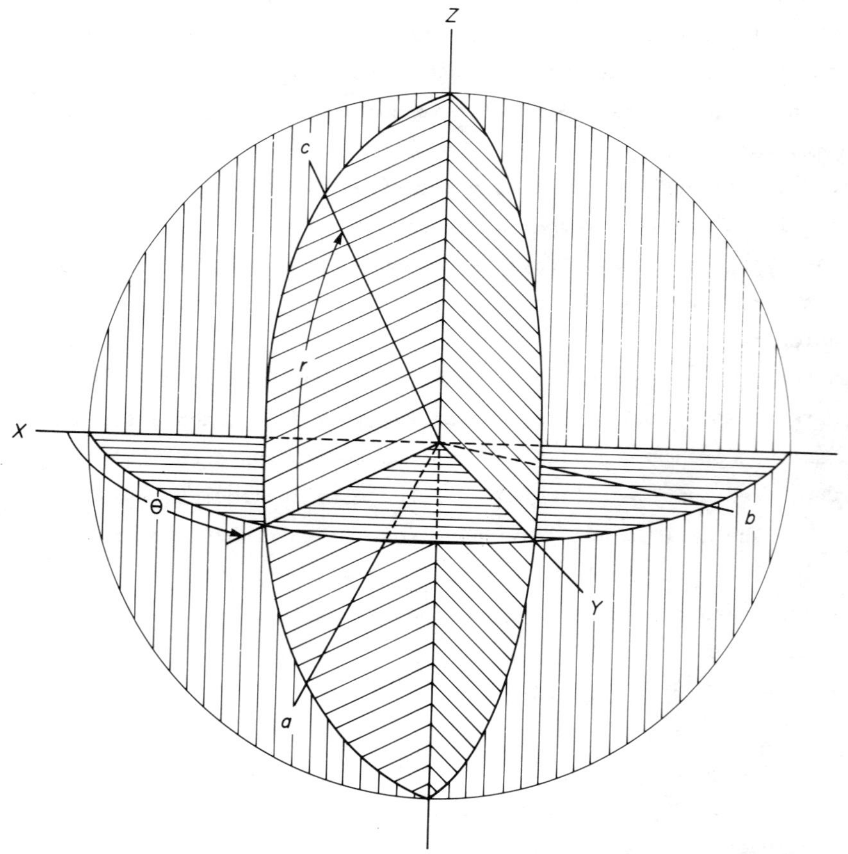

Caption at top of facing page.

Fig. 4.3 The laboratory Cartesian coordinate frame *XYZ* and the film coordinate frame *abc* (reproduced from reference [34] by permission of the publishers Gakujutsu Bunken Fukyu-Kai, Tokyo).

set along the *X*-axis of a system of rectangular cartesian coordinates (Fig. 4.3) and $I_{||}$ and $I_{\perp}$, the parallel and perpendicular components of the fluorescence intensity measured by aligning the polarization vector of the analyser parallel to *X* or *Z*, respectively. The oriented polymer film has its rectangular coordinate system *abc*, with the *ac* plane making an angle θ to the *XZ* plane of the apparatus. The *c* axis of the sample is then rotated in the *ac* plane (described by a further angle ϕ) to give the angular distribution of $I_{||}$ and $I_{\perp}$. A quantity termed the degree of polarization $p = (I_{||} - I_{\perp}) / (I_{||} + I_{\perp})$ is also used as a measure of orientation.

Polar plots of $I_{||}$, $I_{\perp}$ and p are constructed for idealized distributions of molecular orientation. For perfect uniaxial orientation, where the fluorescent probes are aligned along the *c*-axis, a set of such polar plots for $I_{||}$ and $I_{\perp}$ is shown in Fig. 4.4. The essence of Nishijima's work is brought out by comparing such plots for different distributions. In Fig. 4.5 we compare the predicted polar plots for perfect uniaxial orientation with perfect

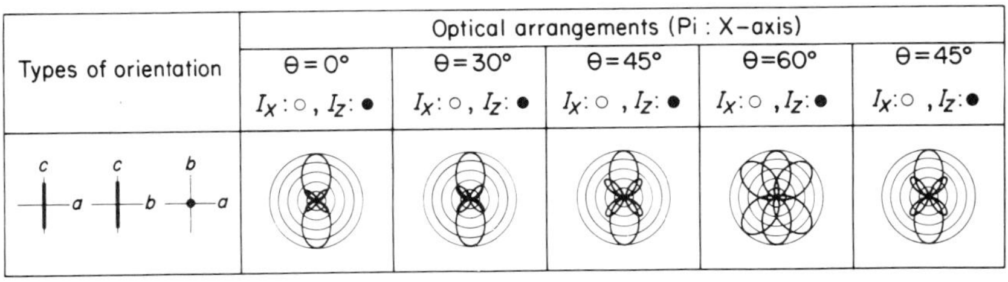

Types of orientation	Optical arrangements (Pi : X-axis)				
	θ=0° I_X: ○, I_Z: ●	θ=30° I_X: ○, I_Z: ●	θ=45° I_X: ○, I_Z: ●	θ=60° I_X: ○, I_Z: ●	θ=45° I_X: ○, I_Z: ●

Fig. 4.4 Theoretical angular distributions of polarized fluorescence for perfect uniaxial orientation. ○($I_{||}$), ●($I_{\perp}$) (reproduced from reference [34] by permission of the publishers Gakujutsu Bunken Fukyu-Kai, Tokyo).

biaxial orientation, where the fluorescent probes are aligned perfectly (and with equal probability) along the a and c axes [30].

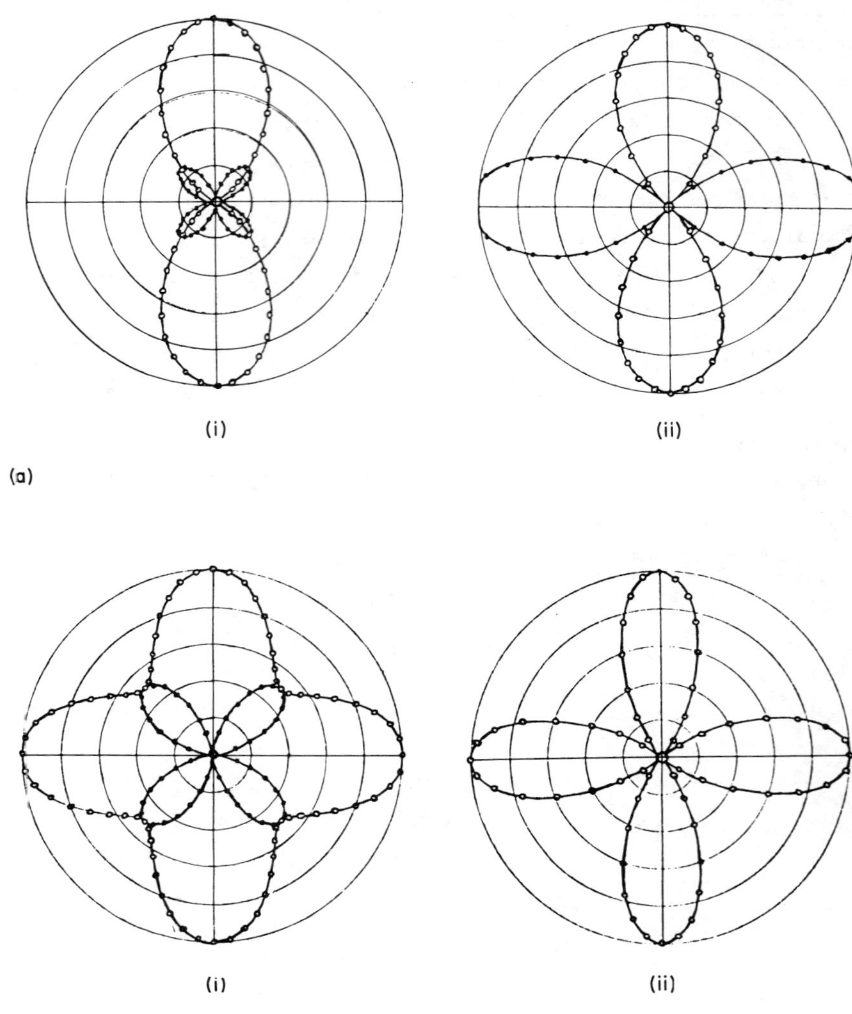

Fig. 4.5 (a) Perfect uniaxial orientation: (i) Polar plots of ○($I_{||}$) and ●($I_{\perp}$); (ii) Polar plots of the degree of polarization for ○(positive) and ●(negative) values.
(b) Perfect biaxial orientation: (i) Polar plots of ○($I_{||}$) and ●($I_{\perp}$); (ii) Polar plots of the degree of

polarization for ○(positive) and ●(negative) values (reproduced from reference [30] by permission of the publishers Gakujutsu Bunken Fukyu-Kai, Tokyo).

Fig. 4.5 shows plots of $I_{||}$, $I_{\perp}$ and p for $\theta = 0$, i.e. the plane of the sheet (*ac* plane) coincides with the *XZ* plane of the apparatus. This is the most straightforward experimental arrangement where the sample is rotated about the normal to the plane of the sheet. The appearance of these polar plots is intuitively appealing, and Nishijima and his colleagues showed that there was reasonable qualitative agreement between observation and their theoretical expectations. Two examples have been selected for comparison with the results shown in Figs 4.4 and 4.5. First, there is a study of uniaxially oriented polyvinyl-alcohol (PVA) films [51]. The fluorescent probe, Whitex RP was introduced at a concentration of 10^{-4} mole ℓ^{-1} into a solution of PVA in water. A cast polymer film was then drawn uniaxially at 100°C at an extension rate of 100% min^{-1}. The drawing process was carried out in two ways, either at constant width or allowing free lateral contraction. Typical results are shown in Fig. 4.6, from which a good qualitative resemblence to the predicted polar plots of Fig. 4.4 can be seen. There is also good qualitative agreement between results for two stage biaxial drawing of polypropylene [51] (Fig. 4.7) and the predicted idealized polar plots of Fig. 4.5. In this case the fluorescent probe was the fluorescent whitening agent:

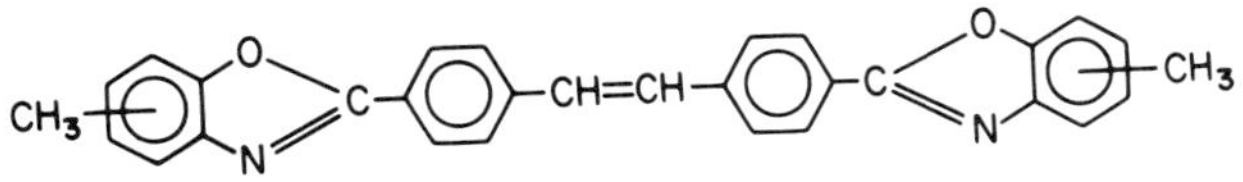

which was dispersed in the polymer melt before extrusion of an unoriented film at a concentration of 10^{-4} mole ℓ^{-1}. The fluoresence results for the biaxially drawing process show clearly the gradual realignment of the probe molecules as the draw ratio of the

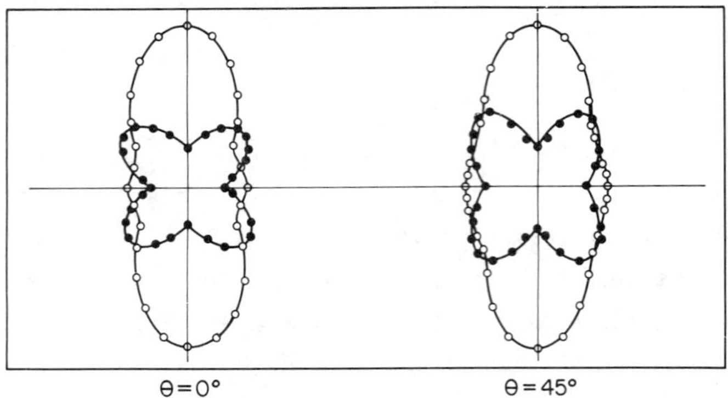

50% elongation

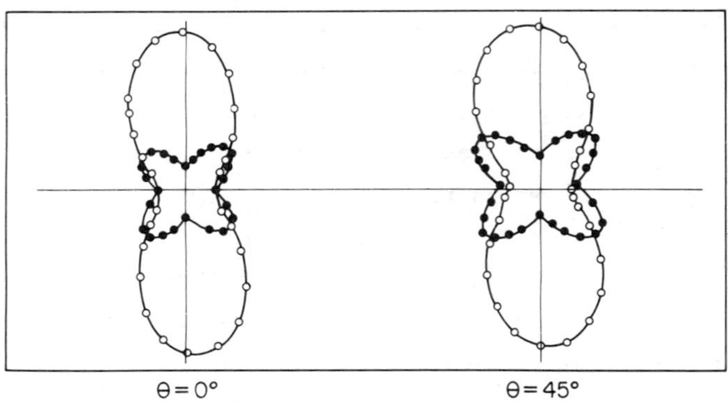

130% elongation

Fig. 4.6 Observed fluorescence orientation pattern for uniaxially drawn PVA films containing Whitex RP(reproduced from reference [51] by permission of the publishers Gakujutsu Bunken Fukyu-Kai, Tokyo).

second stage is increased.

For the uniaxially drawn PVA film there is one particularly noticeable difference between the predicted and measured results, in that the experimental intensity $I_{||}$ in Fig. 4.6 does not show its minimum at 90°, but at an intermediate angle. This was interpreted by Nishijima *et al.* as indicating a degree of biaxial orientation at

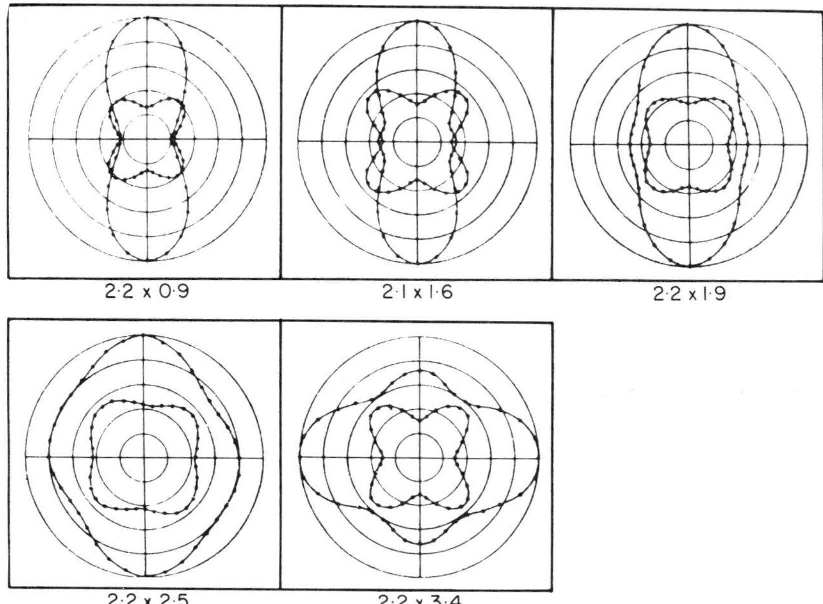

Fig. 4.7 Observed fluorescence orientation pattern for two-stage biaxially drawn polypropylene (reproduced from reference [52] by permission of the publishers Gakujutsu Bunken Fukyu-Kai, Tokyo).

low draw ratios. Although this is a possible explanation, and there is other evidence for such behaviour in some polymers, it is also possible, as shown by Nobbs *et al.* [27], that birefringence effects can lower the intensity observed at 45°. This means that in a uniaxially oriented sample of low molecular orientation, polar plots of the type shown in Fig. 4.6 can be observed, where the minimum in intensity at 45° is entirely due to the birefringence effects.

There is a further important point regarding Nishijima's approach of comparing the predicted and measured polar plots of intensity. As the theoretical analysis shows, the fluorescence intensities determine $\langle P_2(\xi)\rangle$ and $\langle P_4(\xi)\rangle$ only. Knowledge of the values of $\langle P_2(\xi)\rangle$ and $\langle P_4(\xi)\rangle$ enables us to test the validity of possible orientation distributions, but the polar plots for different distributions with the same values for $\langle P_2(\xi)\rangle$ and $\langle P_4(\xi)\rangle$ would be identical. Although as we have noted, Nishijima's comparisons of the predicted and measured polar plots are intuitively appealing, the fluorescence results cannot define the orientation distribution, only the second- and fourth-order orientation averages. A more realistic procedure is therefore to calculate the quantities $\langle P_2(\xi)\rangle$ and $\langle P_4(\xi)\rangle$ from the experimental data, as described above, and use these to discuss the orientation and mechanisms of deformation.

A further limitation of the analyses presented by Nishijima and coworkers was the assumption that the emission and absorption axes of the fluorescent probes were parallel, although inspection of the results for isotropic samples indicates that this is not correct.

Nishijima and his colleagues carried out a wide range of investigations on many polymers, and these will now be reviewed, bearing in mind the reservations noted above.

In PVA, results were presented for the uniaxial stretching of fibres that contain fluorescent polyene segments prepared by partial dehydration of the polymer [53]. The conjugated double bond of the polyene segments lies along the polymer chain, and it was claimed that the emission and absorption axes also lay in this direction. In fact, the fluorescence results on the drawn films were identical to those described above where the fluorescence probe was dispersed in the polymer before drawing. It was considered in both cases that the fluorescence technique indicates the orientation of polymer chains in the non-crystalline regions. In support of this view the optical dichroism of the fluorescent probes was compared with the optical birefringence for a series of drawn films. After a linear correlation between dichroism and birefringence at low draw ratios, there was a break in the curve and the dichroism rose less steeply with increasing draw ratio than the birefringence. This was interpreted as implying that at low draw ratios the amorphous regions orient more rapidly than the crystalline regions. The results are, however, reminiscent of those obtained by Patterson and Ward [54] for poly(ethylene/terephthalate) (PET). Here, it was concluded that at low draw ratio, dichroism and birefringence reflected the overall orientation, but the break occurred because the dichroism related only to the amorphous regions, whereas the birefringence related to both amorphous and crystalline orientation, with the latter making an important contribution at high draw ratios.

A further study by Nishijima *et al.* [30] introduced uranine as the fluorescent probe into PVA, and compared results from polarized fluorescence with those from birefringence and dichroism. Significant differences were again observed, which were interpreted qualitatively.

In polypropylene, the fluorescent probes were either added to the film at the melt extrusion stage [52], as already discussed, or introduced by placing isotropic film in a saturated xylene solution of the fluorescent compound [55]. Although the results are of qualitative interest, especially in the case of biaxial drawing, conclusions regarding the details of the chain orientation are open to question, owing to the reservations mentioned above. In further papers, results are presented for low-density polyethylene and poly (vinyl chloride) [56]. In polyethylene, the redrawing behaviour of strips cut from a tubular film was examined [57]. Samples cut at 0, 45 and 90° to the machine draw direction were redrawn and the fluorescence polarization measured simultaneously. The gradual development of molecular reorientation in the redrawing direction was monitored and made good sense at a qualitative level of interpretation.

Following this pioneering research on polarized fluorescence, and in some cases contemporaneously, several other workers undertook similar investigations on several oriented polymers in the period ca. 1969-71. The interpretation of the data was in all cases at a similar level of sophistication to that proposed by Nishijima, either plotting polar plots of the polarized fluorescence intensity or the parameter p. Although the results must therefore be regarded as to some extent only semi-quantitative, the attempts at correlation with other structural measurements and with physical properties are of some interest. It seems appropriate to review these papers as a group before discussing more recent investigations that have followed a more detailed theoretical treatment of the fluorescence method.

Seki [44] developed an apparatus for studying fibres as well as films, and which permitted the simultaneous measurement of stress and fluorescence. Results were presented for polyethylene, nylon-6, cotton and polyacrylonitrile. In polyethylene, the fluorescent probe Whitex SNR was introduced by mixing it with the polymer before preparation of the isotropic film. The polar plot of $I_{||}$ for a film drawn to draw ratio of five was very close indeed to the idealized polar plot of Fig. 4.5. $I_{\perp}$ shows some discrepancies that were attributed to some depolarization of the fluorescent radiation due to scattering. Polar plots were also presented for cotton fibre and nylon-6 fibres drawn to draw ratio 4, where the Whitex RP was introduced into the oriented fibre by dyeing. Results were also shown for polyacrylonitrile fibre, drawn to $\lambda = 11$ and dyed with diamino stilbene in the aqua-gel state and dried at 20°C. In these cases the plots were consistent with uniaxial orientation, but there was some degree of disorder. This was interpreted by comparison with theoretical polar plots for conical distributions of molecular orientations or by the superposition of a conical distribution and a random distribution. As has been discussed, such idealized distributions cannot be inferred on the basis of the fluorescence data, which depend only on $\langle P_2(\xi)\rangle$ and $\langle P_4(\xi)\rangle$.

In a more detailed study of nylon fibres, Seki [58] studied the effect of draw ratio and of heat treatment, combining polarized fluorescence with X-ray diffraction and birefringence measurements. The spun nylon fibres were dyed before drawing by immersion in an aqueous solution of Whitex RP for several hours at a temperature below 70°C. Most of the results were obtained for nylon-6 fibres, but some measurements were also made on a very wide range of different nylons, including nylon 6.6, nylon 6.12, and nylon 7, 9 and 11. Plots of $I_{||}$ were obtained as a function of draw ratio at constant temperature and as a function of temperature at constant draw ratio. With increasing draw ratio at constant temperature the patterns approach the theoretical patterns for uniaxial orientation, but even at the highest draw ratio there is still appreciable disorientation. An orientation parameter $R = I_{||}(0)/\ I_{||}(90)$ was defined where the polar angle is either parallel $I_{||}(0)$ or perpendicular $I_{||}(90)$ to the draw direction, respectively. It was shown that R falls at the higher draw temperatures for a constant draw ratio of three, and more interestingly, that the development of molecular orientation appeared to be greatest for drawing at about 100°C. It was also shown that increasing the relative humidity produced a similar effect to an increase in temperature, which suggests that the degree of molecular mobility is important.

The effect of heat treatment, either at constant length or with free shrinkage, was also studied as a function of temperature. The fluorescence measurements were combined with measurements of birefringence and density. An important result was that the birefringence was almost unaffected by heat treatment, whereas the polarized fluorescence showed a marked reduction in molecular orientation, which was greatest for the fibres that were heat treated without constraint. There was also a marked increase in density which, for fibres treated at constant length, amounted to an increase in crystallinity from 20% to 50%. Moreover, the X-ray diffraction patterns indicated a considerable increase in crystalline orientation as well as an increase in crystallinity. These results are consistent with the view that the fluorescent probes are primarily in the non-crystalline regions and that heat treatment is associated with disorientation of these regions. Seki also concluded that heat treatment produced planar orientation where the probe molecules became aligned perpendicular to the fibre axis. This conclusion is however open to the objections that have been raised regarding the interpretation along the lines proposed by Nishijima and his colleagues. In particular, Seki [44] concluded that birefringence effects are only important for samples of low orientation, which we will see is at variance with the conclusions of Nobbs *et al.* [27].

Seki's research on nylons was paralleled by a very similar study of poly(ethylene/terephthalate) fibre by McGraw [59]. Fibres containing 200ppm of 2-2' (vinylenedi-*p*-phenylene) bisbenzoxazole (VPBO) were melt spun and drawn in steam at 140°C to a range of draw ratios. In addition to examining the drawn fibres, McGraw also prepared shrunk fibres, by unconstrained heat treatment in an oil bath for 10 min at several temperatures up to 120°C. McGraw was particularly careful to choose a needle-shaped fluorescent probe molecule, and calculations suggested that the transition moment was aligned at 2° to the long axis. No corrections were made for birefringence effects or refraction at the fibre surface.

The fluorescent orientation, described by the factor p with the incident beam vertically polarized ($\phi = 0$) appeared to pass through a maximum with increasing draw ratio. This is a very unusual result, which is contrary to experience in all other polymers so far described, and at variance with later works by Nobbs *et al.* [27], which is to be discussed later. A possible explanation is that the drawing process at 140°C in steam is equivalent to drawing followed by heat treatment, which in other studies by McGraw [60] and by Nobbs *et al.* [61] produces a fall in amorphous orientation. In this respect, there was a very satisfactory correlation between the polarized fluorescence and the shrinkage, which was interpreted as a loss of amorphous orientation. This is consistent with the optical dichroism studies of Patterson and Ward [54], and the further studies by Nobbs *et al.* to be discussed.

In a second paper, McGraw [60] described the changes in the polar plots for shrinkage of spun fibres, where good qualitative results were obtained that correlated with the severity of the heat treatment. It was also shown that for a series of drawn yarns subjected to shrinkage treatments, exactly equivalent fluorescence data could be obtained by introducing the fluorescent probes after drawing, by dyeing from an aqueous dispersion at 50°C. This result is important because it provides further support for the view that the probe molecules are labelling the amorphous regions. Although the probe molecules are much longer than a monomer unit of PET and would therefore produce a very large discontinuity in a crystalline region, incorporation from the melt is open to the possibility that the probes form centres for nucleation in the crystallization process.

McGraw [59] also measured the fluorescence polarization for fibres that had been spun to comparatively high degrees of orientation, by spinning at high wind-up speeds where the tension in the threadline extends the molecular network, as discussed by Pinnock and Ward [62]. These spun fibres showed very high apparent orientation, with p values greater than those for drawn yarns. As pointed out by Bower [4], this remarkable result could be an artifact. Bower has shown that for this situation where the wavelength of the radiation $\lambda \sim d\Delta n$, where d is the sample thickness and Δn the birefringence, the observed intensities are not independent of Δn. For the drawn fibres Δn is very much higher again, and the results are most probably correct in qualitative terms.

A further polymer where semi-quantitative measurements have been undertaken is poly (vinyl alcohol), where results obtained by Nishijima and his colleagues have already been described. Gulrajani and Padhye [63] made a comparative study of orientation using birefringence, infra-red and visible dichroism as well as polarized fluorescence. Films were cast from aqueous solution and drawn to a series of draw ratios up to $\lambda = 5$. In two separate experiments, a reactive and non-reactive optical brightening agent was introduced as the fluorescence probe after drawing by soaking the film in a solution containing the brightening agent. A fluorescence parameter p (for $\phi = 0$) obtained using unpolarized incident radiation was compared with the infra-red dichroic ratio, the birefringence and the visible dichroism. The fluorescence results were then described by an orientation average $f_p = p/(1-p)$, and if this parameter was scaled so that its value at $\lambda = 5$ agreed with that obtained from infra-red dichroism, good qualitative agreement was obtained between the different measures of molecular orientation.

Another investigation by Kryszewski [64] used a parameter $p(\phi=0)/p_0(\phi=0)$ which is the ratio of the parameter p for the polarization direction of the incident light parallel to the draw direction, for drawn and undrawn samples, respectively. Assuming that the fluorescent molecules were in the amorphous regions of the polymer, the results indicated that these regions developed their orientation before the orientation of the crystalline regions.

In the investigations discussed so far, the full potentialities of the fluorescence method have not been realized. As will be appreciated from the theoretical section above, the best procedure is to obtain quantitative measures of the molecular orientation in terms of $\langle P_2(\xi)\rangle$ and $\langle P_4(\xi)\rangle$. However, this requires the more detailed theory presented above, which takes into account the effects of birefringence and absorption, and the key factor that the emission and absorption axes of the fluorescent probe molecule are not in general coincident.

In a series of papers, Bower and Ward and their collaborators describe detailed investigations of poly(ethylene/terephthalate) [27,48,61].

As in the work of McGraw [59,60] and Pinnock and Ward [62] described above, the fluorescent probe was VPBO. The VPBO was mixed with bis(2-hydroxyethyl terephthalate) so that after polycondensation a series of batches of PET were produced with probe concentrations of 200, 150, 100 or 50 ppm by weight. These polymers were then melt spun to give amorphous tapes of PET with cross-section approximately 1.5×10^{-3} m x 1.0×10^{-4} m. These tapes were subsequently drawn around a smooth cylinder (the pin) that was heated to temperatures in the range 65-90°C and located between feed and wind-up rollers rotating at different rates. Samples were produced to a constant draw ratio of 2.66 for different pin temperatures, and for the most widely varying experiments, for a range of draw ratios from 1.38 to 5.85 at a pin temperature of 80 ± 1°C.

The results for an isotropic sample showed clearly that the emission and absorption axes cannot be assumed to be coincident. After detailed consideration of the possible situations, it was assumed that there was no reason to suppose any correlation between the azimuthal angles of the emission and absorption axes. It was shown that the average of the square of the cosine of the angle between the absorption and emission axes of the fluorescent molecule, $\overline{\cos^2\alpha''} = 0.83$, implying a mean angle $\alpha'' = 24°$, and that the angle between the absorption axis and the fluorescent probe axis was 17.5°.

Nobbs *et al.* [27] showed that birefringence effects and absorption effects can dramatically change the observed polar plots of the fluorescence intensity. It was stressed that only $\langle P_2(\xi)\rangle$ and $\langle P_4(\xi)\rangle$ can be obtained from these measurements and that all the information required for calculating these quantities is contained in the quantity $R = I_{||}/I_{\perp}$ for an isotropic sample together with the four intensities $I_{||}(0)$, $I_{||}(\pi/2)$, $I_{\perp}(\pi/2)$ for the uniaxially oriented samples. These four intensities are not affected by birefringence effects, but they are affected by dichroic absorption, for which an appropiate correction must be made.

Nobbs *et al.* [61] used the polarized fluorescence method in conjunction with birefringence and X-ray diffraction measurements to study molecular orientation in drawn PET tapes, and drawn tapes that were subsequently shrunk and crystallized in air under various conditions.

The fluorescence measurements enable the determination of $\langle P_2(\xi_M)\rangle$ and $\langle P_4(\xi_M)\rangle$ where $\xi_M = \cos\theta_M$, and θ_M is the angle between a unique axis M in a typical molecule of the fluorescent additive and the symmetry axis, i.e. the draw direction, of the sample. Values for $\langle P_2(\xi_M)\rangle$ were compared with the optical orientation averages $\langle P_2(\xi)\rangle_{opt}$ that were determined from measurements of refractive index, and relate to the chain axis orientation averaged over all chains, to a good approximation. $\langle P_2(\xi)\rangle_{opt}$ obtained from refractive index data (not simply birefringence) using the relationship:

$$\frac{\Delta\alpha}{3\alpha_o}\langle P_2(\xi)\rangle_{opt} = \frac{\phi_z^e - \phi_x^e}{\phi_z^e - 2\phi_x^e}$$

where $\phi_i^e = (n_i^2-1)/(n_i^2+2)$, $\Delta\alpha$ is the difference between the electronic polarizabilities of a polymer repeat unit parallel and perpendicular to the chain axis, and α_o is the isotropic polarizability. Full details of this procedure are described elsewhere [65].

The X-ray orientation measurements were undertaken on the $(\bar{1}05)$ reflection, and the results corrected using the Legendre addition theorem [47] to allow for the angle of $9^\circ 46'$ between the $(\bar{1}05)$ plane normal and the c-axis. The orientation averages $\langle P_2(\xi)\rangle_c$ and $\langle P_4(\xi)\rangle_c$ that characterize the orientation of the crystallographic c-axis in the specimen were then calculated from the measured diffraction intensity as a function of the angle χ between the draw direction and the bisector of the angle between the incident and diffracted beam:

i.e. $\langle P_2(\cos\chi)\rangle = \langle P_2(\xi)\rangle_c \; P_2(\cos(9^\circ 46'))$

and $\langle P_4(\cos\chi)\rangle = \langle P_4(\xi)\rangle_c \; P_4(\cos(9^\circ 46'))$.

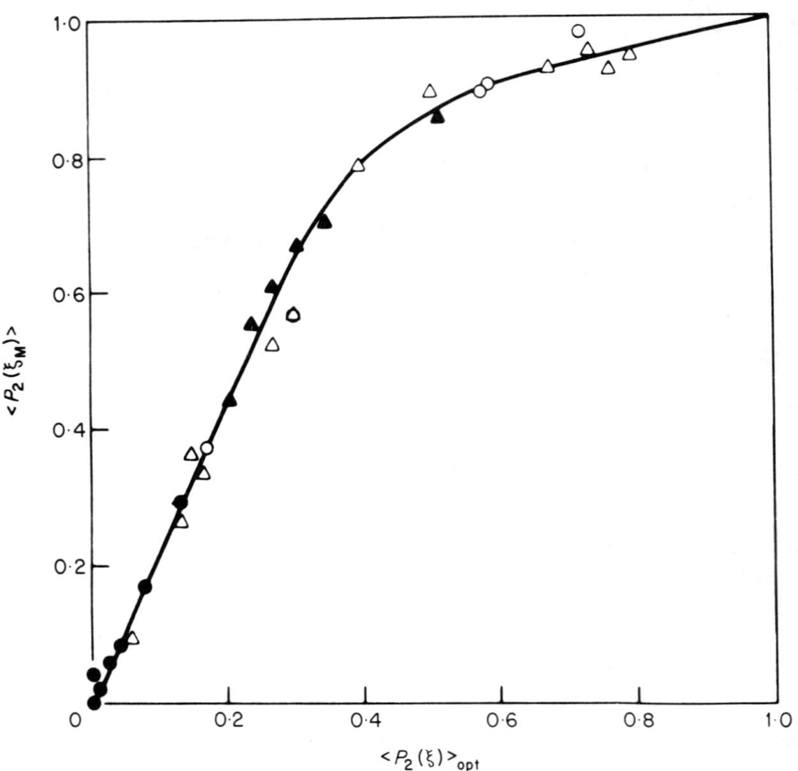

Fig. 4.8 $\langle P_2(\xi_M)\rangle$ compared with $\langle P_2(\xi)\rangle_{opt}$ for samples of low crystallinity (reproduced from *Polymer* 17, 25, 1976 by permission of the publishers, Butterworth Scientific Ltd).

The first comparison was made between $\langle P_2(\xi_M)\rangle$ for all the drawn samples with $\langle P_2(\xi)\rangle_{opt}$. This is shown in Fig. 4.8, and although there is a unique relationship between these two orientation averages, this is by no means linear. For low and intermediate degrees of orientation $\langle P_2(\xi_M)\rangle \sim 2\langle P_2(\xi)\rangle_{opt}$. A possible explanation of the greater orientation of the probe molecules than the overall chain orientation will be discussed later. At this stage it will be assumed that this unique relationship between $\langle P_2(\xi_M)\rangle$ and the overall orientation determined from birefringence in this series of samples of low crystallinity can be used to determine the amorphous orientation averages $\langle P_2(\xi)\rangle_a$ from the measured $\langle P_2(\xi_M)\rangle$. This assumes that the relationship between the distribution of orientation of the fluorescent molecules and the polymer chains in the amorphous regions of semi-crystalline PET is the same as that in completely amorphous PET. If this assumption is correct, then for all types of PET sample the three orientation averages $\langle P_2(\xi)\rangle_a$, $\langle P_2(\xi)\rangle_c$ and $\langle P_2(\xi)\rangle_{opt}$ should be related by:

$$\langle P_2(\xi)\rangle_{opt} = f_c \langle P_2(\xi)\rangle_c + (1-f_c) \langle P_2(\xi)\rangle_a \qquad (4.56)$$

where f_c is the crystalline fraction. In these experiments f_c was estimated from the density ρ, $f_c = (\rho_c/\rho)(\rho-\rho_a)/(\rho_c-\rho_a)$ where ρ_c is the density of the crystalline regions and ρ_a the density of the amorphous regions. ρ_a was assumed to depend on $\langle P_2(\xi)\rangle_a$ and a simple model was used to determine the increase in amorphous density with $\langle P_2(\xi)\rangle_a$.

Some key results are shown in Fig. 4.9 , where the three orientation averages $\langle P_2(\xi)\rangle$ are compared with the shrinkage and crystallinity of samples originally drawn to a draw ratio $\lambda = 3.44$ and subsequently shrunk freely for 10 mins in an air oven at temperatures from 60° to 180°C. The solid points denote the direct experimental data and the open points denote the best fit to Equation (4.56). There is reasonable quantitative agreement between the two sets of points, giving good support to the procedures adopted. It is important to note that for this series of samples, the shrinkage increases rapidly from zero to 15%, as the oven temperature rises from 60° to 90°C and then much less rapidly with

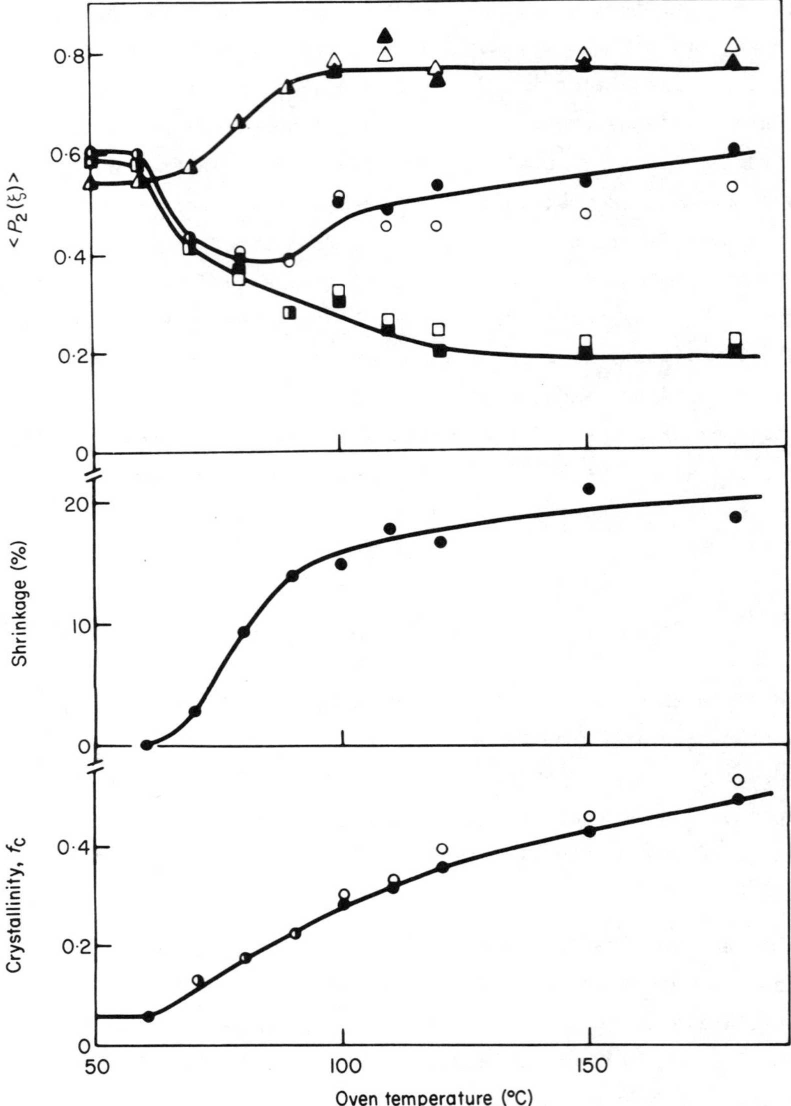

Fig. 4.9 $\langle P_2(\xi)\rangle_c$ △ ▲ , $\langle P_2(\xi)\rangle_{opt}$ ○,●, $\langle P_2(\xi)\rangle_a$ □ ■, f_c and shrinkage plotted against oven temperature for PET samples which were initially drawn to λ = 3.44 and subsequently shrunk freely for 10 min in an air oven. Solid points denote experimental data; open points denote best fit to Equation

(4.56) (reproduced from *Polymer* 17, 25, 1976 by permission of the publishers, Butterworth Scientific Ltd).

increasing temperature. The crystallinity increases gradually with increasing temperature over the whole temperature range. It is of particular interest that $\langle P_2(\xi)\rangle_{opt}$ shows a clear minimum for oven temperatures near 80°C. It can be seen that this minimum is consistent with a monotonic fall in $\langle P_2(\xi)\rangle_a$ and a monotonic rise in $\langle P_2(\xi)\rangle_c$ with increasing temperature. It was concluded from these results that the shrinkage is associated with a decrease in molecular orientation and correlates with the decrease in $\langle P_2(\xi)\rangle_a$, whereas crystallization produces an increase in molecular orientation as reflected by the increase in $\langle P_2(\xi)\rangle_c$ and that this is the major effect above 100°C. For samples shrunk at 80°C that do not crystallize, it is clear that shrinkage is primarily associated with disorientation of the amorphous regions, as had previously been concluded from the optical dichroism studies of Patterson and Ward [54]. It has been proposed, notably by Statton *et al.* [66], that shrinkage should be attributed to the refolding of chains that have been pulled out in the drawing process. These results, however, support the dichroism studies and the view of Wilson [67] that the shrinkage of PET is essentially associated with the disorientation of the amorphous regions and that crystallization occurs at a later stage.

For heat treatment at high temperatures the separation between the shrinkage and crystallization processes is not so clear cut, and there is data to suggest that the rates of these processes are roughly comparable at 120°C, although the possibility of a very fast component of shrinkage taking place on a time scale of a few minutes, as found by Wilson, cannot be ruled out. Both molecular disorientation and crystallization could occur simultaneously in some cases.

These results confirm the previous more qualitative studies of McGraw [59,60] and Seki [44,58] on the heat treament of synthetic fibres and are at a much more quantitative level. A further sophistication made possible by the quantitative determination the second- and fourth- moment orientation averages, is the analysis of the mechanism of deformation. From a comparison of PET samples drawn to a ratio of 2 and subsequently shrunk in a controlled way at 80°C to a series of final effective draw ratios, with samples drawn to draw ratios less than about 2.5, it was shown that for these low draw ratios the deformation is reversible. The relation for $\langle P_2(\xi)\rangle$ derived by Roe and Krigbaum [68] using the inverse Langevin approximation to the rubber model, is approximated by [61]:

$$\langle P_2(\xi)\rangle = \frac{1}{5}B + \frac{36}{875}B^2 + \frac{108}{6125}B^3 + \ldots \qquad (4.57)$$

where $B = \frac{1}{N}(\lambda^2 - 1/\lambda)$ and N is the number of freely jointed links between cross-link points.

By plotting $\langle P_2(\xi)\rangle$ against B from Equation (4.57) a value of B can be read off for each experimental value of $\langle P_2(\xi)\rangle$ obtained from the fluorescence measurements. Fig. 4.10 shows the results obtained

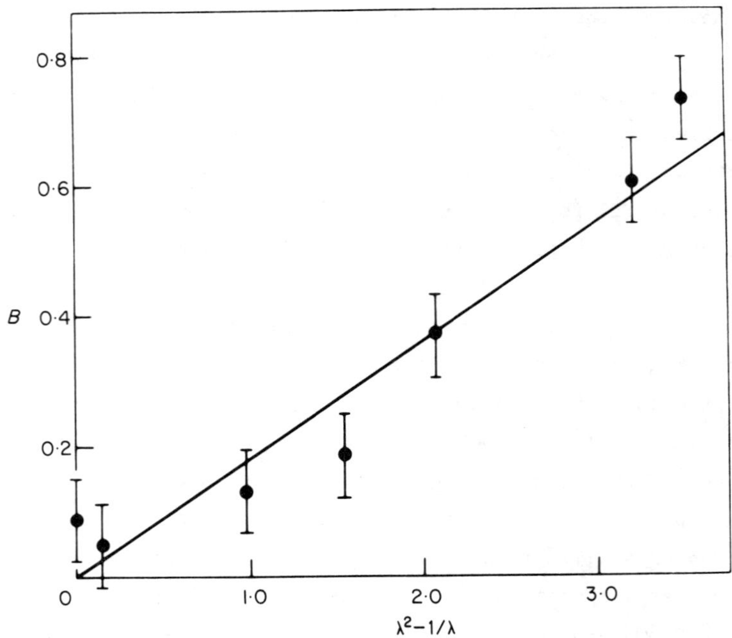

Fig. 4.10 B plotted against (λ^2-1/λ) for PET samples drawn and shrunk at 80°C. (reproduced from *Polymer* 17, 25, 1976 by permission of the publishers, Butterworth Scientific Ltd).

by Nobbs *et al.* for the PET samples drawn and shrunk at 80°C. The plot of B against $(\lambda^2 - 1/\lambda)$ can be fitted very well to a straight line of slope 1/5.6, corresponding to a rubber network with 5.6 freely jointed links between the cross-link points. If we attempt to include samples drawn to much higher draw ratios, it appears that for values of $(\lambda^2 - 1/\lambda)$ greater than 6, the results deviate from a straight line. This value of $(\lambda^2 - 1/\lambda)$ corresponds to a draw ratio of 2.53, which is in the region of maximum extension for a rubber network with $N = 5.6$, since $\lambda_{max} \approx \sqrt{N}$. This result is in good agreement with values deduced previously from stress-optical [62] and infra-red measurements [69].

In a further consideration of the data, Nobbs and Bower [70] considered the possible extension of the rubber network model to predict $\langle P_2(\xi)\rangle$ and also $\langle P_4(\xi)\rangle$ over the whole range of draw ratios. The calculations were carried out both using Treloar's [71] expression for the inverse Langevin function and by making use of numerical methods. It was shown that these two approaches gave essentially identical results for $\langle P_2(\xi)\rangle$ and $\langle P_4(\xi)\rangle$ as functions of λ and N. Fig. 4.11 is a plot of those orientation averages obtained from fluorescence, Raman and refractive index measurements for many PET samples drawn at 80°C. The solid lines are curves predicted by numerical calculation of the modified rubber model with 5.12 equivalent freely jointed links between crosslink points. The value of 5.12 for N is slightly lower that the value of 5.6 obtained from the fit to the low draw ratio samples only, but now data have been fitted to the entire draw ratio range.

In physical terms this calculation assumes that once any chain has been fully extended it rotates like a rigid rod towards the draw direction. This is like the floating-rod model [72] for orientation of crystallites in a crystalline polymer on drawing, first proposed by Kratky [73]. This has been termed pseudo-affine deformation to distinguish it from the affine deformation of a rubber network, in which the segments of chain between the cross-link points are free to rotate so that the chain has maximum entropy.

Finally, Nobbs *et al.* [48] gave consideration to the observation

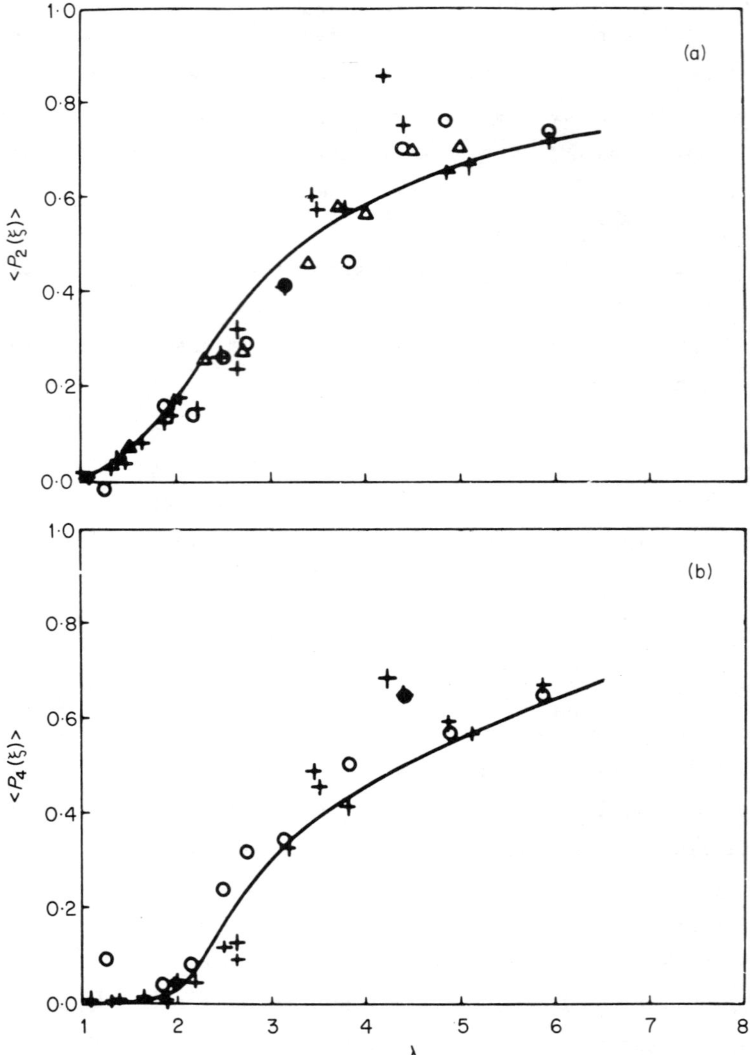

Fig. 4.11 (a) $\langle P_2(\xi)\rangle$ and (b) $\langle P_4(\xi)\rangle$ plotted against draw ratio λ for PET samples drawn at 80°C. The curves are the relationships predicted by the numerical calculation for the modified rubber model with N = 5.12. +, fluorescence measurements; o, Raman measurements; Δ, refractive index measurements (reproduced from *J. Polym. Sci., Polym. Phys. Ed.,*

17 , 259, 1979 by permission of the publishers, John Wiley and Sons, Inc).

that the fluorescent molecules appear to be more highly oriented than the polymer chains viz. the comparison between $\langle P_2(\xi_M)\rangle$ and $\langle P_2(\xi)\rangle_{opt}$ in Fig. 4.8. Two possible explanations were examined. First, there is the possibility that each fluorescent molecule takes up the mean of the orientation of the immediately adjacent polymer chains, which are themselves assumed to be distributed in orientation according to the overall distribution function and moreover that there is no correlation between orientation and position. Calculations were carried out for average chain direction for the case where $\langle P_2(\xi)\rangle = 0.41$, $\langle P_4(\xi)\rangle = 0.35$, assuming the most probable distribution, averaging over either 4, 6 or 10 chains. It was shown that it was not possible in this way to obtain values for the averages $\langle P_2(\xi_M)\rangle$ and $\langle P_4(\xi_M)\rangle$ that were consistent with experimentally observed values for $\langle P_2(\xi_M)\rangle_f$ and $\langle P_4(\xi_M)\rangle_f$. For averaging over 10 chains a sufficiently high value of $\langle P_2(\xi)\rangle_f$ was obtained, but $\langle P_4(\xi_M)\rangle_f$ was much too low. If the averaging were carried out over more chains, a sufficiently high value of $\langle P_4(\xi_M)\rangle$ could be obtained but $\langle P_2(\xi_M)\rangle$ would then be greater than $\langle P_2(\xi_M)\rangle_f$. It is possible that if the correlation that almost certainly exists between the positions and orientations of the polymer chains were taken into account, an averaging scheme might predict the observed fluorescence results, but there is at present no information on which to base such a calculation.

A second possibility stems from a comparison of the orientation of the trans conformations of the PET chains, determined by infra-red measurements with the overall orientation, determined from refractive index measurements. The relationship between $\langle P_2(\xi)\rangle_{IR}^{trans}$ and $\langle P_2(\xi)\rangle_{opt}$ obtained by Cunningham *et al.* [69] is shown in Fig. 4.12, and was found to be independent of the draw temperature. There is a very clear similarity between Fig. 4.12 and Fig. 4.8 where $\langle P_2(\xi_M)\rangle_f$ is plotted against $\langle P_2(\xi)\rangle_{opt}$. In fact the maximum difference between $\langle P_2(\xi)\rangle_{IR}^{trans}$ and $\langle P_2(\xi_M)\rangle_f$ is ± 0.06.

It has therefore been proposed that the axes of the fluorescent molecules align themselves parallel to the chain axis direction of those sections of the chain in the trans conformation. It is also possible to use infra-red measurements of the proportion of chains in the trans conformation, as distinct from the gauche conformation, to obtain values for $\langle P_4(\xi)\rangle_f$ from $\langle P_4(\xi_M)\rangle_f$.

The orientation averages over all chains $\langle P_n(\xi)\rangle$ are given by:

$$\langle P_n(\xi)\rangle = F\langle P_n(\xi)\rangle^{\text{trans}} + (1-F)\langle P_n(\xi)\rangle^{\text{gauche}} \qquad (4.58)$$

where F is the fraction of material in the trans conformation. Cunningham *et al.* showed that the optical measurements that give

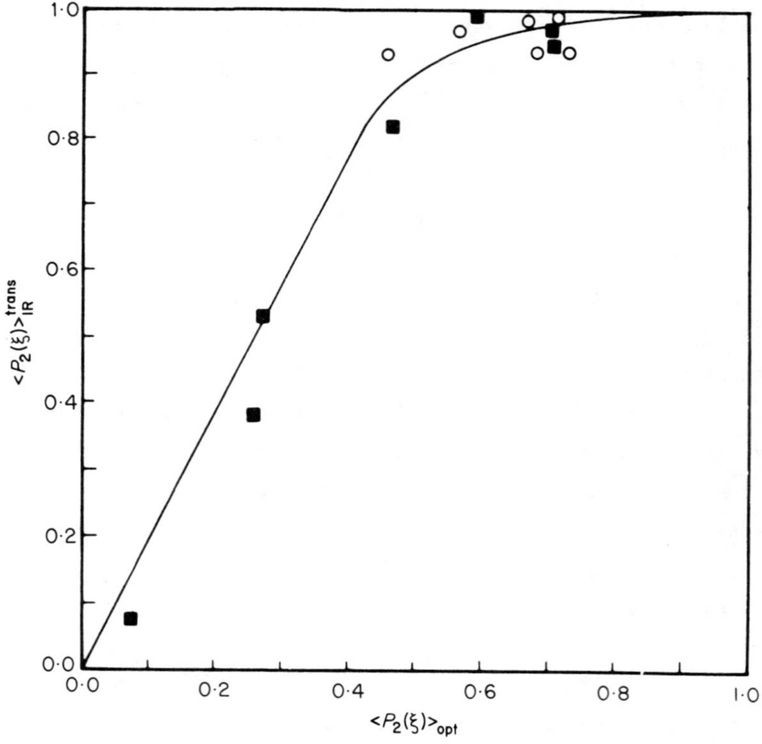

Fig. 4.12 $\langle P_2(\xi)\rangle^{\text{trans}}_{\text{IR}}$ plotted against $\langle P_2(\xi)\rangle_{\text{opt}}$: ○, draw ratio 4 at various temperatures; ■ , various draw ratios at 80°C (reproduced from *J. Polym. Sci., Polym. Phys. Ed.*, 17, 259, 1979.

by permission of the publishers, John Wiley and Sons, Inc).

$\langle P_2(\xi)\rangle$ were consistent to a good approximation with the orientation of the gauche conformations remaining random i.e. $\langle P_2(\xi)\rangle^{\text{gauche}} = 0$, from which it follows that:

$$\langle P_n(\xi)\rangle_f = F\,\langle P_n(\xi_M)\rangle_f\ . \tag{4.59}$$

Cunnungham *et al.* [69] also observed that there was a unique relationship between F and $\langle P_2(\xi)\rangle_{\text{opt}}$. Hence, a value of F can be obtained from $\langle P_2(\xi)\rangle_{\text{opt}}$ for any sample and $\langle P_4(\xi)\rangle_f$ calculated from the relationship:

$$\langle P_4(\xi)\rangle_f = F\langle P_4(\xi_M)\rangle_f. \tag{4.60}$$

The validity of this exercise was confirmed by comparison of values of $\langle P_4(\xi)\rangle_f$ calculated in this way with $\langle P_4(\xi)\rangle_{\text{Raman}}$, the fourth-order average obtained from Raman spectroscopy [74,75]. Very good agreement was obtained, except for the samples of very low orientation where it appeared that the gauche conformations might make a small but significant contribution to the overall orientation.

It was therefore concluded that it is valid to assume that the fluorescent molecules are preferentially aligned parallel to the chain axis direction of the polymer chains in the trans conformation. Thus $\langle P_2(\xi)\rangle_f$, which has been assumed to be equivalent to $\langle P_2(\xi)\rangle$ can be obtained from Fig. 4.12 and $\langle P_4(\xi)\rangle_f$ from Equation (4.60) using a value of F calculated from $\langle P_2(\xi)\rangle_{\text{opt}}$. Values for $\langle P_2(\xi)\rangle_f$ and $\langle P_4(\xi)\rangle_f$ obtained in this way were used for the comparisons discussed above, which are shown in Fig. 4.11.

In the next series of investigations to be described, which are by Maeda and his colleagues, and by Monnerie and his collaborators, the fluorescent probes were chemically incorporated in the polymer chain.

Three papers by Hibi, Maeda and their colleagues [76-78] describe studies of orientation in uniaxially drawn poly(vinyl chloride) films, using polarized fluorescence and refractive index measurements to determine orientation averages, which were then used to predict the mechanical anisotropy on the basis of the aggregate model. Hibi *et al.* [77] measured the fluorescence of polyene and carbonyl group produced in the polymer chain by heat treatment. The ratio of polyene to carbonyl, n, alters according to the details of the heat treatment. It was considered, on the basis of other information, that the value of n lay between 1.0 and 1.5. The results were therefore evaluated for different values of n in this range and $\langle\cos^2\theta\rangle_f$ values compared with $\langle\cos^2\theta\rangle_{opt}$ obtained from birefringence Δn assuming a maximum value $\Delta n_{max} = 10.4 \times 10^{-3}$, on the basis of theoretical calculations. This comparison assumes that the polarized fluorescence is related to the chain orientation. The angles between the polymer chain and the fluorescent probes θ_1 and θ_2 were assumed to be $\theta_1 = 20^{\circ}$ for polyene and $\theta_2 = 90^{\circ}$ for carbonyl, again on the basis of calculation. Hibi *et al.* performed their quantitative calculations on results obtained for principal values of the fluorescence e.g. $I_{||}(0)$, $I_{\perp}(90)$, etc., where there are no birefringence effects. The results were expressed in terms of $\langle\cos^2\theta\rangle$ and $\langle\cos^4\theta\rangle$, which as we have seen is not the best procedure.

Excellent linear correlations were obtained between $\langle\cos^2\theta\rangle_f$ and $\langle\cos^2\theta\rangle_{opt}$, but the value of n depended on the temperature of heat treatment. $\langle\cos^2\theta\rangle_f$ values were compared with values of $\langle\cos^4\theta\rangle_f$. It was proposed that these results, taken together with results for $\langle\cos^2\theta\rangle$ obtained from either fluorescence or birefringence, considered as a function of draw ratio, were consistent with the following models. For unplasticized film drawn below the glass transistion temperature the development of orientation follows the pseudo-affine deformation scheme where the chain axes are considered to rotate as rigid rods towards the draw direction. For all plasticized films, and for unplasticized film drawn above T_g, a composite distribution function of molecular orientations is assumed, in which one fraction of the polymer deforms according to the pseudo-affine deformation scheme and the other fraction according to a rubber-like network. The key parameter for the deformation of a rubber-like network is N, the number of equivalent random links per chain (i.e. between network junction points). The results were shown to be consistent with values for N in the range 6-9, and for the proportions of pseudo-affine to rubber-like deformation in the range 60:40 to 50:50, the actual values depending on the draw temperature and proportion of plasticizer.

In the final paper of this series Hibi *et al.* [78] considered the mechanical anisotropy of uniaxially oriented unplasticized PVC films in the light of the aggregate model. It is considered that the polymer consists of an aggregate of anisotropic units that are gradually aligned during the drawing process. The elastic constants of the anisotropic units were estimated by extrapolation of the mechanical data to a very high draw ratio. From the experimental data it was clear that the measured Young's moduli depended only on the birefringence of the film, irrespective of the draw temperature. Predictions for the Young's moduli, in the directions parallel and perpendicular to the initial draw direction were made on the basis of the compliance averaging scheme only (the Reuss average). Fair agreement with the experimental data was observed. Maeda *et al.* suggested that the unique dependence of the Young's moduli on birefringence justified the approximation $\langle P_2(\xi)\rangle = \langle P_4(\xi)\rangle$, and that with this approximation a reasonable fit to the experimental data could also be obtained.

In two rather similar investigations of polyisoprene [79] and polystyrene [80], respectively, Monnerie and his collaborators have studied the orientation behaviour by introducing the fluorescence probe into the middle of the polymer chain.

High molecular weight polyisoprene of high molecular weight ($M_n \sim 5 \times 10^5$) was prepared by anionic polymerization. Labelled polymer was prepared by joining two chains of molecular weight 3×10^5 by a dimethylanthracene fluorescent group. One advantage of this method of labelling is that the fluorescent transition lies along the chain axis. The labelled polymer of molecular weight 6×10^5 was mixed in solution at a concentration of 1% with the unlabelled polymer. After the solvent had been removed, the polymer was then moulded and cross-linked with dicumyl peroxide at 147°C.

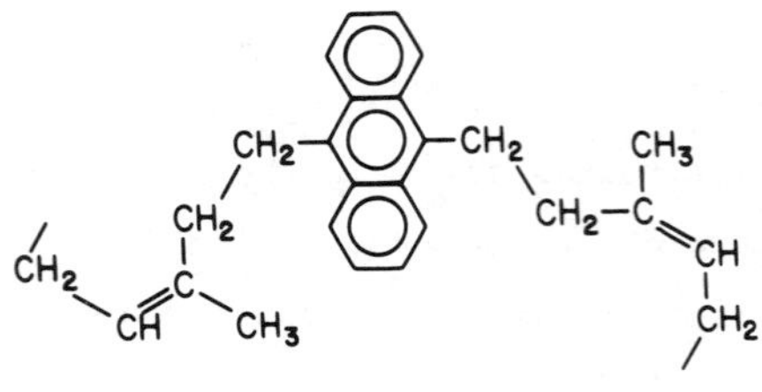

The load - extension curves for the polyisoprene network were measured over the temperature range from -56°C to +85°C. At the lowest temperature of -56°C, the curve showed a yield point before drawing to a ratio $\lambda = 5$. At the higher temperatures homogeneous rubber-like deformation was observed with an increase in the stress level from -30°C to +85°C, which is typical of a rubber network. The plots of $\langle P_2(\xi)\rangle$ versus the elongation λ were determined over the same temperature range, the fluorescence apparatus permitting simultaneous measurements of stress and orientation during stretching. It was found that for temperatures above -30°C, the plots of $\langle P_2(\xi)\rangle$ versus λ were independent of temperature. For temperatures below -30°C, the plots of $\langle P_2(\xi)\rangle$ versus λ rose more steeply as the temperature was reduced, the general shape at -56°C being reminiscent of the pseudo-affine deformation scheme.

Jarry and Monnerie [79] examined the effects of swelling the polyisoprene network. The stress-strain data were described in terms of the Mooney - Rivlin equation:

$$\sigma = (C_1 + C_2/\lambda)\ (\lambda^2 - 1/\lambda). \tag{4.61}$$

It was found that C_1 remained constant on swelling, but C_2 decreases, which would be expected. The values of $\langle P_2(\xi)\rangle$ were reduced on swelling, which the reviewer suggests is consistent with an increase in the number of random links per chain.

Jarry and Monnerie [79] considered that there were two types of deviation from the classical theory of rubber elasticity:

1. At low and moderate elongations, extra orientation occurs in the dry state which is strongly reduced on swelling.
2. At high elongations, saturations of orientation occurs.

It is argued that the extra orientation cannot be attributed to crystallization, because increases are observed in the glass transition region, and not at $-20^{\circ}C$ where maximum crystallization occurs. The existence of anisotropic intermolecular interactions, favouring the alignment of neighbouring segments, is proposed. This is treated formally by a mean-field approximation leading to the equation:

$$\langle P_2(\xi)\rangle = \left(\frac{1}{(1-V)}\right) \frac{1}{5N} (\lambda^2 - 1/\lambda) \qquad (4.62)$$

where the parameter V expresses the extent of intermolecular interaction; $V = 0$ corresponds to the classical theory and $V = 1$ to the critical temperature of spontaneous nematic ordering. Typical values are $V = 0.5$ for temperatures above $-30^{\circ}C$ and $V = 0.65$ at $-47^{\circ}C$.

The saturation of orientation at high elongation ratios is, of course, related to the limiting extensibility of the molecular network, as discussed above for the case of PET. In these experiments the fluorescence orientation average $\langle P_2(\xi)\rangle_f$ is related to the chain orientation average $\langle P_2(\xi)\rangle$ by the linear relationship $\langle P_2(\xi)\rangle = f\langle P_2(\xi)\rangle$, where f is a constant, with a value between 1 and 2. Jarry and Monnerie showed that this linear relationship followed from elementary considerations based on the Kuhn and Grün [81] theory of rubber elasticity, provided that $\langle P_2(\xi)\rangle$ is not too large (typically < 0.1). The fluorescence measurements therefore determine $\langle P_2(\xi)\rangle_f = \frac{1}{5N'}(\lambda^2 - 1/\lambda)$, where N' is the apparent number of random links per chain. Swelling experiments were used to estimate a value for N, the actual number of random links per chain. As discussed above for PET, the maximum elongation ratio $\lambda_{max} \sim \sqrt{N}$ and the network may be expected to deform in a Gaussian manner up to $\lambda_o \simeq 0.7\lambda_{max} = 0.7\sqrt{N}$.

A combination of swelling and fluorescence experiments on the isoprene networks confirmed that the saturation of orientation occurred when the chains approached the limit of Gaussian behaviour, with the constant $f \sim 2$. It was also suggested that beyond the Gaussian range are effects associated with the stretching of the random links. We could envisage this as occurring owing to an increase in the number of monomers per random link (because of gauche -trans conformational changes on stretching). This would reduce N, the number of random links per chain, and hence give a rise in the development of the orientation average $\langle P_2(\xi)\rangle$, through Equation (4.63).

In a further publication [80], Monnerie and his colleagues describe a study of orientation in polystyrene. Using a similar technique to that adopted for polyisoprene, anthracene

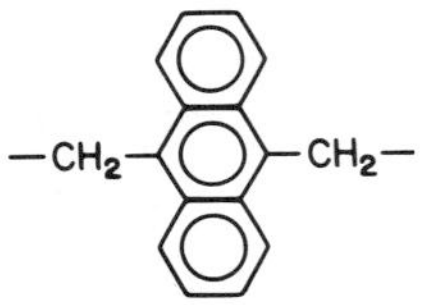

groups were covalently bonded in the middle of anionically polymerized polystyrene chains. Between 0.5% and 1% of the labelled chains (M_n = 287 000) were dispersed in normal narrow dispersion polystyrene ($\overline{M}_w$ = 207 000, M_n = 191 000). After being mixed in solution, the mixture was dried and annealed to give a transparent isotropic film. Previous experiments showed that no molecular motion occurs during the lifetime of the excited state. The un-crosslinked polymer was therefore stretched above T_g, and simultaneous measurements of stress and polarized fluorescence undertaken to give values for $\langle P_2(\xi)\rangle_f$. Subsidiary shrinkage measurements showed that in all cases there was complete recovery to a very good approximation.

Over a wide range of temperatures, the load-extension curves suggested elastic - plastic behaviour, the load rising sharply initially and then less steeply as drawing took place. The results for $\langle P_2(\xi)\rangle_f$ showed that there was always continuous development of orientation with elongation, but the actual values for $\langle P_2(\xi)\rangle_f$ were much smaller at higher temperatures. At low temperatures, the development of orientation with elongation reaches a limiting curve (again reminiscent of pseudo-affine deformation behaviour as observed in PET [82] and polymethylmethacrylate [83]), although the load - elongation curves are not independent of temperature and strain rate. At higher temperatures both the load - elongation curves and the orientation-elongation curves are affected by strain rate.

Monnerie and his collaborators concluded that the orientation determined by the fluorescent probe does not relate to the complete true stress, i.e. the total stress during deformation, but only the equilibrium stress, i.e. the plateau rubber network stress. It is suggested that the fluorescence measurements are sensitive only to the central part of the chain, and that relaxations occur during stretching that affect the central part of the chains (perhaps by the reptation mechanisms preposed by de Gennes [84] and Doi and Edwards [85] These experiments offer the prospect of gaining information on the dynamics of the deformation process, which is a very exciting prospect.

4.10 CONCLUSIONS

Recent developments have provided a firm theoretical understanding of the polarized fluorescence technique, so that it is now established alongside X-ray diffraction, infra-red and Raman spectroscopy and NMR as a technique capable in principle of producing quantitative measurements of orientation in polymers.

In spite of the essential problem of identifying exactly where the fluorescent probe is located with respect to the polymer chain, it is clear that some very useful information has been obtained regarding the deformation of polymers. The polarized fluorescence technique is of most value where it is used in conjunction with other measurements of orientation.

A final point in favour of the technique is that it is comparatively straightforward in experimental terms. The intensity of the scattered radiation can be large, and because it is in the visible range conventional detecting devices are available. For this reason polarized fluorescence offers the prospect of monitoring deformation processes in polymers on a continuous basis with comparatively inexpensive equipment.

REFERENCES

1. GHIGGINO, K.P ., ROBERTS, A.J. and PHILLIPS, D,(1981) *Adv. Polym. Sci.*, 40, 69.
2. MONNERIE, L., This volume.
3. SOUTAR, I.(1982) in *Development in Polymer Photochemistry* - 3 (ed. N.S. Allen), Applied Science Publishers, London.
4. BOWER, D.I. (1975) in *Structure and Properties of Oriented Polymers.* (ed. I.M. Ward), Applied Science Publishers, London.
5. CHAPOY,L.L. and DUPRE,J. (1980) in *Methods in Experimental Physics*, 16A, 404.
6. KIMURA, I., KAGIYAMA,M., NOMURA,S. and KAWAI, H.(1969) *J. Polym. Sci.*, A-2, 7, 709.
7. BADLEY, R.A., MARTIN, W.G. and SCHNEIDER, H.(1973)*Biochemistry*, 12, 208.
8. FREHLAND, E.(1975) *Z.Naturforsch*, 30a, 1241.
9. JARRY, J.P. and MONNERIE, L.(1978) *J. Polym.Sci.*, A-2, 16, 443.
10. ZANNONI, C. (1979) *Mol. Phys.*, 38, 1813.
11. MARGULIES, L., and YOGEV, A. (1978) *Chem. Phys.* 27, 89.
12. MICHL, J. and THULSTRUP, E.W.(1980) *J. Chem. Phys.*, 72,3999.
13. BECKER, R.S.(1969) *Theory and Interpretation of Fluorescence and Phosphorescence*, Wiley-Interscience, New York.
14. PESCE, A. J., ROSEN, C.G. and PASBY, T.L.(Eds.) (1971) *Fluorescence Spectroscopy*, Dekker, New York.
15. HENNECKE, M. and FUHRMANN, J.(1980) *Colloid and Polym. Sci.*,258, 219.
16. ROE, R-J. and KRIGBAUM, W.R.(1964) *J. Chem. Phys.*,40, 2608.
17. ROE, R-J.(1965) *J. Appl. Phys.*, 36,2024.
18. ABRAMOWITZ, M. and STEGUN, I.A.(Eds)(1968) *Handbook of Mathematical Functions*, Dover, New York, Chapter 8.
19. DESPER, C.R. and KIMURA, I.(1967)*J.Appl.Phys.*, 38,4225.
20. NOMURA, S., KAWAI, H., KIMURA, I. and KAGIYAMA,M.(1970) *J. Polym. Sci.*, A-2, 8, 383.

21. ROE, R-J.(1970) *J. Polym. Sci.*, *A*-2,8, 1187.

22. BOWER, D.I.(1972) *J.Polym. Sci.*, A-2, 10, 2135.

23. BOWER, D.I.(1981) *J.Polym. Sci.*, A-2, 19, 93.

24. WARD, I.M. (Ed.) (1975) *Structure and Properties of Orientated Polymers*, Applied Science Publishers, London, Chapters 1 and 5.

25. WARD. I.M. (1962) *Proc. Phys. Soc.*, 80, 1176,

26. ANDERSEN, L. and NORDEN, B.(1980) *Chem. Phys. Lett.*, 75, 398.

27. NOBBS, J.H., BOWER, D.I., WARD, I.M. and PATTERSON,D.(1974) *Polymer*,15, 287.

28. BOWER, D.I. (1982) *Polymer*,23, 1251.

29. NISHIJIMA,Y., ONOGI,Y. and ASAI, T. (1965) *Rep. Prog. Polym. Phys. Jap.*, 8, 131.

30. NISHIJIMA, Y., ONOGI, T. and ASAI, T,(1966) *J. Polym. Sci.*,*C*.15, 237.

31. NISHIJIMA, Y., FUJIMOTO,T. and ONOGI, Y.(1966) *Rep. Prog. Polym. Phys. Jap.*,9,457.

32. NISHIJIMA, Y., ONOGI, Y., and ASAI, T. (1966) *The International Symposium on Macromolecular Chemistry*,7, 161.

33. NISHIJIMA,Y., ONOGI, Y., and ASAI, T. (1967) *Rep. Prog. Polym. Phys. Jap.*, 10,461.

34. NISHIJIMA,Y., ONOGI, Y., ASAI, T., and YAMAZAKI, R.(1967) *Rep. Prog. Polym. Phys. Jap.*, 10, 465.

35. NISHIJIMA,Y., and ONOGI,Y. (1968) *Rep. Prog. Polym. Phys. Jap.*, 11, 395.

36. NISHIJIMA,Y., ONOGI,Y., ASAI, T. and YAMAZAKI, R.(1968) *Rep. Prog. Polym. Phys. Jap.*, 11, 399, and 403.

37. NISHIJIMA,Y.,and ASAI, T. (1968) *Rep. Prog. Polym. Phys. Jap.*, 11, 419 and 423.

38. RUPPRECHT,A., RIGLER, R., FORSLIND, B. and SWANBECK, G.,(1969) *Euro. J. Biochem.*, 10, 291,

39. WEBER, G., (1954) *Trans. Faraday Soc.*,50, 552.

40, PINAUD, F., JARRY, J.P., SERGOT,Ph. and MONNERIE, L., (1982) *Polymer*, 23, 1575.

41. HENNECKE, M. and FUHRMANN, J. (1982) *Polymer*, 23, 797.

42. TSCHANZ, H.P. and BINKERT,Th. (1976) *J. Phys-E, Sci. Instrum.*, 9, 1131.

43. ZINSLI, P.E. (1978) *J. Phys-E.Sci. Instrum.*, 11, 17.

44. SEKI,J. (1969) *Sen - I Gakkashr*, 25, 16.

45. ONOGI,Y. and NISHIJIMA,Y. (1971) *Rep. Prog. Polym.Phys.Jap.* 14, 533, 537 and 541.

46. NOBBS, J.H. (1976) Ph.D. Thesis, University of Leeds.

47. IRVING,J. and MULLINEUX, N.(1959) *Mathematics in Physics and Engineering*, Academic Press, London p.197.

48. NOBBS, J.H., BOWER,D.I. and WARD,I.M. (1979) *J. Polym. Sci.*, *A-2*, 17, 25,

49. MOREY, D.R.(1933) *Text. Res. J.*, 3, 325; (1934) *Text. Res. J.*, 4, 491;(1935) *Text. Res. J.*, 5, 483.

50. MOREY, D.R.(1935) *Text. Res. J.*, 5, 538.

51. NISHIJIMA,Y., ONOGI, Y., YAMAZAKI, R. and KAWAKAMI, K.(1968) *Rep. Prog. Polym. Phys. Jap.*, 11, 407.

52. YAMAZAKI, R., ONOGI, T. and NISHIJIMA, Y. (1969) *Rep. Prog. Polym. Phys. Jap.*, 12, 439.

53. NISHIJIMA, Y. (1970) *J. Polym. Sci. C.*, 31, 353.

54. PATTERSON,D. and WARD, I.M. (1957) *Trans. Faraday Soc.*, 53, 1516.

55. NISHIJIMA,Y., ONOGI, Y. and YAMAZAKI, R. (1968) *Rep. Prog. Polym. Phys. Jap.*, 11, 415.

56. ASAI, T., ONOGI, Y. and NISHIJIMA, Y., (1969) *Rep. Prog. Polym. Phys. Jap.*, 12, 433,

57. ASAI, T. and NISHIJIMA, Y. (1969) *Rep. Prog. Polym. Phys. Jap.*, 12, 437.

58. SEKI, J., (1969) *Sen - I Gakkashi*, 25, 24.

59. McGRAW, G.E. (1970) *J. Polym. Sci.*, *A-2*, 8, 1323.

60. McGRAW, G.E. (1972) In *Structure and Properties of Polymer Films*. (ed. R.W. Lenz and R.S. Stein), Plenum, New York, p.97.

61. NOBBS, J. H., BOWER, D.I. and WARD, I.M. (1976) *Polymer*, 17, 25.

62. PINNOCK, P.R. and WARD, I.M. (1966) *Trans. Faraday Soc.*, 62. 1308.

63. GULRAJANI, M.L. and PADHYE, M.R. (1971) *Ind. J. Technol.*, 9, 211.

64. KRYSZEWSKI, M. (1967) *Faserforsch. Textiltech.* 18, 189.

65. CUNNINGHAM, A., DAVIES, G.R. and WARD, I.M. (1974) *Polymer*, 15, 743.

66. STATTON, W.O., KOENIG, J.L. and HANNON, M.J. (1970)*J. Appl. Phys.* 41, 4290.

67. WILSON, M.P.W. (1974) *Polymer*, 15, 277.

68. ROE, R-J, and KRIGBAUM, W.R. (1964) *J. Appl. Phys.*, 35, 2215.

69. CUNNINGHAM, A., WARD, I.M. WILLIS, H.A. and ZICHY, V. (1974), *Polymer*, 15, 749.

70. NOBBS, J.H. and BOWER, D.I.(1978), *Polymer*, 19, 1100.

71. TRELOAR, L.R.G. (1954) *Trans. Faraday Soc.*, 50, 881.

72. WARD, I.M. (1962) *Proc. Phys. Soc.*, 80, 1176.

73. KRATKY, O. (1933) *Kolloid-Z.*, 64, 213.

74. PURVIS, J., BOWER, D.I. and WARD, I.M. (1973) *Polymer*, 14, 398.

75. PURVIS, J. and BOWER, D.I. (1976) *J. Polym. Sci. Polym. Phys. Ed.*, 14, 1461.

76. HIBI, S., MAEDA,M., KUBOTA, H. and MIURA, T. (1977) *Polymer*, 18, 137.

77. HIBI, S., MAEDA, M.,KUBOTA, H. and MIURA, T. (1977) *Polymer*, 18, 143.

78. HIBI, S., MAEDA, M. and KUBOTA, H. (1977)*Polymer*, 18, 801.

79. JARRY, J.P. and MONNERIE, L. (1980) *J. Polym. Sci., Polym. Phys. Ed.*, 18, 1879.

80. FAJOLLE, R., TASSIN, J.F., SERGOT, P., PAMBRUN,C and MONNERIE, L. (1984) *Polymer* (in press).

81. KUHN, W. and GRÜN, F., (1942) *Kolloid Z. Z. Polym.*,101, 248,

82. ALLISON, S. W. and WARD, I. M. (1967) *Brit. J. Appl. Phys.*, 18, 1151.

83. KAHAR, N., DUCKETT, R.A. and WARD, I.M. (1978) *Polymer*, 19, 136.

84. DE GENNES, P.G. (1971) *J. Chem. Phys.*, 55, 572.

85. DOI, M. and EDWARDS, S.F.(1978) *J. Chem. Soc. Faraday Trans.*, 74, 1789, 1802, 1818.

CHAPTER FIVE

Luminescence of macromolecules in fluid solution

5.1 INTRODUCTION

Interest in the photophysical behaviour of macromolecules has prompted a vast expansion in the relevant literature in recent years. The reasons for this dramatic increase in activity are manifold and range from interests in the use of photophysical techniques in the study of dynamic physical properties of macromolecules to interests in energy transfer and trapping phenomena. The experience gained by polymer photophysicists is directly relevant to considerations of polymer photochemistry and control, energy storage and transmission and to the synthetic modelling of more complex biological systems such as those implicated in photosynthesis.

Consequently, reports involving luminescence characterization of fluid solutions of polymers have encompassed emission depolarization studies of relaxation processes, determination of energy-transfer rates and processes and the study of exciplex or excimer formation. This chapter is concerned primarily with the discussion of intramolecular excimer formation in macromolecules, an area of considerable current interest and topical debate. Aspects of the applications of emission anisotropy measurements and energy transfer phenomena in polymer science are presented in Chapters 6 and 2 and Chapter 3, respectively.

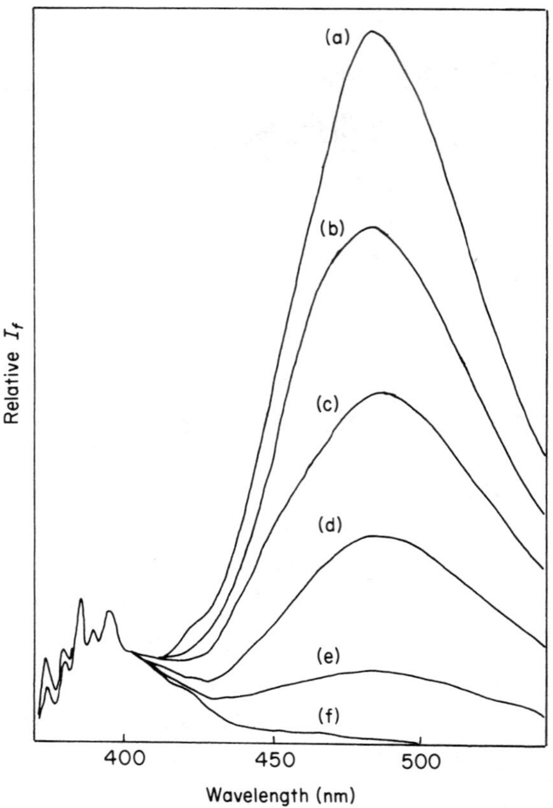

Fig. 5.1 Typical results. Fluorescence spectra of pyrene solutions in cyclohexane: (a) 10^{-2}M; (b) 7.75 x 10^{-3}M; (c) 5.5 x 10^{-3}M; (d) 3.25 x 10^{-3}M; (f) 10^{-4}M (after Birks [3]).

5.2 EXCIMER FORMATION

Excimer formation in fluid solutions was discussed fully in Chapter 1, and is merely referred to here [1-7]. The kinetics are explicable in terms of the kinetic scheme after Birks, shown in Scheme 5.1. Typical results are shown in Fig. 5.1.

5.3 INTRAMOLECULAR EXCIMER FORMATION

The first published fluorescence spectrum that contained a contribution from an intramolecular excimer species is that of polystyrene in 1962 [8], although the origins of the emission were not recognised until 1963 by Yanari *et al* [9]. Informative studies on low molar mass systems appeared shortly thereafter. Hirayama [10] demonstrated the existence of intramolecular excimer species in diphenyl and triphenyl alkanes, and Vala *et al* [11] produced similar evidnece of the effect in paracyclophanes and vinyl polymers. Subsequent studies upon diaryl propanes containing carbazolyl [12] and naphthyl [13] chromophores have demonstrated the generality of the phenomenon.

These pioneering studies, through examination of the geometric and structural requirements for excimer formation in paracyclophanes and diaryl alkanes, have led to the general conception of an excimer being a 'sandwich-type' dimeric structure in which the aromatic

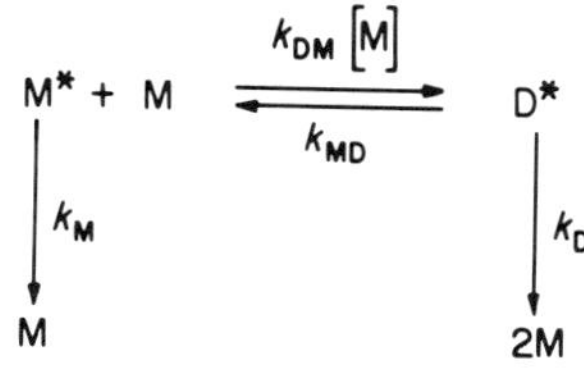

Scheme 5.1.

components are alligned in a periplanar configuration and separated by a distance of the order of 300 pm. These rigorous requirements result in the generalization first propounded by Hirayama [10], known as the n=3 rule, which states that excimer formation is expected betw.en aromatic substituents attached to an alkane chain only when the chromophores are separated by 3 carbon atoms. Although not rigorously obeyed (see subsequent discussion) this rule of thumb has been of considerable value in guiding the throughts of investigators interested in the study of macromolecular photophysics, particularly in the case of vinyl aromatic polymers in which nearest-neighbour chromophores satisfy the ideal conditions for excimer formation.

5.4 STEADY-STATE FLUORESCENCE STUDIES OF INTRAMOLECULAR EXCIMER FORMATION IN POLYMERS

In order to gain information about the parameters governing excimer formation in macromolecules, several approaches have been adopted. Early investigations were concerned primarily with considerations of the thermodynamic characteristics of the interaction through a study of the temperature and solvent dependence of the steady-state emission intensity ratio. In addition, some kinetic information was derived from transient excitation experiments which assumed a validity of the Forster [2] mechanism and direct applicability of Birks [3] kinetic derivations to the intramolecular phenomenon in macromolecules. The extension of these methods to encompass studies of copolymeric species introduced a new dimension to the investigations, namely the ability to examine the intramolecular concentration dependence of the photophysical behaviour. More recently, developments in sophisticated instrumental techniques and the advent of stable, subnanosecond excitation sources has resulted in doubts regarding the validity of direct application of the Förster/Birks [2,3] kinetic scheme to macromolecular photophysics.

5.4.1 *Studies of the temperature dependence of excimer formation*

In the main, studies involving the perturbation of the excited monomer-excimer population through temperature variation have analysed the data on the assumption that the kinetic analysis of Birks [3] is applicable to macromolecular photophysics. The pertinence of the derived thermodynamic or kinetic data is consequently questionable and subject to additional uncertainty by use of the treatment of Stevens and Ban [14] and Birks *et al* [15], which adopt assumptions of dubious validity in the case of polymers. The basis of the approach is discussed below.

The Forster [2]/Birks [3] kinetic scheme results in an expression for the ratio of excimer to monomer intensity ratio as defined in Equation (5.1) as:

$$\frac{I_D}{I_M} \simeq \frac{\phi_D}{\phi_M} = \frac{k_{FD}k_{DM}[M]}{k_{FM}(k_D+k_{MD})} \quad . \tag{5.1}$$

The observed dependence of I_D/I_M as a function of temperature, exemplified in Fig. 5.2, can be explained if several assumptions regarding the thermal behaviour of the rate coefficients are made. The existence of two regions of photophysical behaviour may be rationalized as follows.

(a) Low-temperature regime

At low temperatures, if the rate of excimer dissociation is sufficiently low relative to alternative means of excimer state depopulation ($k_{MD} << k_D$) *and* provided it may be assumed that k_{FM}, k_{FD} and k_D are invariant with temperature, I_D/I_M is represented by:

$$I_D/I_M = kk_{DM}[M] . \tag{5.2}$$

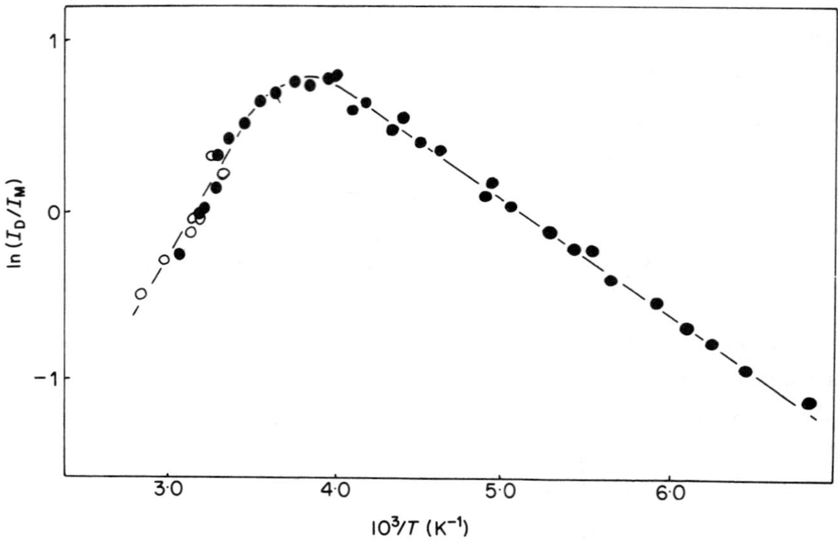

Fig. 5.2 Variation of I_D/I_M as a function of temperature for poly(2-phenyl-5-(*p*-vinyl)phenyl oxazole dissolved in 2-methyltetrahydrofuran (•) and toluene (o) (after McInally [25]).

If it is further assumed that the concentration term [M] (which in the case of a polymer represents a local chromophore concentration, remains constant over the temperature range it is apparent that a plot of $\ln(I_D/I_M)$ as a function of T^{-1} should be linear and yield a value of the activation energy of excimer formation, E_{DM}, from its slope.

(b) High temperature regime

If the temperature dependence of the rate coefficient for excimer dissociation, k_{MD}, is sufficiently greater than that representative of the thermal dependence of k_D, it is to be expected that at sufficiently elevated temperatures k_D may be regarded as negligible relative to k_{MD} and I_D/I_M may be expressed as:

$$\frac{I_D}{I_M} = \frac{k_{FD}}{k_{FM}} \cdot \frac{k_{DM}[M]}{k_{MD}} = k\frac{k_{DM}}{k_{MD}} . \tag{5.3}$$

where k is a constant provided k_{FD}, k_{FM} and [M] are regarded as temperature invariant. Given the validity of these assumptions, a state of equilibrium is obtained in the high-temperature region and the dependence of $\ln(I_D/I_M)$ on T^{-1} should yield an estimate of the binding energy of the excimer. Table 5.1 summarizes the literature results obtained in such a manner.

Since most of the experiments cited in Table 5.1 were performed when the vast majority of the work on macromolecular luminescence was confined to steady-state excitation, the aspirations of the investigators seem reasonable. However, the validity of the data in representation of the functions E_{DM} and $-\Delta H$ is open to considerable doubt, not only in view of the assumptions discussed above in derivation of the low- and high-temperature relationships from the Birks [3] scheme, but also as a result of doubts recently expressed [26-30] regarding the applicability of the basic kinetic scheme itself to macromolecules.

Evidence for the failure of the assumptions inherent in the treatment discussed above to withstand empirical evaluation may be found both from steady-state excitation and transient data. For example Nishijima [31] and David *et al* [32] have presented data (for poly 2- and 1-vinylnaphthalenes, respectively) that indicate that I_M does not increase in the manner expected in the high-temperature region if excimer dissociation were dominant. More emphatically, Holden and Guillet [33] have tabulated individual rate parameters derived from transient decay data for poly(1-vinyl-naphthalene) [33,34] and poly(1-naphthylmethacrylate) [22] in dichloromethane [22,34] or THF [33] at 298K. (Under these conditions, steady-state dependence of I_D/I_M as a function of T^{-1} would be classified as being above the region characterized by 'low-temperature behaviour'). In no case was $k_{MD} >> k_D$, as required by Birks *et al* [15] for high-temperature behaviour. (It is worth noting, however, that the latter data have been derived

Table 5.1 Activation energies and binding energies of excimer formation

Polymer	Solvent	E_{DM} (kJ mol^{-1})	$-\Delta H$ (kJ mol^{-1})	Reference
Polystyrene	THF/E	5.9	34.7	16
	DCE	12.8	-	17
	DCM	15.9	-	18
isotactic Polystyrene	DCM	8.4	-	18
Poly(α-methyl styrene)				
90% syndiotactic	DCM	10.5	-	18
70% syndiotactic	DCM	7.5	-	18
Poly(1-v nyl-naphthalene)	THF/E	8.8	28.9	16
	MTHF	11.3	15.1	19
Poly(2-vinyl-naphthalene)	THF/E	8.4	27.2	16
	PB	15.1	30.9	20
	MTHF	5.4	-	21
Poly(acenaphthylene)	MTHF	3.4	8.8	19
Poly(1-naphthyl methacrylate)	DCM	-	11.7	22
	C	-	15.5	22
	EA	-	19.2	22
Poly(2-naphthyl methyl methacrylate)	THF	-	11.3	23
	EA	-	14.2	23
Poly(2-naphthyl ethyl methacrylate)	THF	-	10.5	23
	EA	-	13.8	23
Poly(2-naphthyl propyl methacrylate)	THF	-	11.7	23
	EA	-	16.3	23
Poly(4-vinyl biphenyl)	PB	8.0	-	24
Poly(vinyl PPO)	MTHF	5.9	17.0	25

THF=tetrahydrofuran; E=ether; DCE=1,2 dichloroethane; DCM=dichloromethane; MTHF=2-methyltetrahydrofuran; PB=n-propylbenzene; C=chloroform; EA=ethyl acetate.

from kinetic data obtained at a single chromophore concentration - that dictated by the equivalent local concentration within the polymer coil dimensions - from inter-relationships between the rate parameters that assume validity of the basic Birks [3] scheme).

Very recently Phillips *et al* [35] have shown on the basis of a revised kinetic scheme [27,36] that for poly(1-vinylnaphthalene) solutions in THF, at no temperature in the region defined as 'low temperature' by steady-state observations is the tenet $k_{MD} << k_D$ valid. Furthermore, values of αk_{DM} obtained as a function of temperature in a manner [27,36] such as to eliminate the dependence upon intramolecular concentration yield an estimate of the 'true' activation energy E_{DM} for this system as 4.8 kJ mol^{-1}.

5.4.2 *Investigations of the influence of microstructure upon excimer formation*

Intramolecular concentration dependence:

Studies of homopolymer photophysics are subject to the limitations imposed by dealing with a fixed but unknown single chromophore concentration. The flexibility enjoyed in low molar mass systems through studies of the concentration dependence of excimer formation may be realized, in part, by the use of series of copolymers containing varying amounts of the chromophoric species of interest and a spectroscopically inert comonomer. Consequently, copolymers present interesting possibilities in attempts to elucidate the nature of the excimer formation mechanisms in macromolecules. The effective local chromophore concentration may also be considered to be altered as the thermodynamic nature of the solvent is varied and as the molar mass of the polymer is changed.

These parameters could affect both the efficiency of energy migration within the polymer and the concentration of potential excimer sites.

Issues that intrigued workers following the early observation of intramolecular excimer emissions in polymers included:

1. The extent to which Hirayama's [10] n=3 rule was obeyed and the fraction of excimer sites generated by longer-range interactions.
2. The role of energy migration in population of potential excimer sites.
3. The influence of chain flexibility upon excimer formation.

It soon became apparent that the n=3 rule of Hirayama is not rigidly applicable to polymer systems even in cases of chromophores that might be expected to follow the generalization closely in low molecular weight materials. For example, excimer emission can be observed in poly(acenaphthylene) [16,26,37-39] and poly(naphthyl methacrylate)s [40-42]. In the latter case even if excimer formation occurs between nearest neighbour-chromophores on the polymer chain, n is equal to 7. In the case of poly(acenaphthylene) it is not possible for adjacent chromophores to adopt the required configuration for excimer formation.

The possibility of excimer formation through interactions of chromophores that are distant upon the polymer chain but are brought into juxtaposition through coiling of the polymer chain is demonstrated in polymers bearing pyrene labels that were designed to measure cyclization kinetics and chain segment encounter frequencies [43-46]. The existence of 'long range' interactions has been observed between naphthyl groups in the backbone of polyamides [47]. The question arises as to the prominence of such long-range interactions in situations where excimer formation between adjacent chromophores is feasible. In this context, solvent dependence or independence of I_D/I_M has been cited as evidence for the presence or absence, respectively, of long-range interactions [16,22,40]. However, such criteria must be regarded with caution, since solvent effects have been observed in polymers [48,49] in which nearest-neighbour interactions are believed to dominate the photophysical behaviour. Furthermore, Abuin *et al* [42] have disputed the contention that the solvent dependence of excimer formation in poly(1-naphthyl methacrylate) arises as a result of long-range effects, and suggest that the effect is due to enhanced energy migration rather than an increased population of potential excimer sites.

Evidence for the dominance of interactions between adjacent chromophores in several vinylaromatic polymers has been provided by studies of copolymers in which the intramolecular chromophore concentration is varied as discussed below, and in alternating copolymers in which no excimer formation was observable [50,51]. In addition, studies of the photophysical behaviour of head-to-head polystyrene in which nearest-neighbour excimer formation is inhibited have revealed no evidence for long-range interactions between phenyl chromophores in this polymer [52,53]. Furthermore, the enhanced probability of energy-trap population in thermodynamically poorer solvents was demonstrated [52], which emphasizes the caution that must be applied in the interpretation of solvent effects upon excimer formation.

Elucidation of the complementary roles played by energy migration and the form of chromophore interactions and consequent potential excimer site concentration is considerably enhanced through observation of the dependence of excimer formation upon intramolecular chromophore concentration, which may be varied over a considerable range by use of copolymeric species. The influence of the intramolecular distribution of chromophores upon the spectral characteristics of polymer luminescence is evident in the works of Nishijima [54] and David *et al* [55]. The latter workers were the first to attempt to characterize the extent of excimer formation in polymers in terms of the microcomposition of the macromolecule. It was argued that in styrene-methyl methacrylate copolymers excimer formation required energy migration from the site of absorption to that suitable for excimer formation. It was proposed that excimer formation was dominated by interactions between adjacent chromophores upon the polymer chain and that the excimer site concentration was consequently proportional to the fraction of pairs of styrene residues in the copolymer. A similar conclusion was reached by Alexandru and Somersall [56], who reported that the occurrence of excimer formation in the styrene-acrylonitrile system (which shows a tendency towards alternation during copolymerization) may be described in terms of the probability of the juxtaposition of aromatic species in the chain.

Reid and Soutar [57] adopted a more general approach and suggested that in polymers in which energy migration occurs to a limited extent *and* serves to populate potential excimer sites; description of the excimer-forming capabilities requires adoption of a function characteristic of the extent to which energy migration may populate the potential excimer sites in addition to that quantifying the concentration of such sites. Accordingly in copolymers incorporating vinylnaphthalenes and styrenes with methyl methacrylate [58] it was shown that the intensity ratio I_D/I_M was adequately described by a function of the form:

$$I_D/I_M = k\,\bar{l}_a f_{aa} \tag{5.4}$$

in which f_{aa} is the fraction of bonds between aromatic species and $\bar{l}_a$ is the mean sequence length of chromophore. The data are reproduced in Fig. 5.3. In development of this model it was considered that:

1. The steady-state expression derived from the Birks scheme (Equation (5.1)) is capable, with suitable modification of the concentration term, of describing intramolecular excimer formation in macromolecules.
2. Excimer formation is resultant upon interactions between nearest-neighbour chromophores.
3. The function descriptive of energy migration as sensed by emission depolarization measurements is capable of quantifying the extent to which potential excimer traps may be populated by this mechanism. This general approach was shown to produce functions descriptive of the I_D/I_M ratio in the 1-vinylnaphthalene-methyl methacrylate system as both the thermodynamic nature of the solvent and the temperature of the system is varied [49].

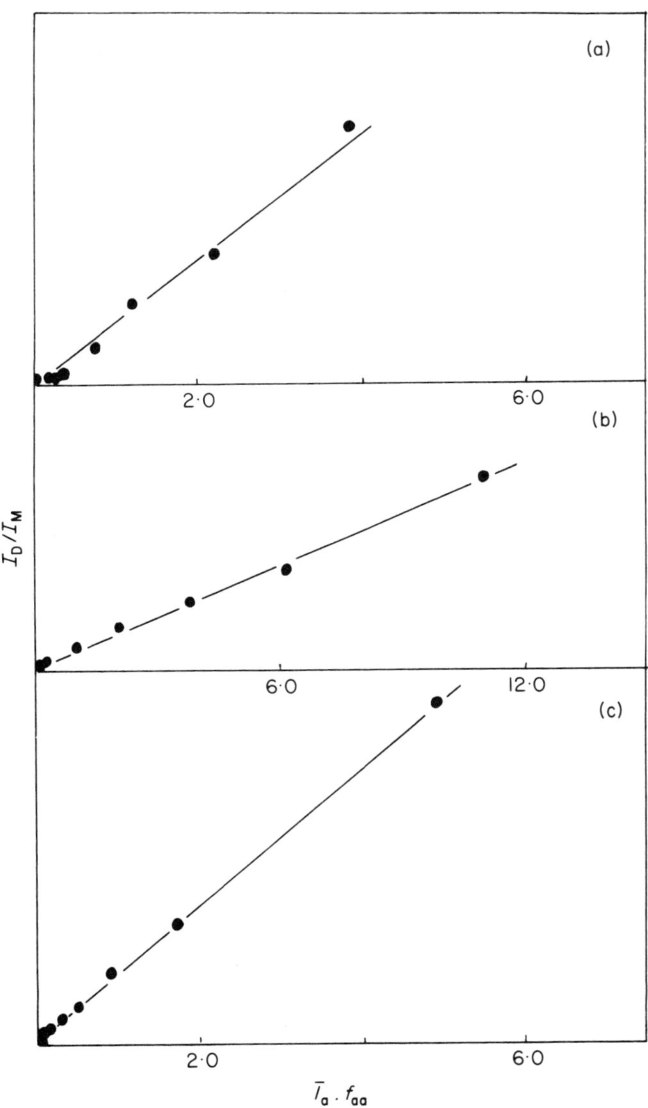

Fig. 5.3 I_D/I_M as a function of $\bar{l}_a \cdot f_{aa}$: (a) 1-vinylnaphthalene-methyl methacrylate; (b) 2-vinylnaphthalene-methyl methacrylate; (c) styrene-methyl methacrylate, copolymers in dichloromethane at 298K (after Reid and Soutar [58]).

The general approach outlined above has been shown to produce adequate fits of stationary-state data as the flexibility of the polymer is varied by variation of comonomer [59] and in copolymers in which chromophores result from acenaphthylene residues [39,60]. In this case, excimer formation can not result from nearest-neighbour interactions and the emission depolarization is not described by the mean sequence length of aromatic. The resultant expression for acenaphthylene copolymers containing methyl methacrylate [39] or methyl acrylate [60] 'spacers' is:

$$I_D/I_M = kf_a\Sigma \tag{5.5}$$

where f_a, the mole fraction of aromatic in the copolymer, is the most appropriate function in description of emission depolarization, and Σ characterizes the concentration of excimer sites in terms of the pentad sequence distribution in the macromolecule. Generation of the function Σ is based upon the assumption that excimer formation in acenaphthylene sequences arises from interactions between next-to-nearest-neighbours in the polymer chain and assumes differential efficiencies of excimer-site population within sites of the type AMA and AAA, where A and M represent copolymer units derived from acenaphthylene and methyl methacrylate, respectively. The suggestion that next-to-nearest-neighbour interactions might dominate the excimer formation in poly(acenaphthylene) was originally mooted by David *et al* [19] and is supported by the work of Wang and Morawetz [37] on copolymers showing high degrees of alternation in addition to the results discussed above.

It is relevant to stress that the implementation of two functions, one descriptive of energy migration and one characteristic of excimer site concentration, is anticipated to be necessary only in cases where energy migration is a determinant in excimer site generation. Two extremes in which this will not be the case are those of:

1. negligible energy migration, and,
2. migration sufficiently extensive to ensure complete statistical sampling of the distribution of excimer sites.

It should be stressed that it is important to generate functions descriptive of excimer formation from a given type of chromophore if it is intended that the influence of factors such as chain mobility and comonomer geometry are to be assessed by studying the emission characteristics of a series of copolymers of a fluorescent chromophore with a series of comonomers of varying structure. It is not sufficient to merely compare the spectral data obtained from copolymers of equal aromatic content [59].

Further studies that report on the relationship between the luminescence properties of copolymers and the chromophore micro-composition include those of Majumdar *et al* [61] and Nakahira *et al* [62,63]. Majumdar *et al* [61] studied copolymers of 1-vinylnaphthalene and the chiral comonomers (-) menthyl acrylate and (-) menthyl methacrylate. The function $\bar{l}_a f_{aa}$ was considered appropriate in description of the fluorescence behaviour consistent with the findings of Reid and Soutar [58] and Anderson *et al* [59] for methyl acrylate and methyl methacrylate copolymers of 1-vinylnaphthalene.

In contast, Nakahira *et al* [62] have reported that the ratio I_D/I_M describing the excimer formation of naphthalene chromophores in copolymers of 2-vinylnaphthalene and 2-t-butyl-6-vinylnaphthalene with styrene vary linearly with f_{aa} across the entire composition range. These results are intriguing. Since for each chromophoric group the linear fit is defined by the coordinates $\{I_D/I_M, f_{aa}\}$ at $f_{aa} = 1$ and $f_{aa} = 0$, it must be concluded either that the trend is peculiar to copolymers of these chromophoric monomers with styrene alone, or that the spectroscopic properties of the copolymers are independent of the nature of the comonomer. Such a situation as the latter would imply that the structure of the comonomer and its influence upon chain flexibility did not affect the energy profiles relevant to the pre-excimeric and excimer state nor the potential energy barrier for excimer formation.

Nakahira *et al* [63] have also examined excimer formation in 2-vinylnaphthalene - vinylferrocene copolymers. This copolymer series is interesting in that the comonomer is not spectroscopically inactive but rather acts as a quencher of naphthalene fluorescence. The data have been analysed in terms of a kinetic scheme involving both static and dynamic quenching of emission. Consistent with this mechanism a linear dependence of I_D/I_M on the aromatic tetrad distribution function f_{aaaa} is reported.

Influence of conformation and configuration:
The importance of conformational considerations in bichromophoric systems has been recognized by several workers [20,30,64-69]. The influence of ground-state conformational control, which has been invoked in explanation of intramolecular exciplex photophysics in α-aryl-ω-*N*-alkylakanes [70,71] and excimer formation in 1,1'-di-2-naphthyldiethylether [30], has been suggested to be of importance too in macromolecular photophysics [30,69].

Configurational aspects are apparent in publications concerning the fluorescence characteristics of polymers of differing tacticity and of model compounds. It has been shown that the intensity ratio of excimer to monomer fluorescence is greater in fluid solutions of isotactic polystyrene [72,73] and poly(*p*-methylstyrene) [74,75] relative to that of the atactic polymers. This phenomenon has been attributed in the case of poly(*p*-methylstyrene) to the existence of a lesser energy barrier to excimer formation in meso dyads compared to the racemic dyad [75]. Similar conclusions of direct relevance to excimer formation in polystyrene were made by Bokobza *et al* [65] in studies of intramolecular excimer formation in model compounds.

A dependence of I_D/I_M upon tacticity in fluid solutions of poly(1- and 2-naphthyl methacrylate)s has also been reported [41]. The implications of these observations are less apparent, since it has been suggested [22,40] that excimer formation in such poly(methacrylate)s is dominated by long-range interactions.

5.4.3 *Summary*

It is convenient at this stage to consider the general conceptions regarding the nature of intramolecular excimer formation that emerged from studies of essentially steady state fluorescence measurements complemented by a few early experiments involving time resolved spectroscopy.

1. A variant of the Birks kinetic scheme was directly applicable to the phenomenon of intramolecular excimer formation. It is important to note that this notion was accepted virtually without question in studies before 1980.
2. In situations in which excimer interactions between chromophores located in close proximity on the polymer chain are sterically viable, such interactions appeared to dominate the photophysical behaviour.
3. The concept of energy migration and consequent sampling of potential excimer sites was cited and discussed by most authors. This situation was somewhat acroamatic, in that little concrete or undisputed experimental evidence for the effect was available except for observations based upon emission depolarization [76,77].

These precepts have prompted interests in the application of time-resolved fluorimetry to the study of excimer formation in macromolecules. Consequently in the remainder of this chapter we shall concentrate our discussions on the application of these techniques to polymer photophysics. It will be shown that the current picture has been considerably altered relative to the situation described above by these developments in the last few years.

5.5 TIME-RESOLVED FLUORESCENCE STUDIES OF INTRAMOLECULAR EXCIMER FORMATION IN POLYMERS

Despite the considerable potential of time-resolved spectroscopy for generating information that would augment and complement that available from steady-state analysis, restrictions upon instrument availability in polymer research groups resulted in a paucity of publications dealing with transient excitation conditions before 1978. Since that time, several publications have appeared which produced a growing awareness that the classical Birks [3] analysis was not adequate in description of the empirical data.

Initial reporters were content to analyse polymer emission data in terms of dual exponential functions consistent with Birks' expressions detailed in Equations (1.64) and (1.65). Particular attention was paid to decays in the monomer region of the total emission spectrum. Table 5.2 presents a selection of decay times characteristic of aromatic macromolecular fluorescences analysed in terms of dual exponential decay functions of the form of Equation (1.65). In these earlier publications it was not common practice to quote statistical parameters such as the reduced chi-square [84], χ^2, which might inform interested parties as to the adequacy of the fitted function in description of the decay data. It is probably fair to state that poor statistical evidence that the fitted function was adequate was frequently accepted on the grounds that polymeric samples never produced 'ideal' results. However, increasing disquiet recently resulted in the expression of serious doubts concerning the validity of the predictions of the Birks [3] analysis as applied to intramolecular excimer formation in polymers [26-30]. The foundation of these doubts was laid upon observations of the following type:

1. The recognition that the inability to describe emission decays in certain polymers in terms of dual exponential functions was photophysically significant [26,27,30].

Table 5.2 Life-time data for excimer forming aromatic polymers in solution

Polymer	Solvent	τ_1 (ns)	τ_2 (ns)	Reference
Polystyrene	DCM	0.76	13.9	34,78
	DCM	<1	15.5	79
	DCE	1.85	17.5	79
	p-Dioxane	<1	22	79
Poly(α-methyl styrene)	DCM	1.0	8.40	78
Poly(1-vinylnaphthalene)	DCM	7.43	43.1	34,78,80
	THF	1.49	62.5	81
Poly(1-naphthyl methacrylate)	EA	1.95	8.48	22
	DCM	1.38	7.34	22
	C	1.03	9.28	22
	THF	1.9	7.9	82
Poly(2-naphthyl methacrylate)	THF	6.0	20.5	82
Poly(1-naphthyl acrylate)	THF	1.4	9.0	82
Poly(2-naphthyl acrylate)	THF	2.7	18.6	82
Poly(2-[9-ethyl]-carbazolylmethyl-methacrylate)	DCM	3.0	8.4	83

DCM=dichloromethane; DCE=1,2-dichloroethane; C=chloroform; THF=tetrahydrofuran.

2. The observation that in dual exponential fitting to the decay data obtained in the spectral region of excimer emission, pre-exponential factors of equal magnitude but opposite sign were not obtained if all four fitting parameters were allowed to vary freely [30,85].
3. The observation that the decay times evaluated in the regions of monomer and excimer emission, respectively, do not coincide [28,85].

The models that have emerged in attempts to explain the empirical photophysical data and details of the experimental evidence leading to their development are discussed below. In the main, the kinetic schemes that have been devised are based on the premise that the observation of multiexponential fluorescence decays from polymers is consequent upon the existence of more than two types of photoemissive sites within the macromolecule. Such an approach, invoking the principle of Occam's Razor, is reasonable and has led to the development of extrapolative procedures that, within the limits set by the adopted mechanism, allow the determination of individual photophysical parameters within the kinetic scheme. Very recently two publications have appeared that question the validity of this approach [86,87].

5.5.1 Modelling of photophysical behaviour of polymers

Holden *et al* [28] have proposed the following kinetic scheme (5.2) to describe the photophysical behaviour of polymers containing naphthalenes whose chromophores are separated by more than three atoms in fluid solutions, and to all naphthalene containing polymers in rigid matrices. Kinetic scheme (5.2) resulted from the study of the transient fluorescence properties of poly(naphthylmethyl methacrylate) (PNMMA) and poly(naphthylethyl methacrylate) (PNEMA) and their model compounds 1-naphthylmethyl pivalate (NMP) and 2-(1-naphthyl)ethyl pivalate (NEP) in deoxygenated THF and toluene and in polystyrene or poly(methyl methacrylate) matrices at 298K.

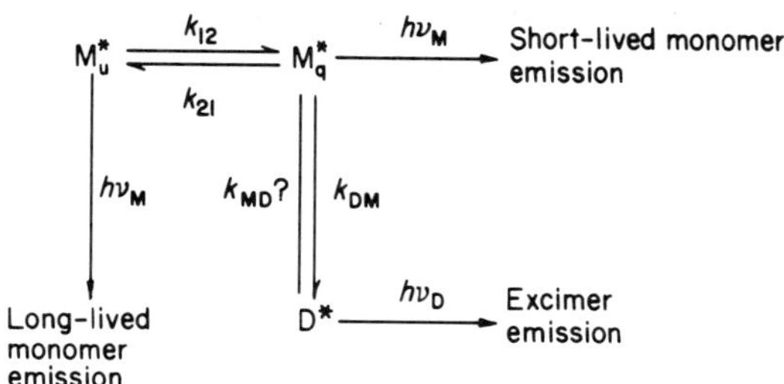

Scheme 5.2

The decay curves were analysed in terms of two exponentials in the regions of monomer and excimer, respectively. The lifetimes of the longer lived components, $\tau_1 = 1/\lambda_1$, of the decays in the regions of monomeric and excimeric emission are compared in Table 5.3 with those obtained from the model compounds.

Table 5.3 Comparison of the longer lived fluorescence decay time in regions of monomer and excimer emission with those of model compounds [28]

Polymer	*Model Compound*	*Solvent*	*Monomer* τ_1/(ns)	*Excimer* τ_1/(ns)	*Model* τ/(ns)
PNMMA	NMP	THF	36(±2)	48(±2)	34(±1)
PNMMA	NMP	Toluene	40	63	39
PNEMA	NEP	THF	66	60	63
PNEMA	NEP	Toluene	58	76	60

Comparison of the data reveals that the value of τ_1 obtained in the region of monomer fluorescence is very different from that characteristic of the decay of excimer fluorescence. Within experimental error, the monomer τ_1 values are equal to those of the model compounds. Consequently, it was proposed that the slowly decaying component of the monomer fluorescence arises not through dissociation of the excimer but as a result of the existence of unquenched monomer species (M_u^* in scheme (5.2)) supposed to exist in configurations too far removed from excimer to be quenched within its lifetime. Further support for the existence of isolated monomer species was gained from a study of the fluorescence of PNMMA and PNEMA dispersed in glassy polymer matrices in which the monomer decays exhibited a remarkable similarity to those in fluid solution while again being recognisably different from the long-lived excimer decay.

A somewhat similar kinetic scheme to that of Holden *et al* [28] was proposed independently by Phillips, Roberts and Soutar as a consequence of studies of polymers and copolymers of the vinylnaphthalenes [27,35,85,88], naphthyl methacrylate [85], acenaphthylene [26] and styrene [89,90] in fluid solutions. Variation of the intramolecular chromophore concentration in series of copolymers and observation of the consequences in the decay kinetics has resulted in the proposal of scheme (5.3) for application to macromolecules containing the chromophores listed above.

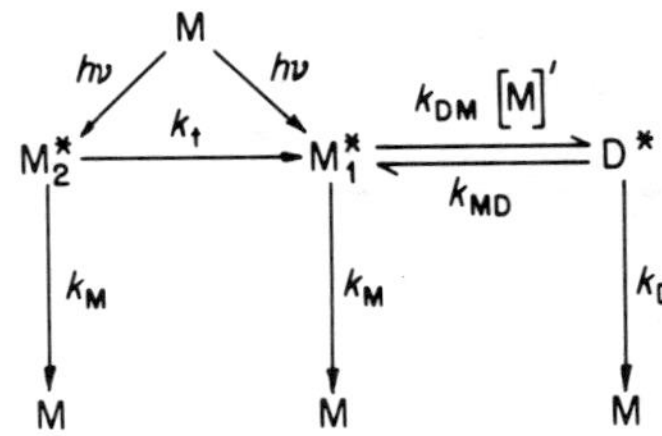

Scheme 5.3

This scheme was adopted as a result of the following general observations:

1. The Birks scheme is generally inapplicable to description of intramolecular excimer formation in polymers.
2. For the copolymer series studied, decay in the monomer region could only be described adequately if three

exponential terms are adopted, as in Equation (5.6):

$$i_m(t) = A_1\exp(-\lambda_1 t) + A_2\exp(-\lambda_2 t) + A_3\exp(-\lambda_3 t). \qquad (5.6)$$

3. Time-resolved emission spectra reveal the existence of only two spectrally distinct species as illustrated in Fig. 5.4 for polystyrene in degassed dichloromethane [89].

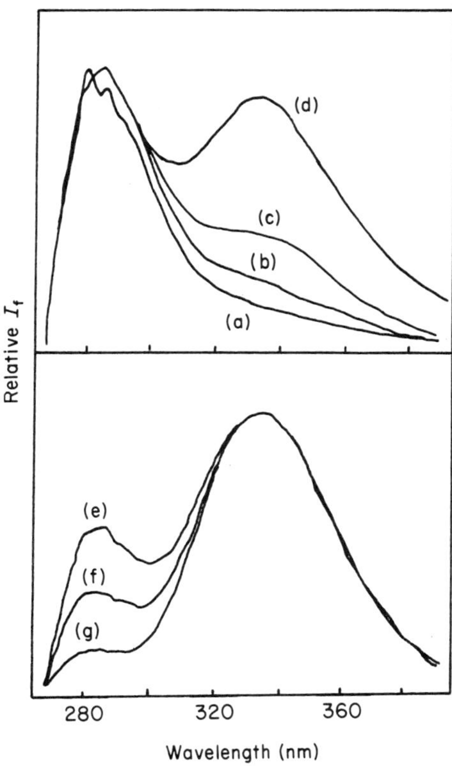

Fig. 5.4 Time-resolved fluorescence spectra of a styrene-methyl methacrylate copolymer (75 mole % styrene; dichloromethane solution; 298K) recorded at delays of: (a) 0; (b) 3.8; (c) 7.7; (d) 11.5; (e) 15.4; (f) 19.3; (g) 28.8 ns. Excitation at 257.3 nm using the frequency doubled, cavity dumped output of a 4W argon ion laser (FWHM 7 ns). Gate width 3.2 ns. (After Soutar *et al* [89]).

4. The decay parameters, λ_i, assigned as resultant upon the existence of three kinetically distinct species, vary with the aromatic content of the polymers.
5. Time-resolved emission spectra suggest that dissociation of excimers to create excited state monomer does occur in the copolymers studied including polymers containing styrene [89] in contrast to previous reports [34,78].

In scheme (5.3), it is envisaged that D*, the excimer, is populated from M*, by an 'exciton sampling' mechanism; M_2^* is a chromophore that enjoys a certain degree of kinetic isolation from the normal distribution of quenched excited monomer. The M_2^* sites are not unquenched as supposed in the kinetic scheme of Holden *et al* [28], but suffer an intramolecular concentration dependent quenching that is supposed to originate from energy transfer into the reservoir of M_1^* sites.

The relative importance of M_2^* in the photophysical behaviour of macromolecules appears to depend upon the type of chromophore and macromolecular structure involved. For example, there is no detectable influence from isolated monomeric groups in the decays of the monomeric emission band in polystyrene in fluid solution [89]; the decay is adequately described by a dual exponential function. In contrast the photophysical behaviour of the poly(vinylnaphthalene)s [85], poly(1-naphthyl methacrylate) [85] and poly(acenaphthylene) [85] requires a third exponential term in description of monomer decay. Consequently it may be inferred that M_2^* sites in copolymers containing styrene are associated with chromophores that are physically isolated within the macromolecular microcomposition. In naphthalene polymers, M_2^* sites are evident in the absence of 'spectroscopic spacers' and can be associated with species whose kinetic isolation is not solely consequent upon separation from similar chromophores.

An alternative kinetic scheme, represented as scheme (5.4), was considered by Phillips *et al* [27,36,88] in which the longer lived monomeric species M_2^* was considered to be quenched by excimer

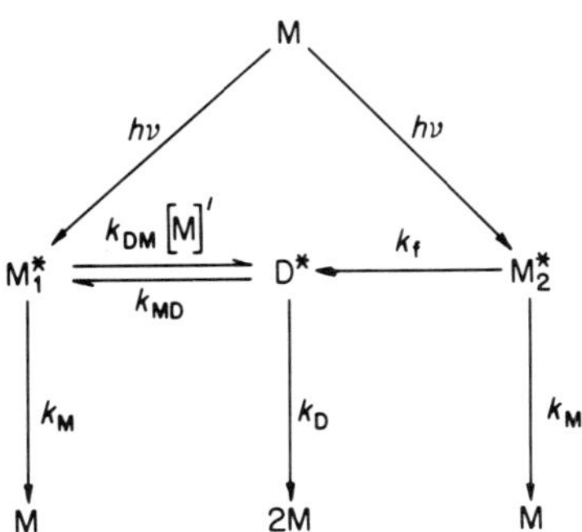

Scheme 5.4

formation, largely as a result of long-range segmental diffusion acting to align two chromophores distant along the chain backbone into spatial proximity by chain coiling. This model was discarded following studies of the temperature dependence of intramolecular excimer formation in copolymers of 1-vinylnaphthalene and methyl methacrylate [35]. Consequently it was considered that of the two schemes (5.3) and (5.4) that arose from the study of homopolymers and copolymers, only scheme (5.3) remains a reasonable proposition.

The application of scheme (5.3) has allowed the determination of rate parameters within the photophysical scheme to be evaluated [27,35,36,88]. Furthermore, as will be described later, the study of copolymers has allowed the removal of the influence of intramolecular concentration effects upon the calculated rate coefficient k_{DM} to an extent not hitherto undertaken.

De Schryver *et al* [30] have undertaken a painstaking study of painstaking study of the comparison of the photophysics of polymer systems with those of low molar mass bichromophoric analogues. In an early report the inapplicability of the Birks scheme [3] to the description of excimer formation in poly(2-vinyl naphthalene) (P2VN) was expressed [29]. Comparison of the emission behaviour of P2VN with that of 1,3-di(2-naphthyl)propane combined with an observed dependence of the emission characteristics of P2VN upon excitation wavelength led to the formulation of scheme 5.5 [29].

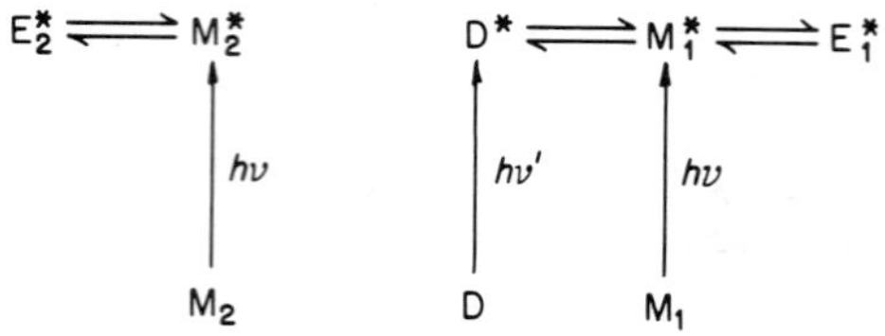

Scheme 5.5

The scheme invokes the existence of different local environments that can not interconvert on the lifetime of the excited states (M_1^* and M_2^*). It is further supposed that all emission in the monomer region is due to one type of localized excited state (M_1^*) whereas excimer formation occurs very rapidly from (M_2^*). Deviations in the excitation spectra and excimer decay kinetics are explained by the existence of a subset (D) of M_1 responsible for the 'deviation at 304 nm' constituting an excited-state rapid equilibrium between unperturbed (M_1^*) and perturbed (D*) alkylnaphthalene species. The temperature dependence of the decay of monomer emission was rationalized in terms of two excimeric species E_1^* and E_2^*. At low temperatures, decay from M_1^* is biexponential, the longer-living component resulting from dissociation of E_1^* whereas the greater complexity of the decay kinetics at higher temperatures is a consequence of the influence of the excimer E_2^*.

Amplification of the concepts involved in scheme 5.5 resulted from studies of the photophysics of the meso and racemic isomers of 1,1'-di-2-naphthyl diethyl ether (D2NEE) [30]. These isomers serve as models for the isotactic and syndiotactic sequences in P2VN respectively. It was shown that:

1. The ratio I_D/I_M of excimer to monomer fluorescence in both P2VN and meso D2NEE is dependent upon excitation wavelength, whereas no such dependence is evident in the racemic isomer of D2NEE. It is therefore supposed that the observed excitation wavelength dependence, P2VN emission characteristics

is due to isotactic sequences in the polymer.

2. The decay kinetics of P2VN are temperature dependent and complex (compared, for example, to the predictions of the Birks scheme). Comparison with the kinetic behaviour of meso D2NEE led to the proposal that isotactic sequences in the polymer were responsible for two exponential decay terms as contribution to the total excimer emission in P2VN.
3. The decay processes in P2VN are considered to be a superposition of excimer forming processes from different conformational sites. Further, the efficiency of excimer formation at different conformational sites differs substantially. The dual decay process in P2VN, if indicative of the existence of two excimer sites, is supposed to result from ground state conformational control of excimer formation.

Further information relevant to the configurational and conformational influences upon excimer formation has been obtained through time-resolved studies of 2,4-diphenylpentanes as models for polystyrene, [91]. The rate constant k_{DM} for excimer formation in the racemic diastereoisomer in iso-octane at 293K was estimated as $9 \times 10^7\ s^{-1}$ whereas that for the meso compound was quoted as $2 \times 10^9\ s^{-1}$. In this respect the value of k_{DM} for the racemic isomer [91] was reported to be 30 time smaller that that reported for atactic polystyrene [92] and that for the meso isomer comparable to k_{DM} for the isotactic polymer [75]. Similarly, differences in k_{DM} have been reported between the meso and racemic steroisomers of 2,4-dinaphthylpentane [93].

Although such studies on model compounds are extremely useful and pertinent to studies of polymer photophysics, the extent to which the conformational energy surface of high polymers is distorted relative to the 1,3-dimer and the role of energy migration between chromophores in dictating excimer site and pre-excimer site population lends some uncertainty to the direct application of model compound data. A caveat may be sounded on comparison of rate coefficient data between low molar mass analogues and polymeric species with reference to the comparisons of the 'k_{DM}' data for polystyrene and diphenylpentanes cited above. First, the polystyrene data contain a constant (for polymer of given tacticity and sufficiently high molecular weight) but undefined contribution due to the local chromophore concentration and thence can not be directly compared with dimer data. Second, the rate coefficient k_{DM}[M] shows a marked molar mass dependence in polystyrene [90,92] at low molecular weights, such that correspondence between dimer and polymer should not be expected.

Other kinetic schemes:

Following publication of the proposals outlined above, several papers have appeared in the literature that offer alternative variations on kinetically distinct moieties whose existence would lead to deviations from the kinetic behaviour resultant upon adoption of the Birks kinetic scheme. Some of these proposals are discussed briefly below.

MacCallum [76,77] has discussed the possibility of kinetic discrimination within the compositional environment of styrene copolymers in the absence of energy migration being based upon quenching probability of monomer emission in the sites -M$\overset{*}{S}$M-, -M$\overset{*}{S}$S- and -S$\overset{*}{S}$S-. It is proposed that the monomer life-time will be least in the -S$\overset{*}{S}$S- site and greatest in the -M$\overset{*}{S}$M-. This model would lead to decays of monomer emission that would consist of three exponential terms whose relative contributions would vary with copolymer composition. Consequently the predictions of this model would be in broad agreement with the empirical data on styrene - methyl methacrylate copolymers [89]. However, the model can not be held as consistent since:

1. It presumes that dissociation of excimer to form excited-state monomer is negligible, in contradiction with the findings of time-resolved spectroscopy [89].
2. Recent work on block copolymers containing styrene (S) and butadiene (B) has shown that the photophysical behaviour can not be explained in terms of differences in activity of -S$\overset{*}{S}$S- and -B$\overset{*}{S}$S- triads [90].

3. Time resolved fluorescence quenching data on polystyrene in which the aromatic fluorescence was quenched by an intramolecular energy trap [94] is inconsistent with MacCallum's scheme. In particular, dual exponential decays were observed (as for polystyrene [89]) in which the long-lived component is unquenched. This is not feasible in terms of the model proposed by MacCallum [77], since the end of sequence styrene chromophores would be located adjacent to the energy traps and consequently should suffer considerable quenching.

Itagaki *et al* [94] have proposed that the decay kinetics observed for fluid solutions of poly(1-methoxy-4-vinylnaphthalene) (PMVN) may be explained in terms of two excimer species in addition to excited monomer. The proposal is based on variations in steady-state spectra as a function of temperature and, by analogy to studies on 1,3-di(4-methoxy-1-naphthyl)propane (BMNP) [95,96], presumes that the lower energy excimer comprises the normal structure with fully overlapped aromatic rings, whereas the second excimer is thought to comprise a partially overlapped structure. Although the steady-state spectra provide reasonable support for the existence of three excited-state species, it must be stated that spectral deconvolution into three intensity components representing monomer, normal excimer and second excimer appears optimistic in view of the uncertainty of the spectral shapes of the two excimer species and the high degree of overlap between the spectrum of the second excimer and those of monomer and normal excimer. Data analysis of decay curves did not use whole-curve fitting nor deconvolution procedures, and at best represents consistency with the proposed scheme.

A similar proposal of fully overlapping and partially overlapped excimers has been proposed for poly(1-vinylnaphthalene) by Gupta *et al* [97]. However the evidence for such a proposal is somewhat tenuous. Time-resolved data have been fitted by dual exponential or single exponential functions, which mean that these at best must be regarded as averages in which the short lived component corresponding to quenched monomer M_1^* [27] in scheme (5.3) is neglected. The observed decay times, for which rigorous fitting criteria do not appear to have been adopted, vary from those derived on the basis of dual exponential fits at short (325 nm) wavelengths, to single exponential at 375 nm, then dual exponential at longer wavelengths. It is assumed that the decay at 375 nm may be exclusively attributed to the partially overlapped excimer. This assumption must be regarded with suspicion, considering the high degrees of overlap that would be expected to occur between the emission band of such a species with those of normal monomer and excimer. Furthermore, if one inspects the equations resultant upon adoption of the Birks [3] scheme, to which the dual exponential fitting procedures adopted in this work might approximate, it is apparent that, in the presence of spectral overlap of emission bands corresponding solely to monomer and normal excimer, decays analysed at longer times of analysis at intermediate wavelengths could (depending upon relative magnitudes of pre-exponential terms) approximate to single exponential behaviour. Further concern regarding the data generated in this work arises from the fact that it is not evident to these authors how the naphthalene chromophores of the polymer could be directly excited by the output of a nitrogen laser.

5.5.2 *Derivation of rate coefficients*

One of the prime objectives in attempting to derive a kinetic scheme appropriate to the characterization of the photophysical behaviour of a given polymer system is that it allows, in principle, derivation of rate coefficients for the individual processes within the mechanism. In turn, such derivations may yield data through comparison of the photophysics of different macromolecular systems that permit a greater insight to be gained into the nature of the fluorescence behaviour of such complex species. Ultimately it would be hoped that the resultant information might allow enhanced characterization of more complex phenomena, such as energy trapping by sites other than excimers, photochemical processes consequent upon energy migration and trapping, etc.

The derivation of individual rate parameters of photophysical reaction schemes are subject to the limitations imposed both by the reaction scheme deemed appropriate to the system under study and by the system itself. In the latter respect the study of homopolymers is restricted through the adoption of a fixed yet unknown chromophore concentration. This approach results in the determination of a pseudo first-order 'k_{DM}' term in description of excimer formation. Such a single concentration procedure has been adopted by several workers, based on dual exponential fitting of decay data in combination with estimates of unquenched life-time data based upon model compounds [21,22,40,75,92].

Alternatively, the intramolecular concentration of chromophores may be varied by use of suitable copolymers and suitable extrapolation procedures employed as an adaptation of the methods available for the derivation of kinetic parameters for intramolecular excimer formation in low molar mass species [3,98]. The application of such extrapolative procedures requires that:

1. the kinetic scheme of Birks [3] is applicable to the intramolecular phenomenon in polymers or that suitable modification is possible; and
2. a term descriptive of the concentration of chromophores capable of formation of intramolecular excimer formation may be substituted for [M].

Subject to the limitations of adoption of kinetic schemes (5.3) or (5.4) to describe the copolymer photophysics it is apparent that [27,35,36,88]:

$$i_m(t) = A_1\exp(-\lambda_1 t) + A_2\exp(-\lambda_2 t) + A_3\exp(-\lambda_3 t) \qquad (5.6)$$

and

$$i_D(t) = A_4\exp(-\lambda_1 t) + A_5\exp(-\lambda_2 t) + A_6\exp(-\lambda_3 t) \qquad (5.7)$$

where

$$\lambda_{1,2} = \tfrac{1}{2}[(X+Y) \mp \{(Y-X)^2 + 4k_{MD}k_{DM}[M]\}^{\frac{1}{2}}] \tag{5.8}$$

and

$$X = k_M + k_{DM}[M] \quad ; \quad Y = k_D + k_{MD}$$

where

$$k_D = k_{FD} + k_{ID}$$

and

$$\lambda_1 + \lambda_2 = k_M + k_{DM}[M] + k_D + k_{MD} \tag{5.9}$$

and

$$\lambda_3 = k_i + k_M$$

where $k_i = k_t$ or k_f for schemes (5.3) or (5.4), respectively.

Table (5.4) summarizes the procedures for deriving rate constants from the rate parameters λ_i resulting from analysis of $i_m(t)$ according to Equation (5.6). It should be noted that a measure of the self-consistency of the scheme, the extrapolative procedures and the intramolecular concentration functions adopted is afforded by the fact that each of the derived parameters may be checked through comparison of values obtained by alternative extrapolation approaches.

Implementation of this extrapolation method requires appropriate concentration terms to be adopted to describe the intramolecular concentration dependence of the rate parameters λ_i. In this context it has been assumed that the variation in aromatic content in the copolymers dominates the variations in photophysical behaviour across the range of aromatic compositions and that microcompositional terms derived from steady-state analysis of I_D/I_M are sufficient to characterize the dependence of λ_1, λ_2 with aromatic content. These assumptions seem to be justified in terms of the adequate plots obtained for the copolymer systems studied to date. Typical examples

Table 5.4 Procedures deriving rate constants [27].

Method	Procedure	Function derived	Derived parameter
(a)	Measurement of unquenched monomer lifetime	k_M	k_M
(b)	Extrapolation of λ_1 to $[M] = 0$	k_M	k_M
(c)	$\frac{\partial(\lambda_1 + \lambda_2)}{\partial[M]}$	αk_{DM}	αk_{DM}
(d)	$\frac{\partial\lambda_2}{\partial[M]} \rightarrow k_{DM}$ as $[M] \rightarrow \infty$	αk_{DM}	αk_{DM}
(e)	$\lambda_1 + \lambda_2 \rightarrow k_M + k_{MD} + k_D$ as $[M] \rightarrow 0$	$k_M + k_{MD} + k_D$	$k_{MD} + k_D$ [by combination with (a) or (b)]
(f)	$\lambda_1\lambda_2 \rightarrow k_M(k_{MD} + k_D)$ as $[M] \rightarrow 0$	$k_M(k_{MD} + k_D)$	$k_{MD} + k_D$ [by combination with (a) or (b)]
(g)	$\frac{\partial(\lambda_1\lambda_2)}{\partial[M]} = k_{DM}k_D$	$\alpha k_{DM}k_D$	k_D [by combination with (c) or (d)]
(h)	$\lambda_1 \rightarrow k_D$ as $[M] \rightarrow \infty$	k_D	k_D
(i)	(e) or (f) plus (g) or (h)	k_{MD}	k_{MD}
(j)	Extrapolation of λ_3 to $[M]' = 0$	k'_M	k'_M
(k)	$\frac{\partial\lambda_3}{\partial[M]'} = k'_i$	$\beta k'_i$	$\beta k'_i$

are shown in Figs. 5.5 and 5.6 for the copolymer systems 1-vinylnaphthalene-methyl methacrylate and acenaphthylene-methyl methacrylate, respectively.

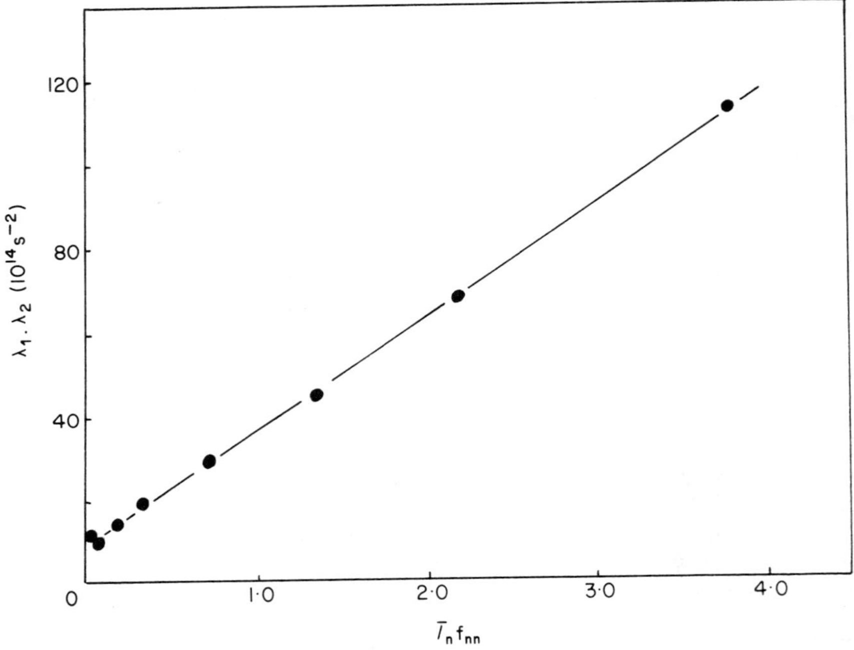

Fig. 5.5 Product of the decay parameters λ_1 and λ_2 as a function of $\bar{l}_n f_{nn}$ for copolymers of 1-vinylnaphthalene and methyl methacrylate in tetrahydrofuran at 298K (after Phillips *et al* [27]).

The procedure has been applied to the copolymer series 1-vinylnaphthalene-methyl methacrylate [27], 1-vinylnaphthalene-methyl acrylate [88], acenaphthylene-methyl methacrylate [36] and styrene-methyl methacrylate [89]. The consistency of the data derived from alternative extrapolations are good as exemplifed for the 1-vinylnaphthalene-methyl methacrylate copolymer series in Table 5.5. A study of the temperature dependence of the rate parameter k_i for the latter copolymer series has resulted in the

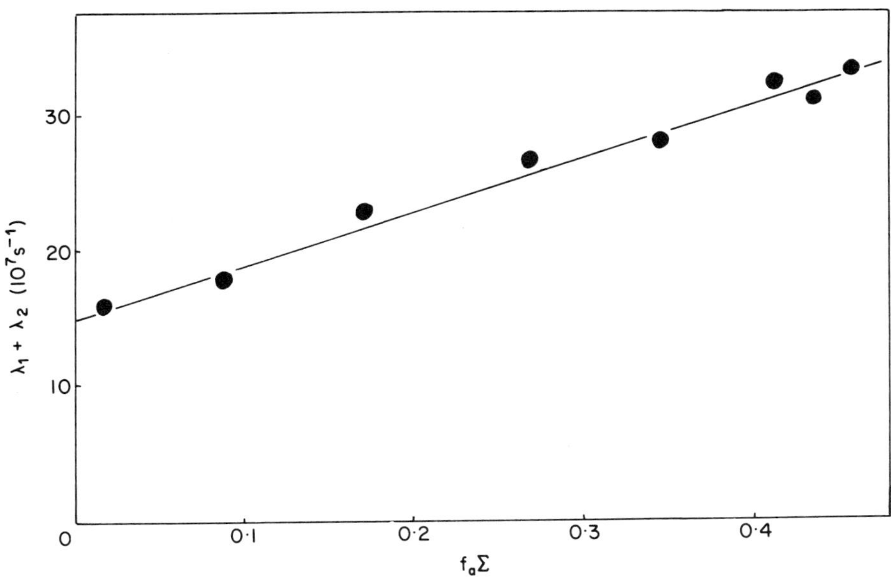

Fig. 5.6 Sum of the decay parameters λ_1 and λ_2 as a function of $f_a\Sigma$ for copolymers of acenaphthylene and methyl methacrylate in tetrahydrofuran at 298K (after Phillips *et al* [36]).

conclusion that kinetic scheme 5.4 that would implicate large scale segmental diffusion of the polymer in the involvement of species M_2^* in the photophysical reaction mechanism is not consistent with the observed low activation energy obtained [35]. Consequently, it has been proposed that, of the two kinetic schemes, only scheme (5.3) remains viable in description of the kinetic behaviour of vinyl-aromatic polymers in fluid media [35].

As a closing remark regarding the importance of establishing kinetic schemes suitable for the description of macromolecular photophysics we should wish to cite the recent application of observations on the decay behaviour of polystyrene [89] in fluid solution to the study of energy trapping by diphenyloxazole (PPO) species incorporated as intramolecular traps in the macromolecule [99].

Table 5.5 Kinetic parameters for 1-vinylnaphthalene-methyl methacrylate copolymers [27]

Kinetic parameter	*Value* (10^7 s^{-1})	*Method (see text)*
k_M	1.72	(a)†
	2.15	(a)‡
	1.3	(b)
αk_{DM}	12.1	(c)
	12.0	(d)
$k_{MD} + k_D$	6.5	(e)
	6.2	(f)
k_D	2.2	(g)
	2.2	(h)
k_{MD}	4.2	(i)

†Determined using 2 x 10^{-5}M 1-methylnaphthalene.

‡Determined using copolymer containing less than 0.5 mole % naphthalene chromophore.

It was concluded that transfer from excited chromophores in the polymer to PPO acceptor occurred from monomeric species and that excimer sites act predominantly as competitive traps in this system. The study also implicated the existence of a degree of energy migration in polystyrene to complement data on fluorescence depolarization [58,100,101] and upon the influence of molar mass on the transient photophysical behaviour of polystyrene and styrene-butadiene block copolymers [90].

5.5.3 Poly(N-vinylcarbazole) and related compounds

Although the photophysical properties of carbazole containing polymers are discussed in Chapter 8 it is appropriate to present a brief account of the studies of excimer formation in poly(*N*-vinylcarbazole) in this review. Poly(*N*-vinylcarbazole) (PNVCz) is distinct from other vinylaromatic polymers studied to date in that no emission from monomeric species is observed [102]. The fluorescence spectrum of PNVCz in fluid solution is supposed to result from radiative energy loss from two spectrally distinct excimers [102-105]. The lower energy excimer is generally accepted to be formed in a totally eclipsed conformation [102], whereas the higher energy excimer is assumed to involve a dimeric conformation in which partial overlap of the aromatic nuclei is attained [102,105]. Time-resolved fluorescence spectroscopic evidence has suggested that these excimers are interconvertible [103,104], although a recent study has suggested that such may not be the case [106].

The fluorescence decay characteristics of PNVCz have been shown to be complex and resolvable in terms of three components [107-110]. The decay data have been analysed in terms of kinetic schemes involving three excited state species which, to varying degrees within each scheme, are interconvertible [107-110]. Ng and Guillet [110] and Tagawa *et al* [109] have analysed their data in terms of excited monomer and two excimers, whereas Roberts *et al* [107,108] favour an analysis involving three excimeric species. The situation may be summarized in terms of scheme (5.6) [111].

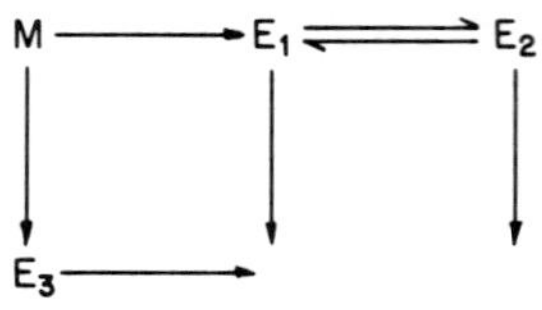

Scheme 5.6

The interpretations of Tagawa *et al* [109] and Roberts *et al* [108] vary in respect of the identity of species E_1. Tagawa *et al* [109] favour a 'relaxed monomer', whereas Roberts *et al* [108] favour a structure similar to the higher energy excimer E_3, but subjected to kinetic isolation due to its configuration. The higher energy excimer (E_1 and E_3) is supposed to constitute (by virtue of its extremely rapid formation; <10 ps) a spectral species corresponding to preformed excimer sites. The lower energy excimer, on the other hand, shows a 'growing in' on the nanosecond scale and is supposed to originate from either relaxed monomer [109] or from a fraction (E_1) of the higher energy species [107,111].

Two studies of model systems containing carbazole chromophores warrant particular mention in this discussion of the fluorescence behaviour of PNVCz:

1. Ng and Guillet [110] have shown that the transient photophysical behaviour of 1,3-di(*N*-carbazolyl)propane in benzene solution can not be described in terms of the Birks [3] kinetic shceme. Adequate description of the decay curves corresponding to monomeric (365 nm) and excimer (420 nm) could only be attained by resort to three exponential functions. A kinetic scheme based on an earlier suggestion of ground state conformational control by Goldenberg *et al* [69] as illustrated in scheme (5.7) was shown to be consistent with the observed decay behaviour in the temperature range 10-40°C.

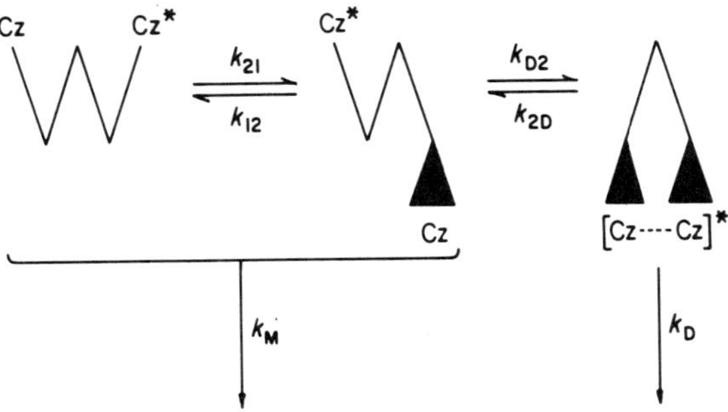

Scheme 5.7

This temperature range corresponds to the 'high-temperature region' of the system in which dynamic equilibrium between various conformations is established. Furthermore it is proposed that in this T region, k_M and k_D are much smaller than those of the internal processes k_{21}, k_{12}, k_{D2}, k_{2D}. On this basis, of the three empirical decay times obtained at a given temperature one is characteristic of the interacting M*, D* system, whereas the other two are resultant upon internal relaxation processes. This interpretation is of obvious relevance to the analysis of the complex data derived from polymer systems, and is interesting in view of recent interpretations of such complexities made by Itagaki *et al* [87].

2. De Schryver *et al* [106] have recently reported some extremely interesting observations on the photophysical behaviour of the two diastereoisomers of 2,4-di(*N*-carbazolyl)-pentane. It was shown that the fluorescence spectrum of the meso isomer in isooctane consists of monomeric emission and that of an excimer at 420 nm (see Fig. 5.7). The racemic diastereoisomer shows, at room temperature, emissions corresponding to monomer and a higher energy excimer at 370 nm the contribution of which, to the spectrum, increases as the temperature is lowered (see Fig. 5.7). Both excimers

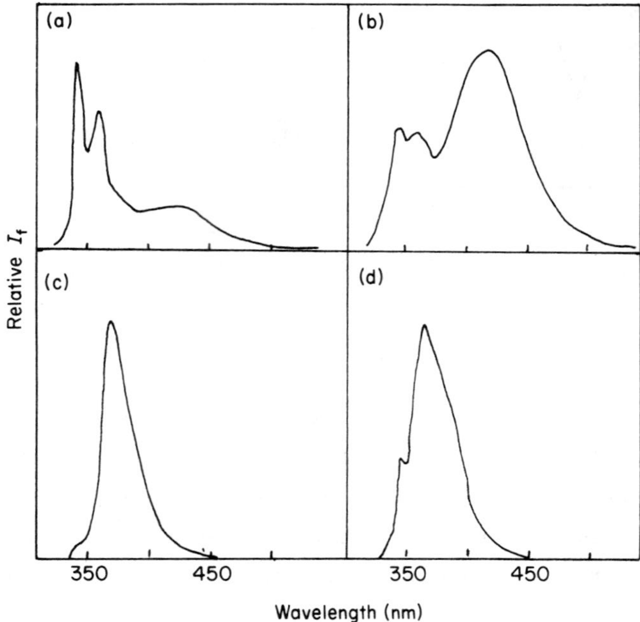

Fig. 5.7 Fluorescence spectra of 2,4-di(*N*-carbazolyl)pentanes:
(a) meso isomer in isooctane at 197K;
(b) meso isomer in isooctane at 293K;
(c) racemic isomer in isopentane at 133K;
(d) racemic isomer in isopentane at 293K;
(after De Schryver *et al* [106]).

have been observed to decay as single exponentials. These observations are clearly of relevance to the photophysics of PNVCz, and implicate isotactic sequences in formation of the lower energy excimer in the polymer, whereas the syndiotactic sequences would appear to be responsible for observation of the higher energy species. The data also substantiate the assumptions of Johnson [102] and Itaya *et al* [105] that the higher energy excimer comprises a partially overlapped structure. Perhaps most significantly, the data suggest that the two excimeric species do not interconvert in an equilibrium via a direct pathway as has been assumed in previous analyses of PNVCz photophysics. During the final stages of manuscript preparation, a report has been published [112] that confirms the observations of De Schryver *et al* [106] on these model compounds.

5.6 CONCLUSIONS AND CONSOLIDATION

The generation of mechanisms descriptive of the complex emission decay processes observed in macromolecules is desirable since, in principle, it affords some means of characterizing the kinetic behaviour in polymers and thereby promote greater understanding of subsequent energy trapping and photochemistry. The approach adopted in the main, as detailed in Section 5.5.1, has assumed that the observed life-times obtained by multi-exponential analyses represent average decay parameters associated with species representative of distributions of chromophores within polymodal distributions of chromophores. In other words it is assumed that groups of chromophores within the total time-resolved distribution may be taken as adequate averages of maxima within the overall chromophore distribution.

Within the limitations of such assumptions the models which have been generated independently and discussed in Section 5.5.1 should not be regarded as mutually exclusive. There does appear to be an influence of tacticity upon polymer photophysical behaviour, supported by kinetic and spectroscopic evidence from low molar mass analogues [30]. These influences of tacticity may be responsible, for example, for the differences in kinetic behaviour observed in homopolymers and may be reflected in the 'kinetic isolation' of monomeric chromophores observed in some types of macromolecule [27,28].

Superimposed on these microcompositional effects are those of intramolecular concentration effects revealed in molecular weight effects upon time-resolved emission data [90,92] and the influence of intramolecular chromophore concentration upon the luminescence characteristics of copolymers as described in Sections 5.4.2, 5.5.1 and 5.5.2. The self-consistency of rate-parameter data derived from time-resolved measurements on series of copolymers (Section 5.5.2) is good evidence for the assignment (in the copolymers studied to date) of observed decay times as averages representative of chromophore distributions centred on species of the type M_1^*, M_2^* and D* in kinetic scheme (5.3).

In some systems, such as those containing carbazoles and sterically hindered naphthalenes, there seems to be good spectroscopic evidence for the involvement of more than one type of excimeric state. There are without doubt systems in which, owing to the sheer complexity of the processes involved, no average lifetimes will be adequate in description of the decay kinetics, i.e. average decay parameters will not be capable of characterizing local chromophore environments within the total emitting distribution. Examples of such situations may be apparent in consideration of polymers containing pyrene or oxadiazole fluors.

In copolymers containing pyrene, decay kinetics are complex [113] as might be expected from long-lived chromophores that exhibit a strong tendency to form excimers. Intramolecular excimer formation is liable to involve both long- and short-range interactions on the polymer chain requiring various diffusive processes of both energy and material with concomitant complexities of associated rates.

At the other extreme, excimer formation has been observed in oxadiazoles constrained to high local concentrations by restriction to the confines of a polymer coil [114,115]. This observation contradicts theoretical proposals that excimer formation is forbidden in diaryloxadiazoles due to mutual repulsion between the ground and excited states [116]. Complexity of decay kinetics in this polychromophoric system may arise as a result of the short fluorescence lifetime of phenylbiphenyloxadiazole (PBD) and its derivatives and the extremely low binding energy of the excimer.

The overall result may be the existence of a continuous distribution of excimer sites corresponding to varying degrees of overlap of the aromatic rings rather than the 'bimodal distributions" envisaged, e.g. in carbazoles or sterically hindered naphthalenes, corresponding to fully overlapped and partially overlapped structures. Evidence for continuous temporal and energy distribution of excimer sites in oxadiazole polymers is afforded by late-gated time-resolved spectra (see Fig. 5.8), which show shifts to lower energy in the excimer band for spectra sampled at longer times following excitation. This behaviour is not observed, for example, in polymers showing excimer formation between vinylnaphthalene [27], acenaphthylene [26] or styrene [89] - derived chromophores (see Fig. 5.4).

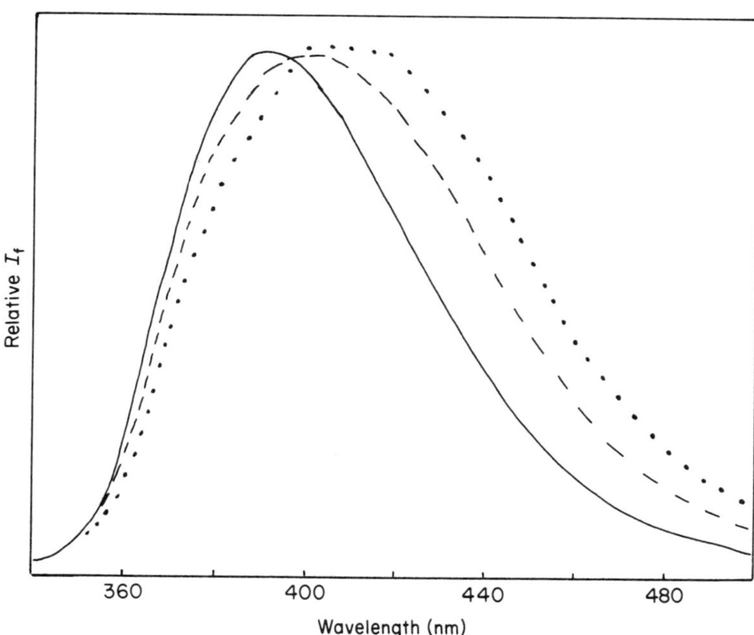

Fig. 5.8 Time-resolved fluorescence spectra of poly 2-(4-*t*-butylphenyl)- 5-(4'vinyl-4-biphenylyl)-1,3,4-oxadiazole in toluene at 298K recorded at delays of 4.5(—) 7.5(- - -) and 10.5 ns (. . .). Gate width 3 ns. (After Birch *et al* [117].)

Rationalization of the kinetic aspects of polymer photophysics is complicated for several reasons. Not the least of these may arise from the efficiency of energy trapping in macromolecules by trace levels of low energy entities. This immediately produces complications in the correlation of information produced in different laboratories on samples of the same chemical type but of different origin. In relation to such considerations are reports of ground-state dimers in polymers containing naphthalene chromophores. Irie *et al* [118] report emission from a ground-state dimer in poly 1- and 2-vinylnaphthalenes whereas Nakahira *et al* [119, 120] report a very similar emission profile for dimers in poly(2-*t*-butyl-6-vinylnaphthalene) [119] and poly(2-napthylalkyl methacrylate) [120] yet do not observe the emission in poly(2-vinylnaphthalene) [119].

It is relevant to note without further comment that spectral features of the type reported in support of ground-state dimers [118-120] are apparently absent in the studies on similar types of polymer by many other groups of workers. Indeed, prolonged efforts to generate spectra representative of ground state dimers in poly(vinylnaphthalenes) or poly(acenaphthalene) proved fruitless in one of the authors' laboratories [121]. However, a spectrum of very similar features, tentatively assigned to an impurity, has been obtained recently from a sample of 1,3-bis(2-naphthyl)propane we have synthesized [122]. The species appears to be photolabile in air-saturated fluid solutions, which would not be consistent with identification as a ground-state dimer. The phenomenon is the subject of continued research [122].

Of further relevance to the differences in result that can arise through the subjection of different polymer samples to scrutiny is the fact that in our laboratories it has not proved possible to reproduce the marked dependence of I_D/I_M upon excitation wavelength observed for poly(2-vinyl-naphthalene) by DeSchryver *et al* [30] upon examination of the fluorescence spectra of poly 1- and 2-vinylnaphthalenes synthesized by various methods [123].

Finally, we should like to note that during the preparation of this manuscript there have appeared [86,87] two independent publications that question the viability of associating time constants derived from decay analyses with the existence of distinct photophysical moieties within macromolecules and subsequent generation of kinetic schemes of the type discussed in Section 5.5. Both groups of workers derive, from a different basis, decay functions of the form:

$$i(t) = A \exp[-(at + bt^{\frac{1}{2}})] + B \exp(-ct) \qquad (5.10)$$

Fredrickson and Frank [86] have argued on the premise of a time-dependent energy-trapping function (in contrast to the k_{DM}[M] term adopted in scheme 5.3) in the absence of excimer dissociation to excited monomer in derivation of Equation (5.10). Itagaki *et al* [87] base their arguments on the existence of non-equilibrium diffusion controlled excimer sampling and make allowance for the existence of excimer dissociation to excited monomer. It has been shown that functions of the form of Equation (5.10) can simulate (through adjustment of the parameters) curves of the form observed in empirical monomer and excimer decays [86]. Alternatively, the functional type $\exp[-(at + bt^{\frac{1}{2}})]$ has been shown to emulate dual exponential components of triple-exponential fits [87].

These publications should stimulate further discussion and experiment designs in polymer photophysics. It will be of interest to note the success that expressions of the type represented in Equation (5.10) achieve in describing systems in which energy migration and/or polymer flexibility are varied. In the latter context it is relevant to note that in the copolymer series studied to date [27,35,36,88,89] that the two shorter decay components obtained in triple-exponential analyses increase as the basic homopolymer repeat units are replaced by comonomers that will induce chain flexibility. It is difficult to rationalize these results with the trend expected in the presence of rotational diffusional as described by Equation (5.10).

Testing of the model of Fredrickson and Frank [86] might be achieved through studies of intramolecular energy transfer especially in polymers in which the incidence of energy trapping by excimeric species might be eliminated or drastically reduced. In a recent study of energy trapping in the presence of excimer sites, good agreement between decay constraints in the emission of the intramolecular oxazole energy trap with those of the polystyrene donor was achieved [99]. This might appear unusual in the presence of a time dependent energy trapping function as postulated by Fredrickson and Frank [86].

It is clear that the origins of the photophysical behaviour of polymers will continue to be the subject of lively debate for some time. Activity in a field that has proved to be rapidly expanding and stimulating during the past decade seems likely to maintain momentum in the next few years. It is hoped that this article will prove useful to those working in the area and that it may provoke continued speculation and debate.

5.7 EXPERIMENTAL METHODS

The discussion above has leaned heavily upon results obtained by the method of time-correlated single-photon counting, and in order to provide background, a full discussion of the technique is included here.

Much of the greater understanding of excimer formation in macromolecules that has been realized in recent years is due to the availability of high resolution time-resolved fluorescence spectroscopic data. Of the methods available for examining the temporal characteristics of excited states 124,125 [1, 124], that of time-correlated single-photon counting has enjoyed the greatest application to the study of synthetic macromolecules for fluorescence life-times in the subnanosecond to microsecond time domain. In this section, the technique is described briefly, and some aspects of data analysis and recent developments in the technique are discussed.

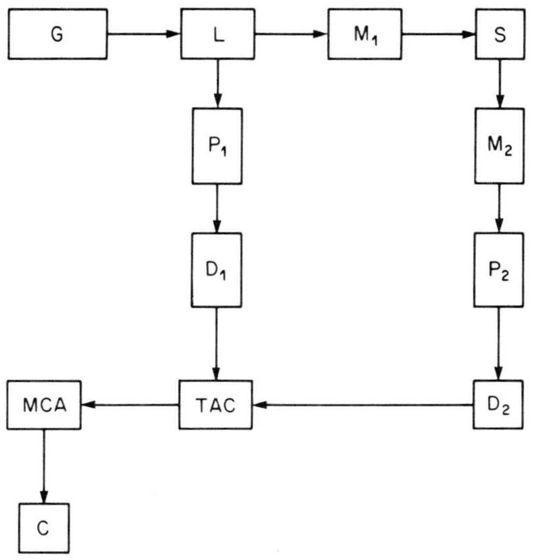

Fig. 5.9 Schematic layout of a time-correlated single photon counter:
G = trigger control; L = nanosecond source;
M = monochromator; S = sample; P = photomultiplier;
D = constant fraction discriminator; TAC = time-to-amplitude converter; MCA = multichannel analyser;
C = computer.

Fig. 5.9 shows a block diagram of a conventional single-photon counting arrangement. The sample is excited with a short pulse of light, typically from a nanosecond spark discharge lamp, and the fluorescence is detected at right angles to the excitation light path by a fast, high-gain photomultiplier. The excitation pulse (monitored for example by a photomultiplier P_1) initiates a voltage ramp in a time-to-amplitude converter (TAC). This voltage ramp is terminated when a single emitted photon is observed by the fast photomultiplier P_2. If, after a certain time has elapsed, a fluorescence photon has not been detected, the TAC is reset in preparation for the next excitation pulse. The time delay between the excitation flash and the arrival of the fluorescence photon is represented by a certain voltage. The event is registered as a single count in the appropriate channel (according to the voltage realized) in a multichannel analyser (MCA) operating in the pulse-height analysis mode. Provided the emission is attenuated such that the probability of detecting more than one fluorescence photon per excitation pulse is negligible (single photon conditions), repetition of the procedure many times results in the accumulation of a histogram in the MCA representative of the probability of emission as a function of time after excitation. In order to attain a high signal-to-noise ratio and a significant statistical data distribution a minimum of 10^4 counts at the channel of maximum population is desirable. A greater number of counts is desirable if multi-exponential decay fitting is to be attempted. Several texts have appeared that describe the technique in greater detail (e.g. [124,125]).

The observed decay curve, $F(t)$, is a convolution of the instrument response function $P(t)$ and the true decay function $G(t)$ of the form:

$$F(t_i) = \int_0^{t_i} G(t')P(t_i - t')\,dt' \,. \tag{5.11}$$

Deconvolution involves solution of Equation (5.11). The method of least squares iterative reconvolution [126-128] is generally adopted in such deconvolutions. $G(t)$ is assumed to be of a specific functional form and is convoluted with $P(t)$ according to Equation (5.11). The calculated curve $C(t)$ is compared with the measured curve $F(t)$ and the reduced chi-square, χ_r^2, is calculated:

$$\chi_r^2 = \frac{\Sigma_{i=1}^{n} W_i^2[F(t_i) - C(t_i)]^2}{n - p} \tag{5.12}$$

W_i is the weighting factor for the ith data point and is equal to $F(t_i)^{-\frac{1}{2}}$, n is the number of channels over which fitting is performed and p is the number of adjustable parameters in the adopted fitting function. The values of the parameters are varied until χ_r^2 is minimized. A value of the minimized χ_r^2 close to unity is indicative of a good fit to the data.

The 'goodness of fit' may also be assessed by examination of the distribution of the residuals $R(t_i)$ and the autocorrelation of the residuals $A(t)$ defined as:

$$R(t_i) = \frac{F(t_i) - C(t_i)}{F(t_i)^{\frac{1}{2}}} , \tag{5.13}$$

$$A(t_i) = \frac{\frac{2}{N} \Sigma_{j=1}^{N/2} R(t_j) R(t_{j+i})}{\frac{1}{N} \Sigma_{j=1}^{N} [R(t_j)]^2} . \tag{5.14}$$

A good fit to the data requires a random distribution of the weighted residuals about zero and the absence of correlation between the residuals.

An additional parameter that may be employed to assess the viability of the trial function is the serial correlation coefficient [129,130], D, calculated according to the expression:

$$D = \frac{\Sigma_{i=2}^{N} [R(t_i) - R(t_{i-1})]^2}{\Sigma_{i=1}^{N} [R(t_i)]^2} \tag{5.15}$$

Rigorous examination of the 'goodness of fit' is necessary in the application of multi-exponential analyses of data. It is also desirable to test the ability of the deconvolution procedures in extraction of multiple decay times through tests on synthetic data [131]. It is apparent that the ability to derive meaningful data by curve fitting will depend upon the degree of distinction between the excited state life-times and the magnitudes of such quantities relative to the instrumental response function in multi-exponential analyses [131].

Although several commercial instruments are available for the routine analysis of simple single and bi-exponential decays, much interest has been shown in the use of high repetition rate, sub-nanosecond light sources (synchrotrons and mode-locked lasers), in place of the conventional nanosecond flashlamp for sample excitation, thus facilitating the analysis of fluorescence decay profiles in terms of more complex functions such as triple exponential schemes. Although synchrotron radiation sources have the advantage of being tunable over a wide range of excitation wavelengths, their disadvantages (high cost, long time-scale for development, inflexibility associated with central research facilities) have detracted from their attraction as excitation sources for photochemistry and, currently, we are aware of only one research report in which these sources have been adopted for an investigation of excimer formation, in polymers [25]. However, mode-locked gas and dye lasers have been shown to offer a relatively inexpensive, high repetition rate picosecond pulsed light source, ideal for single-photon counting applications. Several time-resolved fluorimeters based upon these sources are currently in operation or under construction.

A diagram of one such instrument is shown in Fig. 5.10. In this system a mode-locked argon ion laser is used to excite a jet stream dye laser, whose cavity has been extended to match that of the ion laser. Under these conditions, the output of the dye laser is mode-locked with a pulse width typically less that 6 ps (FWHM). In a typical arrangement the pulse repetition rate (around 80-100 MHz) is too high for total system relaxation between excitation events, and a cavity dumper is employed in order to reduce this rate down to about 1 - 4 MHz. The laser output is frequency-doubled into the ultra-violet region using angle and/or temperature tuned non-linear crystals. Residual, non-doubled light is removed by a filter. Conventional single-photon counting detection methods are employed [124] although in order to realize the full potential of the 1 - 4 MHz excitation rate, it is necessary to operate the time - to - amplitude converter (TAC) in reverse [132]. In this mode of

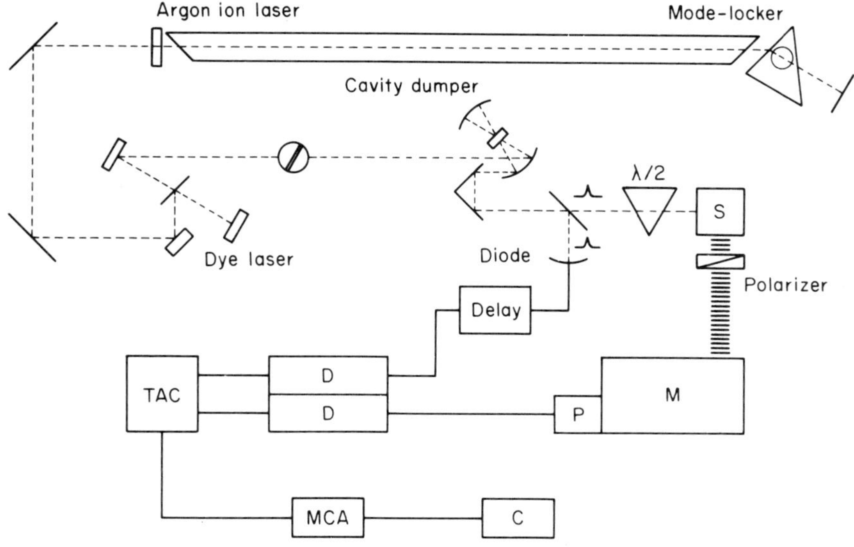

Fig. 5.10 Block diagram of a laser-excited time-resolved spectrofluorimeter: D = discriminator; P = photomultiplier; M = monochromator; S = sample; MCA = multichannel analyser; C = computer (after Ghiggino *et al* [111]).

operation, a pulse corresponding to a detected photon of fluorescence is sent from the photomultiplier tube to the start input of the TAC. A signal corresponding to the next excitation pulse, and derived from a fast photodiode monitoring a fraction of the laser output, terminates the TAC, and the accumulated voltage corresponds to the time interval between the emission of a photon and the arrival of the next excitation pulse. Since the laser interpulse separation is stable to 1 part in 10^5 (or better) it is a simple matter to 'count back' one pulse and derive the true decay profile. Spectrometers of this type have been shown to be capable of recording fluorescence decay data of high quality facilitating the measurement of subnanosecond and heterogeneous decays [131, 133].

5.7.1 *Future developments*

The potential of time-resolved spectroscopic techniques in the study of polymer photophysics will be enhanced with the development of more sophisticated laser technology and improvements in reduction of the instrumental response profile. The latter objective may be realized in part by improved photomultiplier design (provided such improvements in transit time are not at the expense of signal-to-noise discrimination).

It is to be hoped further that rigorous application of criteria of the quality of fits to decay data will become more widely recognized. In many reports in the polymer literature, unacceptably high values of chi-square have been tolerated (when such values have been quoted at all) and little use has been made of the alternative criteria discussed above. Furthermore, many literature reports show evidence of insufficient attention to experimental procedures. Some examples of common faults are listed below:

1. Decay curves may be accumulated to insufficient signal-to-noise resolution for the analysis intended.
2. Attempts are made to analyse short-lived decay components using too coarse a time scale (channel width).
3. Decay curves are analysed solely over the 'decay' portion (i.e. from the maximum to longer times) of the transient profile.
4. Decay curves are not analysed over a minimum of 3 decades of intensity.

It is important that attention be given to such details. Understandably, workers in related fields will accept and quote conlcusions derived from transient decay data without critical evaluation of the basis for the claims. Instances of this type have undoubtedly led to the number of misconceptions commonly accepted among differing factions of polymer photochemistry that have become compounded and convolved with relay into subsequent experiment design and interpretation.

ACKNOWLEDGEMENT

Ian Soutar wishes to acknowledge, with gratitude, funding by the Nuffield Foundation in the form of a Science Research Fellowship during part of the time in which this manuscript was being prepared.

REFERENCES

1. STEVENS, B. and HUTTON, E. (1960), *Nature* 186, 1045.
2. FORSTER, Th. and KASPER, K. (1954), *Z. Phys. Chem. NF* 1, 275.
3. BIRKS, J.B. (1970), *Photophysics of Aromatic Molecules*, Wiley, London.
4. BIRKS, J.B. (1964), *Acta Phys. Polon.* 26, 367.
5. STEVENS, B. (1971), *Adv. Photochem.* 8, 161.
6. BIRKS, J.B. and CHRISTOPHOROU, L.G. (1962), *Nature,* 194, 442.
7. BIRKS, J.B. and CHRISTOPHOROU, L.G. (1965), *Nature,* 197, 1064.
8. BASILE, L.J. (1962), *J. Chem. Phys.* 36, 2204.
9. YANARI, S.S., BOVEY, F.A. and LUMRY, R. (1963), *Nature* 200, 242.
10. HIRAYAMA, F. (1965), *J. Chem. Phys.* 42, 3163.
11. VALA, M.T., HAEBIG, J. and RICE, S.A. (1965), *J. Chem. Phys.* 43, 886.
12. KLOPFFER, W. (1969), *Chem. Phys. Lett.* 4, 193.
13. CHANDROS, E.A. and DEMPSTER, J.C. (1970), *J. Amer. Chem. Soc.* 92, 3586.
14. STEVENS, B. and BAN, M.J. (1964), *Trans. Faraday Soc.* 60, 1515.
15. BIRKS, J.B., LUMB, M.D. and MUNRO, I.H. (1964), *Proc. Roy. Soc. London. Ser. A* 280, 289.
16. FOX, R.B., PRICE, T.R., COZZENS, R.F. and McDONALD, J.R. (1972), *J. Chem. Phys.* 57, 534.
17. WU, S.K., JIANG, Y.C. and RABEK, J.F. (1980), *Polym. Bull.* 3, 319.
18. BOKOBZA, L. and MONNERIE, L. (1981), *Polymer* 22, 235.
19. DAVID, C., PIENS, M. and GEUSKENS, G. (1972), *Eur. Polym. J.* 8, 1019.
20. HARRAH, L.A. (1972), *J. Chem. Phys.* 56, 385.
21. ITO, S., YAMAMOTO, M. and NISHIJIMA, Y. (1976), *Rep. Prog. Polym. Phys. Japan* 19, 421.
22. ASPLER, J.S. and GUILLET, J.E. (1979), *Macromolecules* 12, 1082.
23. NAKAHIRA, T., MARUYAMA, I., IWABUCHI, S. and KOJIMA, K. (1970), *Makromol. Chem.* 180, 1853.
24. FRANK, C.W. (1974), *J. Chem. Phys.* 61, 2015.
25. McINALLY, I., STEEDMAN, W. and SOUTAR, I. (1977), *J. Polym. Sci. Polym. Chem. Ed.* 15, 2511.
26. PHILLIPS, D., ROBERTS, A.J. and SOUTAR, I. (1980), *J. Polym. Sci. Polym. Lett. Ed.* 18, 123.
27. PHILLIPS, D., ROBERTS, A.J. and SOUTAR, I. (1980), *J. Polym. Sci. Polym. Phys. Ed.* 18, 2401.

28. HOLDEN, D.A., WANG, P.Y.-K. and GUILLET, J.E. (1980), *Macromolecules* 13, 295.
29. DEMEYER, K., VAN DER AUWERAER, M., AERTS, L. and DE SCHRYVER, F.C. (1980), *J. Chim. Phys.* 77, 493.
30. DE SCHRYVER, F.C., DEMEYER, K., VAN DER AUWERAER, M. and QUANTEN, E. (1981), *Ann. N.Y. Acad. Sci.* 366, 93.
31. NISHIJIMA, Y. (1973), *Prog. Polym. Sci. Japan* 6, 199.
32. DAVID, C., PIENS, M. and GEUSKENS, G. (1979), *Eur. Polym. J.* 15, 373.
33. HOLDEN, D.A. and GUILLET, J.E. (1981), in *'Developments in Polymer Photochemistry,'* vol. 1 (ed. by N.S. Allen) Applied Science Publishers, London.
34. GHIGGINO, K.P., WRIGHT, R.D. and PHILLIPS, D. (1978), *J. Polym. Sci. Polym. Phys. Ed.* 16, 1499.
35. PHILLIPS, D., ROBERTS, A.J. and SOUTAR, I. (1982), *J. Polym. Sci. Polym. Phys. Ed.* 20, 411.
36. PHILLIPS, D., ROBERTS, A.J. and SOUTAR, I. (1981), *Eur. Polym. J.* 17, 101.
37. WANG, Y.C. and MORAWETZ, H. (1975), *Makromol. Chem., Suppl.* 1, 283.
38. DAVID, C., LEMPEREUR, M. and GEUSKENS, G. (1972), *Eur. Polym. J.* 8, 417.
39. REID, R.F. and SOUTAR, I. (1980), *J. Polym. Sci., Polym. Phys. Ed.* 18, 457.
40. SOMERSALL, A.C. and GUILLET, J.E. (1973), *Macromolecules* 6, 218.
41. BOUDEVSKA, H., BRUTCHKOV, C. and ANSTRUG, A. (1979), *Makromol. Chem.* 180, 1113.
42. ABUIN, E.A., LISSI, E.A., GARGALLO, L. and RADIC, D. (1979), *Eur. Polym. J.* 15, 373.
43. REDPATH, A.E.C. and WINNIK, M.A. (1980), *J. Amer. Chem. Soc.* 102, 6869.
44. REDPATH, A.E.C. and WINNIK, M.A. (1981), *Ann. N.Y. Acad. Sci.* 366, 75.
45. CUNIBERTI, C. and PERICO, A. (1980), *Eur. Polym. J.* 16, 887.
46. CUNIBERTI, C. and PERICO, A. (1981), *Ann. N.Y. Acad. Sci.* 366, 35.
47. IBEMISI, A.B., KINSINGER, J.B. and EL-BAYOUMI, M.A. (1980), *J. Polym. Sci. Polym. Chem. Ed.* 18, 879.
48. NISHIHARA, R. and KANEKO, M. (1969), *Makromol. Chem.* 124, 84.
49. SOUTAR, I. (1981), *Ann. N.Y. Acad. Sci.* 366, 24.
50. FOX, R.B., PRICE, T.R., COZZENS, R.F. and ECHOLS, W.H. (1974), *Macromolecules* 7, 937.
51. YOKOYAMA, M., TAMAMURA, T., ATSUMI, M., YOSHIMURA, M., SHIROTA, Y. and MIKAWA, H. (1975), *Macromolecules* 8, 101.
52. LINDSELL, W.E., ROBERTSON, F.C. and SOUTAR, I. (1981), *Eur. Polym. J.* 17, 203.
53. TORKELSON, J.M., LIPSKY, S. and TIRRELL, M. (1981), *Macromolecules* 14, 1601.
54. NISHIJIMA, Y. (1970), *J. Polym. Sci.* C31, 353.
55. DAVID, C., LEMPEREUR, M. and GEUSKENS, G. (1973), *Eur. Polym. J.* 9, 1315.

56. ALEXANDRU, L. and SOMERSALL, A.C. (1977), *Polym. Sci. Polym. Chem. Ed.* 15, 2013.
57. REID, R.F. and SOUTAR, I. (1977), *J. Polym. Sci. Polym. Lett. Ed.* 15, 153.
58. REID, R.F. and SOUTAR, I. (1978), *J. Polym. Sci. Polym. Phys. Ed.* 16, 231.
59. ANDERSON, R.A., REID, R.F. and SOUTAR, I. (1979), *Eur. Polym. J.* 15, 925.
60. ANDERSON, R.A., REID, R.F. and SOUTAR, I. (1980), *Eur. Polym. J.* 16, 945.
61. MAJUMDAR, R.N., CARLINI, C., ROSATO, N. and HOUBEN, J.L. (1980), *Polymer* 21, 941.
62. NAKAHIRA, T., MINAMI, C., IWABUCHI, S. and KOJIMA, K. (1979), *Makromol. Chem.* 180, 2245.
63. NAKAHIRA, T., SAKUMA, T., IWABUCHI, S. and KOJIMA, K. (1980), *Makromol. Chem. Rapid Commun.* 1, 413.
64. LONGWORTH, J.W. and BOVEY, F.A. (1966), *Biopolymers* 4, 1115.
65. BOKOBZA, L., JASSE, B. and MONNERIE, L. (1977), *Eur. Polym. J.* 13, 921.
66. FRANK, C.W. and HARRAH, L.A. (1974), *J. Chem. Phys.* 61, 1526.
67. DE SCHRYVER, F.C., BOENS, N. and PUT, J. (1977), *Adv. Photochem.* 10, 359.
68. BOKOBZA, L., JASSE, B. and MONNERIE, L. (1980), *Eur. Polym. J.* 16, 715.
69. GOLDENBERG, M., EMERT, J. and MORAWETZ, H. (1978), *J. Amer. Chem. Soc.* 100, 7171.
70. VAN DER AUWERAER, M., GILBERT, A. and DE SCHRYVER, F.C. (1980), *J. Amer. Chem. Soc.* 102, 6107.
71. MEEUS, F., VAN DER AUWERAER, M. and DE SCHRYVER, F.C. (1980), *Chem. Phys. Lett.* 74, 218.
72. LONGWORTH, J.W. (1966), *Biopolymers* 4, 1131.
73. ISHII, T., MATSUSHITA, H. and HANDA, T. (1975), *Kobunshi Ronbunshu* 32, 211.
74. ISHII, T., MATSUNAGA, S. and HANDA, T. (1976), *Makromol. Chem.* 177, 283.
75. ISHII, T., HANDA, T. and MATSUNAGA, S. (1977), *Makromol. Chem.* 178, 2351.
76. MacCALLUM, J.R. (1981), *Eur. Polym. J.* 17, 797.
77. MacCALLUM, J.R. (1982), in *Photophysics of Synthetic Polymers* (ed. D. Phillips and A.J. Roberts) Science Reviews Ltd, Northwood.
78. GHIGGINO, K.P., ROBERTS, A.J. and PHILLIPS, D. (1978), *J. Photochem.* 9, 301.
79. HEISEL, F. and LAUSTRIAT, G. (1969), *J. Chim. Phys.* 66, 1881.
80. GHIGGINO, K.P., WRIGHT, R.D. and PHILLIPS, D. (1978), *Chem. Phys. Lett.* 58, 552.
81. ITO, S., YAMAMOTO, Y. and NISHIJIMA, Y. (1978), *Rep. Prog. Polym. Phys. Japan* 21, 361.
82. MERLE-AUBRY, L., HOLDEN, D.A., MERLE, Y. and GUILLET, J.E. (1980), *Macromolecules* 13, 1138.

83. KEYANPOUR-RAD, M., LEDWITH, A., HALLAM, A., NORTH, A.M., BRETON, M., HOYLE, C.E. and GUILLET, J.E. (1978), *Macromolecules* 11, 1114.
84. BEVINGTON, P.R. (1967), *Data Reduction and Error Analysis for the Physical Sciences*, McGraw-Hill, New York.
85. PHILLIPS, D., ROBERTS, A.J. and SOUTAR, I. (1981), *Polymer* 22, 427.
86. FREDRICKSON, G.H. and FRANK, C.W. (1983), *Macromolecules* 16, 572.
87. ITAGAKI, H., HORIE, K. and MITA, I. (1983), *Macromolecules* 16, 1395.
88. PHILLIPS, D., ROBERTS, A.J. and SOUTAR, I. (1981), *Polymer* 22, 293.
89. SOUTAR, I., PHILLIPS, D., ROBERTS, A.J. and RUMBLES, G. (1982), *J. Polym. Sci. Polym. Phys. Ed.* 20, 1759.
90. PHILLIPS, D., ROBERTS, A.J., RUMBLES, G. and SOUTAR, I. (1983), *Macromolecules* 16, 1597.
91. DE SCHRYVER, F.C., MOENS, L., VAN DER AUWERAER, M., BOENS, N., MONNERIE, L. and BOKOBZA, L. (1982), *Macromolecules* 15, 64.
92. ISHII, T., HANDA, T. and MATSUNAGA, S. (1978), *Macromolecules* 11, 40.
93. NISHIJIMA, Y. and YAMAMOTO, M. (1979), *Polym. Prepr. Amer. Chem. Soc. Div. Polym. Chem.* 20, 391.
94. ITAGAKI, H., OKAMOTO, A., HORIE, K. and MITA, I. (1982), *Eur. Polym. J.* 18, 885.
95. ITAGAKI, H., OBUKATA, N., OKAMOTO, A., HORIE, K. and MITA, I. (1981), *Chem. Phys. Lett.* 78, 143.
96. ITAGAKI, H., OBUKATA, N., OKAMOTO, A., HORIE, K. and MITA, I. (1982), *J. Amer. Chem. Soc.* 104, 4469.
97. GUPTA, A., LIANG, R., MOACANIN, J., KLIGER, D., GOLDBECK, R., HORWITZ, J. and MISKOWSKI, V.M. (1981), *Eur. Polym. J.* 17, 485.
98. O'CONNOR, D.V. and WARE, W.R. (1976), *J. Amer. Chem. Soc.* 98, 4706.
99. PHILLIPS, D., ROBERTS, A.J. and SOUTAR, I. (1983), *Macromolecules* 16, 1593.
100. GUPTA, M.C., GUPTA, A., HORWITZ, J. and KLIGER, D. (1982), *Macromolecules* 15, 1372.
101. McINALLY, I., REID, R.F., RUTHERFORD, H. and SOUTAR, I. (1979), *Eur. Polym. J.* 15, 723.
102. JOHNSON, G.E. (1975), *J. Chem. Phys.* 62, 4697.
103. GHIGGINO, K.P., WRIGHT, R.D. and PHILLIPS, D. (1978). *Eur. Polym. J.* 14, 567.
104. HOYLE, C.E., NEMZEK, T.L., MAR, A. and GUILLET, J.E. (1978), *Macromolecules* 11, 429.
105. ITAYA, A., OKAMOTO, K. and KUSABAYASHI, S. (1976), *Bull. Chem. Soc. Japan* 49, 2092.
106. DE SCHRYVER, F.C., VANDENDRIESSCHE, J., TOPPET, S., DEMEYER, K. and BOENS, N. (1982), *Macromolecules* 15, 406.
107. ROBERTS, A.J., CURETON, C.G. and PHILLIPS, D. (1980), *Chem. Phys. Lett.* 72, 554.
108. ROBERTS, A.J., PHILLIPS, D., ABDUL-RASOUL, F. and LEDWITH, A. (19) *J.C.S. Faraday I.* 77, 2725.

109. TAGAWA, S., WASHIO, M. and TABATA, Y. (1979), *Chem. Phys. Lett.* 68, 276.
110. NG, D. and GUILLET, J.E. (1981), *Macromolecules* 14, 405.
111. GHIGGINO, K.P., ROBERTS, A.J. and PHILLIPS, D. (1981), *Adv. Polym. Sci.* 40, 71.
112. EVERS, F., KOBS, K., MEMMING, R. and TERRELL, D.R. (1983), *J. Amer. Chem. Soc.* 105, 5988.
113. NEVES, M.E.P.J.M. and SOUTAR, I. to be published.
114. SOUTAR, I. (1982), in *Photophysics of Synthetic Polymers* (ed. D. Phillips and A.J. Roberts), Science Reviews Ltd, Northwood.
115. ANDERSON, R.A., BIRCH, D.J.S., DAVIDSON, K., IMHOF, R.E. and SOUTAR, I. (1982), in *Photophysics of Synthetic Polymers* (ed. D. Phillips and A.J. Roberts), Science Reviews Ltd. Northwood.
116. LAMI, H. and LAUSTRIAT, G. (1968), *J. Chem. Phys.* 48, 1832.
117. BIRCH, D.J.S., DAVIDSON, K., IMHOF, R.E. and SOUTAR, I. to be published.
118. IRIE, M., KAMIJO, T., AIKAWA, M., TAKAMURA, T., HAYASHI, K. and BABA, H. (1977), *J. Phys. Chem.* 81, 1571.
119. NAKAHIRA, T., ISHIZUKA, S., IWABUCHI, S. and KOJIMA, K. (1982), *Macromolecules* 15, 1217.
120. NAKAHIRA, T., ISHIZUKA, S., IWABUCHI, S. and KOJIMA, K. (1983), *Macromolecules* 16, 297.
121. GRAHAM, C. and SOUTAR, I. Unpublished data.
122. ARCIERO, G.W., GRAHAM, C., PHILLIPS, D., RUMBLES, G. and SOUTAR, I. to be published.
123. LINDSELL, W.E., SERVICE D.M., KOUTSOKOSTAS, N. and SOUTAR, I. to be published.
124. WARE, W.R. (1971), in *Creation and Detection of the Excited State* (ed. A.A. Lamola) Vol. 1A, Marcel Dekker, New York, p. 213.
125. O'CONNOR, D.V. and PHILLIPS, D. (1984) *Time-correlated Single Photon Counting*, Academic Press, New York.
126. KNIGHT, A.C. and SELINGER, B.K. (1971) *Spectrochim. Acta.* 27A, 1223.
127. McKINNON, A.G., SZABO, A.G. and MILLER, D.R. (1977), *J. Phys. Chem.* 81, 1564.
128. O'CONNOR, D.V., WARE, W.R. and ANDRÉ, J.C. (1979), *J. Phys. Chem.* 83, 1333.
129. DURBIN, J. and WATSON, G.S. (1950), *Biometrika* 37, 409.
130. DURBIN, J. and WATSON, G.S. (1951), *Biometrika* 38, 159.
131. ROBERTS, A.J., O'CONNOR, D.V. and PHILLIPS, D. (1981), *Ann. N.Y. Acad. Sci.* 366, 109.
132. GHIGGINO, K.P., PHILLIPS, D., SALISBURY, K. and SWORDS, M.D. (1977), *J. Photochem.* 7, 141.
133. GHIGGINO, K.P., ROBERTS, A.J. and PHILLIPS, D. (1980), *J. Phys. E : Sci. Instrum.* 13, 446.

CHAPTER SIX

Dynamic depolarization of luminescence

6.1 INTRODUCTION

The first dynamic study of fluorescence polarization was performed in 1929 by Perrin [1] for measuring the fluorescence life-time of small molecules in solution. Much later, in 1948, Weber [2] applied this technique for studying rotational motions of proteins. Since then, fluorescence polarization has been extensively used in molecular biology mainly for looking at conformational changes occuring during denaturation or enzyme-substrate reactions.

In the synthetic polymer field, only a limited number of studies have been performed with the fluorescence polarization technique, in spite of its real potential for analysing polymer dynamics. Such a situation could be due to the greater difficulty of fluorescent labelling polymers (compared to proteins), owing to the lack of chemically reactive groups in most synthetic polymers.

Recently, phosphorescence polarization has been applied to bulk polymers for studying molecular motions.

In this Chapter, we first describe the luminescence polarization phenomenon and the related basic expressions. Then, the labelling methods and the different measuring techniques are reviewed. Finally, typical examples of studies on polymer dynamics are reported, both for polymer solutions and bulk polymers.

6.2 LUMINESCENCE POLARIZATION PHENOMENON

As previously mentioned in Chapters 1 and 7, an 'absorption transition moment', M_A, with fixed orientation with respect to the geometry of the molecule, is associated with the absorption phenomenon. In the same way there is an 'emission transition moment', M_E, with a definite orientation in the molecule geometry. Depending on fluorescence or phosphorescence emission, the transition moments involved are different.

It is worth noticing that the orientation of the absorption and emission transition moments depends on the nature of the electronic transition undergone. Thus, in aromatic hydrocarbons, the allowed transitions $^1(\pi \leftarrow \pi)$ correspond to moments that lie in the plane of the molecule, in such a way that the fluorescence emission is called 'in-plane' polarized. But, in these compounds, the phosphorescence emission is polarized perpendicular to the plane of rings for it corresponds to a forbidden transition. As an example, some transition moments of anthracene are shown in Fig. 6.1. On the contrary, with molecules where $\pi \leftarrow n$ transition occurs, as heterocyclic molecules or carbonyl containing molecules, the $\pi \leftarrow n$ singlet transition is out-of-plane polarized whereas the phosphorescence from n, π triplet states is in-plane polarized. The relationship between the chemical structure of a molecule and the nature of electronic transitions can be found in Becker [3].

As far as luminescence polarization is concerned, luminescent molecules can be represented by their absorption and emission transition moments $\vec{M}_0$ and $\vec{M}$ (Fig. 6.2). When a polarized light is used, the probability of absorption is proportional to $\cos^2 \alpha_A$. In the same way, the luminescence intensity measured through an analyser A is proportional to $\cos^2 \beta$. Thus, for $\vec{P}$ and $\vec{A}$ direction of polarizer and analyser, the observed luminescence intensity is proportional to $\cos^2 \alpha_A \cos^2 \beta$.

Owing to the lack of phase correlation between excitation and emission, luminescence emission can be described as resulting from three independent radiations respectively polarized along OX, OY, OZ

(a) Absorption

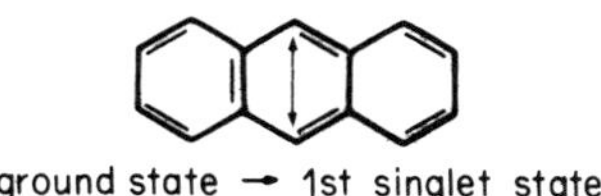

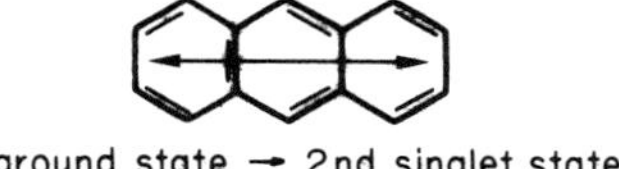

(b) Fluorescence emission

(c) Phosphorescence emission

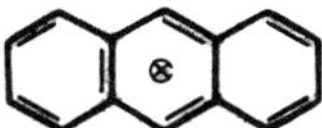

Fig. 6.1 Directions of transition moments for absorption, fluorescence and phosphorescence of anthracene.

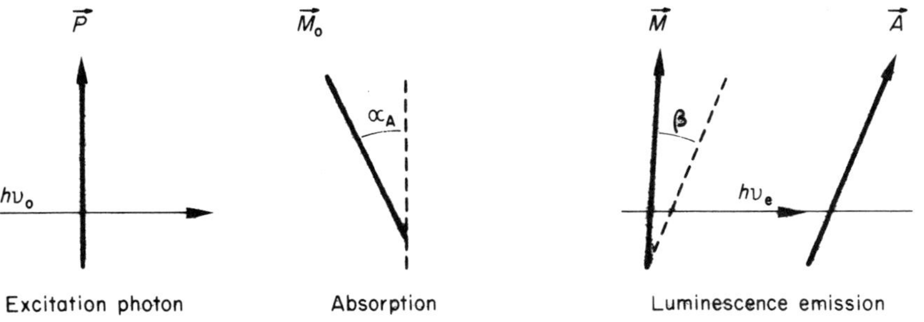

Fig. 6.2 Polarized absorption and luminescence emission, $\vec{P}$ = polarizer, $\vec{A}$ = analyser.

with intensities I_X, I_Y and I_Z (Fig. 6.3).

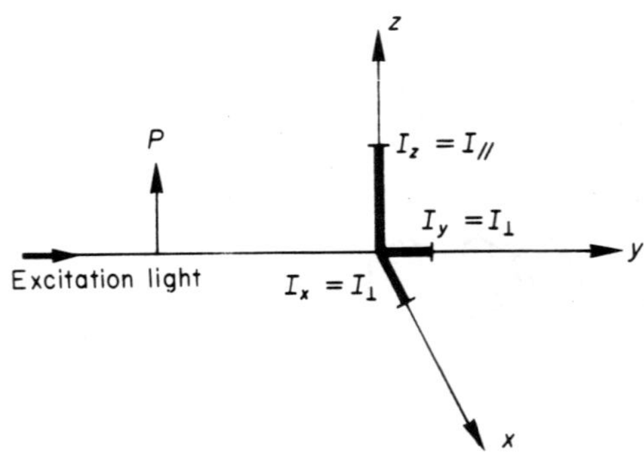

Fig. 6.3 States of polarization of luminescence emission.

The total luminescence intensity I is equal to:

$$I = I_X + I_Y + I_Z.$$

Luminescence emission is commonly observed either at right angles to the excitation direction or along it. To recapitulate, two quantities are used for characterizing the emission polarization:

1. The polarization ratio, p: $p = (I_{||} - I_\perp)/(I_{||} + I_\perp)$.
2. The emission anisotropy, r: $r = (I_{||} - I_\perp)/(I_{||} + 2I_\perp)$.

$I_{||}$ and $I_\perp$ correspond to luminescence intensity obtained with an analyser direction respectively parallel and perpendicular to that of the polarizer. The quantity $(I_{||} + 2I_\perp)$ corresponds to the total luminescence intensity.

The emission anisotropy, r, is more convenient for deriving analytical expressions and will be used hereafter. Nevertheless the two quantitites are related through:

$$r = (2/3)(1/p - 1/3)^{-1}.$$

6.2.1 *Emission anisotropy for motionless molecules isotropically distributed*

Consider a set of molecules, excited at time $t = 0$, by an infinitely short light pulse, vertically polarized (Fig. 6.4 (a)).

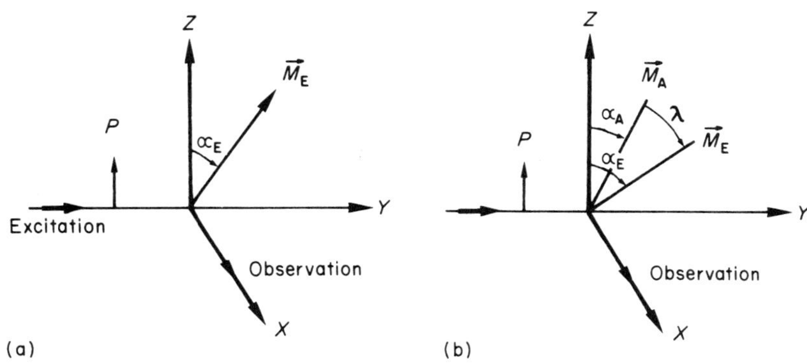

Fig. 6.4 Orientation of transition moments in the reference frame: (a) emission moment; (b) non-parallel absorption and emission moments.

At time t, the emission transition moment of molecule i lies along $\vec{M}_E$, whose components are:

$$(M_E)_a = M\,\ell_{a_i}(t) \qquad a = X,\ Y,\ Z$$

where ℓ_a's are direction cosines related by $\sum_a \ell_{a_i}^2 = 1$.

Owing to the lack of phase correlation between individual emissions, the emission intensity components are:

$$I_a(t) = KM^2 \sum_{i=1}^{N} \ell_{a_i}^2(t) = KM^2\, N\, \overline{\ell_{a_i}^2}(t)$$

where K is a constant and $\overline{\ell_{a_i}^2}$ is the average of $\ell_{a_i}^2$ over the N molecules emitting at time t.

The excitation light being vertically polarized, excited molecules are symmetrically distributed around OZ, in such a way that $\overline{\ell_X^2} = \overline{\ell_Y^2}$ and $\overline{\ell_Z^2} = 1 - 2\,\overline{\ell_X^2}$. This leads to the following expressions for the emission anisotropy:

$$r(t) = (3\,\overline{\ell_Z^2(t)} - 1)/2,$$

$$r(t) = (3\,\overline{\cos^2 \alpha_E(t)} - 1)/2.$$

Now, suppose that the molecules do not change their orientations between excitation and emission. In addition, consider that they are randomly distributed.

Further evaluation requires the calculation of the average value of $\cos^2 \alpha_E(t)$ and two cases have to be considered depending on the relative position of absorption and emission transition moments.

(a) Parallel absorption and emission moments

To evaluate $\overline{\cos^2 \alpha_E}$, it is necessary to consider the excitation probability of those molecules with transition moments at an angle $\alpha_A = \alpha_E = \alpha$ to the polarization direction OZ. If $N_o(\alpha)$ is the total number of molecules with absorption moments characterized by α, the number of molecules in the excited state is $N_o(\alpha)\cos^2 \alpha$. For randomly oriented molecules, $N_o(\alpha)$ is proportional to $\sin \alpha\, d\alpha$, so that the average of $\cos^2 \alpha$ can be expressed by:

$$\overline{\cos^2 \alpha} = \left(\int_o^{\pi/2} \cos^2 \alpha (\cos^2 \alpha \sin\alpha) d\alpha\right) / \left(\int_o^{\pi/2} \cos^2 \alpha \sin \alpha\, d\alpha\right),$$

$$\cos^2 \alpha = 3/5. \tag{6.1}$$

It leads the expression for the emission anisotropy becoming:

$$r_{o,\ell} = 2/5 = 0.4. \tag{6.2}$$

This value is the highest that could be reached for the emission anisotropy; it is called 'limit anisotropy'. Experimental r values are lower than 0.4, as is shown hereafter.

(b) Non-parallel absorption and emission moments

Such a situation first deals with phosphorescence emission, for in this case the emission transition moment is almost perpendicular to that of the absorption.

In the case of fluorescence emission, vibronic coupling between molecular vibrations and electronic motion leads to some electronic delocalization which results in an average angle between transition moments (Fig. 6.1 (a),(b)), even when the same electronic transition is involved, as for excitation to the 1st singlet state. Nevertheless, in this case, small average angles between the moments are observed.

The corresponding situation is represented in Fig. 6.4(b). If Ψ is the angle between the planes (OZ, M_A) and (OZ, M_E), $\cos \alpha_E$ can be written:

$$\cos \alpha_E = \cos \alpha_A \cos \lambda + \sin \alpha_A \sin \lambda \cos \Psi.$$

By squaring both members of this relation and taking into account the random orientation of the molecules, Ψ takes any value between 0 and 2π ($\overline{\cos \Psi} = 0$, $\overline{\cos^2 \Psi} = 1/2$), and after rearrangement becomes:

$$(3 \cos^2 \alpha_E - 1)/2 = ((3 \overline{\cos^2 \alpha_A} - 1)/2)((3 \cos^2 \lambda - 1)/2).$$

It is worthwhile to point out that this 'product rule' is a general property [4] of the quantity $(3 \overline{\cos^2 \beta} - 1)/2$ and it applies to the average value of any angle β corresponding to a direction resulting from 2 orientation changes, provided that this direction have a random azimuth relative to the preceding direction (as $\vec{M}_E$ relatively to $\vec{M}_A$).

As $\cos^2 \alpha_A = 3/5$ (see Equation (6.1)), the expression describing the emission anisotropy is:

$$r_o = 0.4(3 \cos^2 \lambda - 1)/2. \tag{6.3}$$

Notice that this r_o value, called 'fundamental emission anisotropy', depends only upon the angle λ determined by the moments of absorption and emission:

λ	0°	54.73^o	90°
r_o	0.4	0	- 0.2

Fundamental anisotropy measurements yield a determination of $\cos^2 \lambda$. In the case of fluorescence emission, if measurements are performed at different excitation wavelengths (polarization spectrum), it is possible to distinguish between $S_o \rightarrow S_1$ transition ($r_o > 0$) and $S_o \rightarrow S_2$ transition ($r_o < 0$). For phosphorescence emission r_o is always negative.

This loss of anisotropy from the maximum value of 0.4 may be considered as due to intrinsic causes in opposition to the extrinsic causes (excitation migration, Brownian rotations) which yield further changes of the emission anisotropy.

6.2.2 *Emission anisotropy for mobile molecules isotropically distributed*

If we consider a system in which the luminescent molecules undergo a motion during the lifetime τ of the excited state, the polarization state of the luminescence light will be dependent on the nature and amplitude of the motion.

Assume that during time t between absorption and emission, the emission transition moment $\vec{M}_E$ rotates through an angle $\Theta(t)$ (Fig. 6.5).

The application of the 'product rule' to the successive rotations α_A, λ, $\Theta(t)$ yields:

$$r(t) = (3\cos^2\alpha_E(t) - 1)/2 = ((3\cos^2\alpha_A(0) - 1)/2) \times ((3\cos^2\lambda - 1)/2)((3\cos^2\Theta(t) - 1)/2)$$

which can be written:

$$r(t) = r_o(3\,\overline{\cos^2\Theta(t)} - 1)/2. \tag{6.4}$$

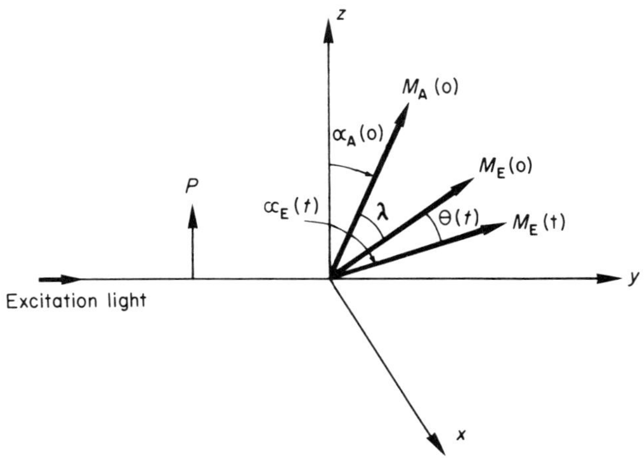

Fig. 6.5 Orientation of transition moments in the reference frame for mobile molecules.

The quantity $(3\ \overline{\cos^2\Theta(t)} - 1)/2$ corresponds to the orientation autocorrelation function of the luminescent molecule and reflects the type of motion undergone (isotropic, anisotropic...). It is frequently designed as $P_2(\cos\Theta(t))$, the second Legendre polynomial.

The various motional models that have been proposed for polymer chains will be described later and the resulting expressions of the emission anisotropy will be given there.

Equation (6.4) gives the emission anisotropy of the luminescence light emitted at a time t after an excitation at $t = 0$ by an infinitely short light pulse.

Under continuous illumination, the measured emission anisotropy yields a mean value, $\bar{r}$. Indeed, since luminescence is an exponential decay phenomenon, molecules will be emitting at various times and, consequently, after different degrees of rotation. Thus, the final average rotation will be the average of the rotations occuring over different intervals of time t, weighted by the contribution $f(t)$ that molecules emitting at time t makes to the total luminescence intensity. Hence:

$$\overline{\cos^2 \Theta} = \int_0^\infty f(t)\ \overline{\cos^2 \Theta(t)} dt$$

and: $$\bar{r} = \int_0^\infty f(t)\ r(t) dt.$$

The luminescence intensity corresponding to molecules emitting after time t, considering an exponential decay is:

$$I(t) = I_0 \exp(-t/\tau)$$

where I_0 is the luminescence intensity at $t = 0$. The total luminescence is:

$$I = \int_0^\infty I_0 \exp(-t/\tau) dt = \tau I_0$$

in such a way that:

$$f(t) = (1/\tau) \exp(-t/\tau)$$

and:

$$\bar{r} = (1/\tau) \int_0^\infty r(t) \exp(-t/\tau) dt$$

$$\bar{r} = (r_0/\tau) \int_0^\infty P_2[\cos \Theta(t)] \exp(-t/\tau) dt. \qquad (6.5)$$

If instead of the emission anisotropy, $\bar{r}$, the degree of polarization, $\bar{p}$, is used, the corresponding expression is:

$$[1/\bar{p} - 1/3]^{-1} = [1/p_0 - 1/3]^{-1}\{(1/\tau) \int_0^\infty P_2[\cos \Theta(t)]\}\exp(-t/\tau)dt. \qquad (6.6)$$

6.3 INSTRUMENTATION FOR STUDYING MOLECULAR MOBILITY OF ISOTROPIC SAMPLES

As previously noticed, fluorescence anisotropy can be studied either under continuous excitation or as a function of time after a light pulse excitation.

Some requirements for convenient measurements are common to the two techniques, in particular those related to the optical arrangement.

In the present part the typical equipment needed in each case is described and the main advantages as well as the limitations are pointed out for each type of measurement.

6.3.1 Continuous-excitation measurements

These are the easiest to perform and several experimental arrangements have been reported in the literature, using one or two analytical arms as shown in Fig. 6.6. Such equipment is presently commercially available from S.L.M. Instruments (810W Anthony Drive, Urbana Illinois, 61801, USA).

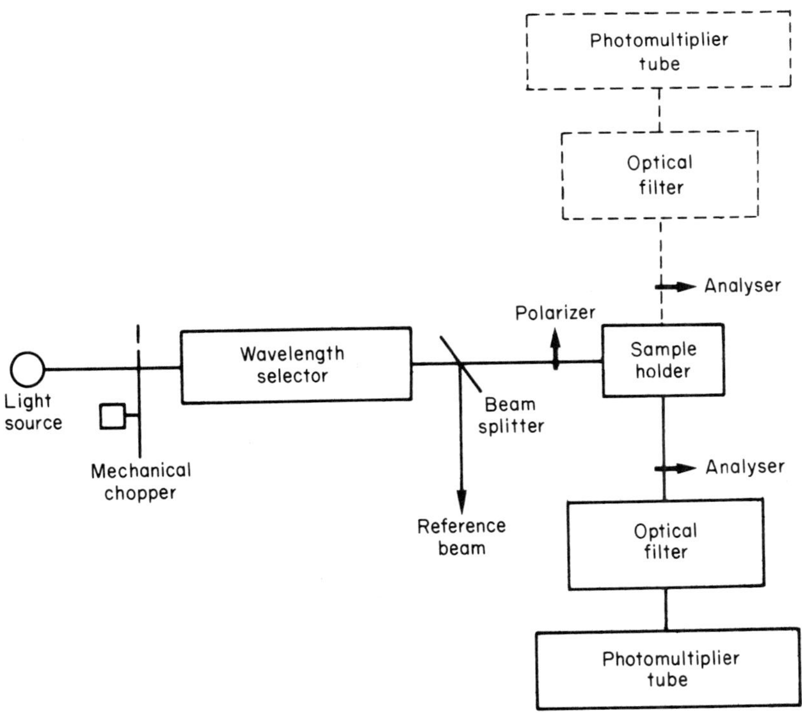

Fig. 6.6 Arrangement for luminescence polarization measurements

Some commercial companies manufacturing spectrofluorimeters have designed attachments for measuring fluorescence polarization; nevertheless the accuracy of measurements using them is not in general actually sufficient for their use for molecular mobility studies.

From the basic diagram shown in Fig. 6.6, some comments can be made about the choice of the various components (further details of components can be found in Parker [5]).

(a) Light source

Hg or Xe arc lamp is commonly used, the former gives high intensity in a limited number of spectral lines, the latter has a continuous emission but a lower intensity. In recent years, some equipment has been designed with a dye-laser as a source; the use of lasers will be described in Section 6.3.

(b) Wavelength selector

Different requirements exist for excitation or emission wavelength selection. With regard to excitation wavelength, either a narrow band width optical filter or a monochromator can be used. For emission, monochromators lead to a change in light polarization owing to reflection on mirrors and gratings. Such effects depend on wavelengths and can be modified by ageing. For correcting the experimental results one has to use a calibration curve, periodically tested. For this reason, optical filters, with broad band width covering the luminescence spectrum, are preferred. It is necessary to check that optical filters do not exhibit intrinsic luminescence excited by the incident wavelengths.

(c) Polarizers

Nicol or equivalent prisms can be used as well as dichroic films. Intrinsic luminescence must be tested. It is worth pointing out that a polarized excitation light is more interesting than a natural one for it leads to larger polarization values (x 1.7) and thus to a better accuracy of measurements.

(d) Electronic devices

In order to get a better accuracy, it is of interest to chop the excitation beam mechanically and to use lock-in amplifiers. Elimination of lamp-intensity fluctuations and long time range decrease can be achieved by means of ratiometers operating either on $I_{||}$ and $I_{\perp}$ and leading to $I_{||}/I_{\perp}$ or on $I_{||}$ and $I_{\perp}$ and a reference intensity I_{ref} in such a way that the two quantities $I_{||}/I_{ref}$ and $I_{\perp}/I_{ref}$ are obtained. The latter procedure is preferable, since total luminescence intensity $(I_{||} + 2I_{\perp})$ can be derived from these measurements. Such information is particularly important when the influence of temperature or plasticizer content is examined. Indeed, these factors can change the life-time of the luminescent excited state, which is reflected in the total luminescence intensity.

(e) Optical arrangement

For studying solutions, the most convenient arrangement is one with right-angle excitation and observation. Indeed, it avoids any trouble from excitation stray light.

For bulk polymer samples, the problem is more difficult to solve. Thus, an arrangement with a straight-through excitation and observation is attractive, for it only requires samples with two parallel smooth faces which can be easily obtained from compression moulding or solution casting and evaporation, but it is very difficult to get a good cut-off of excitation light. The perpendicular arrangement involves polymer samples with two perpendicular faces with smooth surfaces, which are difficult to achieve. The in-front illumination arrangement (Fig. 6.7(a)) commonly used for recording luminescence spectra of solid samples cannot be applied to polarization measurements for some polarization changes occur at the air-polymer interface, which yield wrong anisotropy values. A convenient attachment (Fig. 6.7(b)) has been developed recently in our laboratory in order to reduce such effects. It is based on the fact that air glass interfaces are perpendicular to light beams, whereas depolarization effects decrease at glass-polymer interfaces owing to closer indices of refraction. With such an attachment, only polymer samples with parallel smooth surfaces are required.

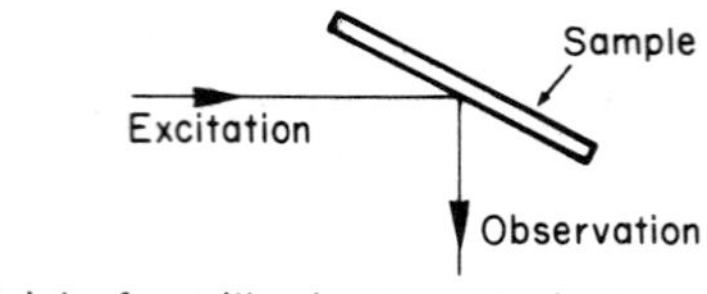

(a) In-front illumination attachment

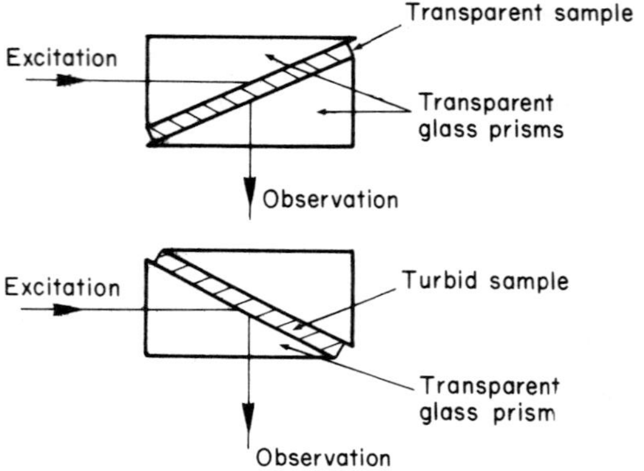

(b) Attachment for polarization measurements on solid samples

Fig. 6.7 Attachments for bulk polymer samples

For kinetics studies, the arrangement with two analytical arms is the more convenient for it leads to simultaneous measurements of $I_{||}$ and $I_{\perp}$. Nevertheless it can only be applied to solutions or transparent samples held with attachment shown Fig. 6.7(b).

Another apparatus has been described recently [6] in which a Wollaston prism is used as analyser. The partially polarized luminescent light entering such a prism is separated in two beams corresponding to $I_{||}$ and $I_{\perp}$, respectively, whose intensities are measured through two photomultiplier tubes. Thus, luminescence polarization can be continuously recorded for any type of sample (transparent or turbid with attachments; Fig. 6.7(b)). This analytical system has been combined with an optical microscope [6] or put on a stretching machine [7] for studying the change of molecular mobility of polymer chains during stretching.

(f) Phosphorescence
In such a case, of course mechanical shutters are required in addition to the above described equipment.

It is interesting to look further into information that can be obtained from continuous excitation measurement.

First, this technique yields the mean emission anisotropy $\overline{r}$:

$$\overline{r} = r_0 \quad \tau^{-1} \int_0^\infty P_2[\cos \Theta(t)] \exp(- t/\tau) dt$$

and requires independent measurements of the life-time, τ, and of the fundamental anisotropy r_0. This latter quantity, r_0, can be determined through measurements in a frozen medium (rigid polymer, organic glass) or a very viscous medium (glycerol or propylene glycol at low temperatures). As r_0 does not seem to depend too much on the surrounding medium, the value determined by such a procedure can be usually used for further experiments in other conditions. Secondly, no information can be obtained on the type of motion undergone by the luminescent molecule (isotropic, anisotropic see Section 6.4). Furthermore, only one characteristic of molecular mobility is available. If isotropic motion or orientation diffusion motional model are considered (see Sections 6.4.1 and 6.4.3), only one correlation time is involved and it can be derived from continuous excitation measurements. For motional models implying several correlation times, as anisotropic or flexible chain motions (see Sections 6.4.2 and 6.4.3), this method is less interesting for it leads only to an average correlation time.

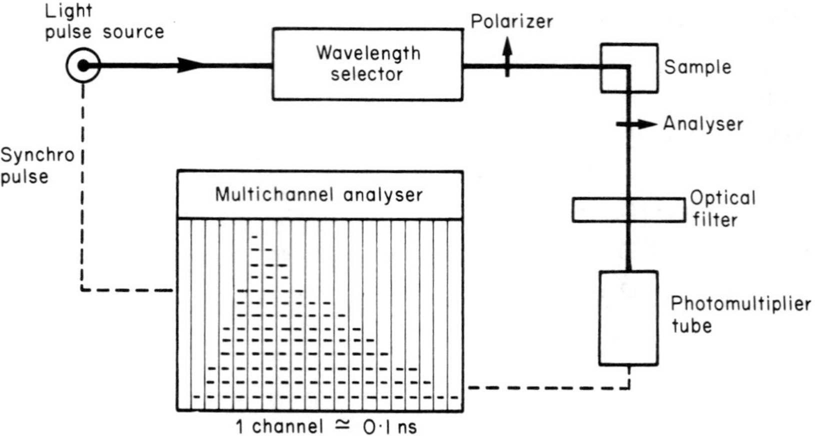

Fig. 6.8 Time-dependent fluorescence polarization equipment

6.3.2 Time-dependent measurements

The second technique deals with time-dependent measurements using the 'single-photon counting' technique of Chapter 2. A schematic diagram is shown in Fig. 6.8 for fluorescence. The spectrometer uses a nanosecond light pulser with a width of 2 ns and a typical flash rate of 20 000 pulses s^{-1}. The photomultiplier tube is used for photon counting and the delays between flash emission and the fluorescence photon arrival are sampled and stored in a multichannel analyser. Data corresponding to $I_{||}$ and $I_{\perp}$ are successively stored by changing the analyser direction at regular time intervals (typically a few minutes) in order to avoid the effect of long-time changes in flash intensity. More details concerning these time-dependent measurements are given in Chapter 2.

The remarks above on monochromators, straight-through or right-angle optical arrangements, sample holder attachments are still valid for time dependent measurements.

The main experimental difficulty in studying fluorescence anisotropy with time-dependent measurements comes from the fact that the time interval between excitation and emission is so short (10^{-10} to 10^{-7}s) that a deconvolution treatment is required for getting the emission anisotropy $r(t)$. Indeed, for an infinitely short flash, measured intensities would directly correspond to the fluorescence intensities sought, $I_{||}(t)$ and $I_{\perp}(t)$ from which $r(t)$ would be derived by:

$$r(t) = (I_{||}(t) - I_{\perp}(t))/(I_{||}(t) + 2\ I_{\perp}(t)) = D(t)/S(t).$$

Because the exciting flash has a width around 2 ns, one measures experimental functions $i_{||}(t)$ and $i_{\perp}(t)$ which lead to:

$$d(t) = i_{||}(t) - i_{\perp}(t),$$
$$s(t) = i_{||}(t) + 2\ i_{\perp}(t).$$

The functions $D(t)$ and $S(t)$ are related to these experimental functions by the following convolution integrals:

$$d(t) = \int_0^t D(t - T)\ g(T)\mathrm{d}T,$$
$$s(t) = \int_0^t S(t - T)\ g(T)\mathrm{d}T,$$

where $g(T)$ is the result of the flash-emission function and the response function of the measuring equipment (photomultiplier tube and electronic devices).

The resolution of above equations is the main difficulty in such a technique, and several deconvolution methods have been proposed.

It should be stressed that in contrast to the continuous-excitation technique, the time-dependent technique represents a very powerful tool for studying molecular mobility.

Indeed, in each experiment first the fluorescence life-time is obtained from the intensity decay ($I_{||}(t) + 2\ I_{\perp}(t)$), and secondly, the fundamental anisotropy r_0 is derived from the $t = 0$ limit of $r(t)$.

Furthermore, the orientation autocorrelation function $P_2[\cos \Theta(t)]$ is directly given by $r(t)$. This provides a unique means of probing the type of motion undergone by the luminescence molecule. Other molecular relaxation techniques as nuclear magnetic relaxation, electron spin relaxation, involve $P_2(t)$, but it is not possible to derive it from measured quantities.

Finally, when several correlation times are involved for describing molecular motions (see Sections 6.4.2 and 6.4.3) they can be determined from a fitting of the anisotropy decay curve.

6.3.3 *Laser excitation*

In order to get short light pulses and a higher light intensity, it is of interest to use a laser as an excitation source in time-dependent measurements. Such apparatus was designed a few years ago [8-10] and their use is becoming more frequent in spite of their very high cost.

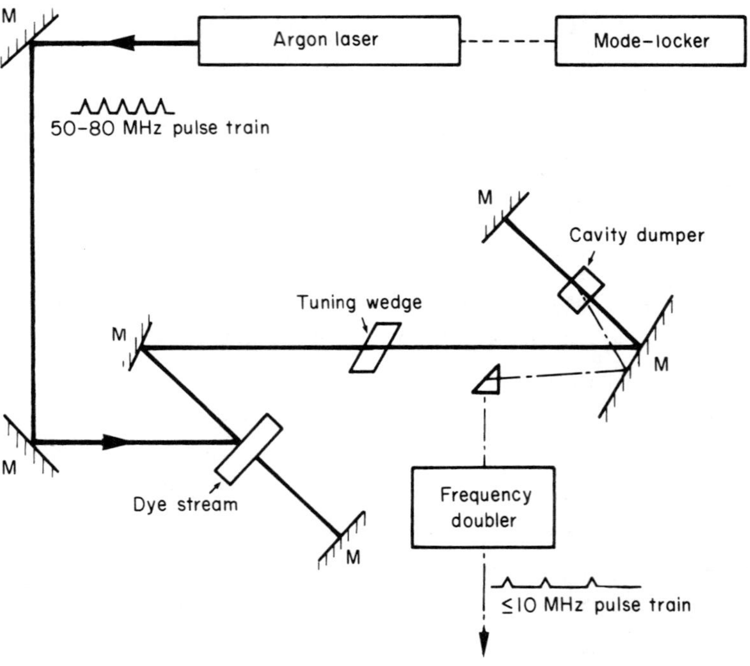

Fig. 6.9 Laser excitation equipment for time-dependent measurements [10].

A typical arrangement is shown schematically in Fig. 6.9. The argon laser combined with a mode-locker device yield a train of light pulses. Depending on the arrangement, the pulse width can vary from 30 to

200 ps with a repetition between 50 and 80 MHz. To get a range of wavelengths convenient for fluorescence excitation it is necessary to combine the argon laser with a tunable dye laser and a frequency doubler. It is possible to use directly this pulse-train excitation [8], but the short time interval between pulses leads to a large overlap of excitation and fluorescence emission. Performing a deconvolution in Fourier space allows the determination of fluorescence life-times as long as 30 ns [8].

To decrease the frequency of the excitation pulses, an additional device can be used, called a cavity dumper. The principle is to diffract, at some time intervals, a part of the light out of the laser cavity. For this purpose an acousto-optic Bragg cell (or a Pockel cell), monitored through a high-frequency acoustic wave (about 400 MHz) accurately synchronized to the mode-locker frequency, can decrease the pulse-train frequency down to less than 10 MHz and to extract till 50% of the instantaneous cavity light energy. Of course, such an arrangement requires very accurate optical alignments and electronic synchronization.

Time-correlated single-photon counting systems are associated to laser excitation equipment to obtain the convoluted fluorescence decay curve.

The main interest of such an excitation system is to lead to satisfactory measurements at very short times. Thus, fluorescence life-times as short as 70 ps have been determined [10].

6.4 MOTIONAL MODELS

The orientation autocorrelation function $P_2[\cos \Theta(t)]$ which is involved in the luminescence polarization technique, depends on the nature of the Brownian motion undergone by the luminescent molecule. Several models have been developed; they are briefly described hereafter, in pointing out the corresponding expressions for emission anisotropy.

6.4.1 Isotropic motion

This is the simplest model and corresponds to the Brownian motion of a rigid sphere, whose diffusion coefficient is D. For such a system:

$$P_2(t) = \exp(-t/\tau_{is}) \tag{6.7}$$

where τ_{is}, is the characteristic correlation time.

For a motion occuring at temperature T in a medium of viscosity η, the rotational diffusion coefficient D of a sphere with a hydrodynamic volume V, is expressed through:

$$D = kT/6 \quad \eta V.$$

The time-dependence of the emission anisotropy, $r(t)$ is a single exponential:

$$r(t) = r_o \exp(-6\,Dt)$$

whereas under continuous excitation, the mean anisotropy, $\bar{r}$, is given by:

$$1/\bar{r} = (1/r_o)(1 + 6\,D\tau) = (1/r_o)(1 + 3\tau/\tau_R) \tag{6.7}$$

and the degree of polarization under polarized excitation light $\bar{p}$ is expressed by:

$$(1/\bar{p} - 1/3) = (1/p_o - 1/3)(1 + 3\tau/\tau_R) \tag{6.8}$$

where τ_R is the time of rotatory diffusion:

$$\tau_R = (2\,D)^{-1}.$$

These expressions were first derived by Perrin [1] and used for measuring the fluorescence life-times τ of small molecules in solution, the hydrodynamic volumes of which being derived from viscosity studies.

It is worth noticing that accurate determinations of the correlation time τ_R requires that τ_R lies in the same range as the lifetime τ (i.e. 10^{-1} to 10^{-7} s). For τ_R>>/<<τ, r tends to r_0 and for τ_R>>/<<τ $\bar{r}$ is close to zero.

6.4.2 *Anisotropic motion*

The simplest model consists of an axis-symmetric ellipsoid. The autocorrelation function is composed of 3 exponential terms weighted by coefficients that depend on the angle β between the transition moment and the ellipsoid symmetry axis:

$$P_2(t) = A \exp(-t/\tau_A) + B \exp(-t/\tau_B) + C \exp(-T/\tau_C) \qquad (6.9)$$

where $A = (3 \cos^2 \beta - 1)^2/4 \qquad \tau_A^{-1} = 6 D_\perp$

$B = 3 \cos^2 \beta(1 - \cos^2 \beta) \qquad \tau_B^{-1} = D_\| + 5 D_\perp$

$C = 3(1 - \cos^2 \alpha)^2/4 \qquad \tau_C^{-1} = 4 D_\| + 2 D_\perp$.

In these expressions $D_\|$ and $D_\perp$ are the coefficients of the rotational diffusion about the symmetry axis and perpendicular to it.

6.4.3 *Motional models for flexible chains*

In the case of polymer chains, luminescence polarization reflects either side chain motions or mostly segmental motions relative to a few monomer units. The latter will occur in polymer solutions or in bulk polymers above their glass transition temperature.

Polymer dynamics are much more difficult than small molecule dynamics, in particular when dealing with segmental motions that occur at very short times. Nevertheless, during the last 10 years Monte Carlo approaches as well as analytical theories have been developed that yield a new insight on this area. The main results will be summarized hereafter.

(a) Description of local motions

In these models, the motion of a chain results from successive local jumps, each of them affecting a very small part of the chain. Successive jumps affect any part of the chain at random.

Such a model was first treated by Monte Carlo simulation on a cubic lattice [11] then on a tetrahedral lattice that corresponds to a more realistic picture [12]. The elementary motions of the chain inside the lattice are such that the bonds always coincide with the lattice edges, the connectivity of the chain bonds is preserved and the back-step is avoided. In the tetrahedral lattice, these conditions imply that within the chain, the smallest groups which can move include 3-bonds and 4-bonds (Fig. 6.10).

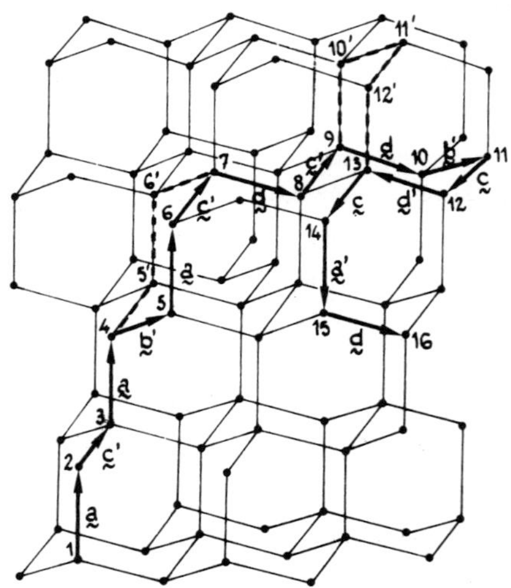

Fig. 6.10 Representation of a chain on a tetrahedral lattice. Three-bond motion: 4-5-6-7 (g^-); 4-5'-6'-7 (g^+). Four-bond motion: 9-10-11-12-13 (g^+g^-); 9-10'-11'-12'-13 (g^-g^+). (From Monnerie and Gény [12]).

It is worthwhile noting that a 3-bond motion does not create any new orientation but that it does make it possible to diffuse an orientation along the chain. On the contrary, in the case of a 4-bond jump, the orientations of the two internal bonds remain unchanged, whereas a new orientation is created for the first and fourth bonds. Thus, the 4-bond motion does not result in the diffusion of an orientation along the chain; rather it leads to a loss of orientation inside the chain.

Other motions involving a larger number of bonds have been proposed in polymers, such as the crankshaft motion [13] or the 'dopplekink' motion [14]. When looking at the effects of these motions on chain bond orientation, it follows that a crankshaft motion yields a diffusion of chain bond orientation along the chain sequence, whereas the 'dopplekink' motion results in a loss of orientation.

The reality of such motions with fixed ends is questionable, and departure from the tetrahedral lattice should be considered. Actually, such discrepancies from the lattice lead to a loss of correlation of orientation along the chain sequence, and are equivalent to a motion yielding a loss of orientation.

The most realistic simulation has been performed by Helfand *et al* [15] on a 200-bond chain, taking into account potential energy contributions from bond lengths, bond angles and internal rotation angles. Simulation results show that most motions correspond to change in the orientation of one bond, leading to a translation of the other part of the chain. Examples of such motions are shown in Fig. 6.11. Furthermore, there seems to be correlation of local motions which leads to a diffusion of bond orientation along the chain, though there is no evidence of direct 3-bond motions. The fluctuations of bond angles and internal rotation angles which are present yield a loss of bond orientation as do the discrepancies from the lattice.

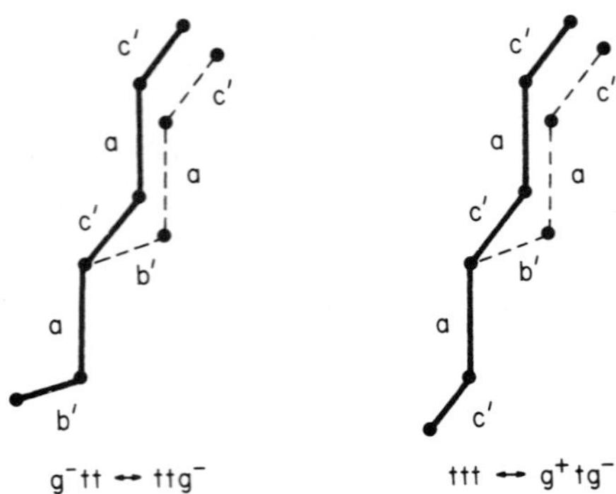

Fig. 6.11 Examples of one-bond motions.

(b) Analytical expressions of the orientation autocorrelation function

The first analytical treatment valid at short times has been derived using the jump model in the tetrahedral lattice [16].

When only local motions leading to a diffusion of bond orientation along the chain are present, the orientation autocorrelation function, $P_2(t)$ is expressed by:

$$P_2(t) = \exp(t/\rho)\ \mathrm{erfc}(t/\rho)^{1/2} \qquad (6.10)$$

where erfc is the complementary error function and ρ is the characteristic correlation time for 3-bond motion.

If additional processes giving a loss of bond orientation occur, they lead to an exponential term, $\exp(-\ t/\Theta)$, multiplying the previous expression. Thus:

$$P_2(t) = \exp(-\ t/\Theta)\ \exp(t/\rho)\ \mathrm{erfc}(t/\rho)^{1/2}. \qquad (6.11)$$

Since this early work other analytical expressions have been derived [17, 18], but they result in quite similar behaviour.

It appears from experiments that the orientation autocorrelation function of a chain bond does not depend on the precise description of local motions. The most important feature comes from the resulting effect of these motions on chain-bond orientation. From this point of view, two classes of motions should be considered:

1. The first class corresponds to any process leading to a diffusion of chain orientation along the chain. They all lead to a correlation function, such as $\exp(t/\rho)\ \mathrm{erfc}(t/\rho)^{1/2}$.
2. The second class deals with processes resulting in a loss of orientation without implying any conformational transitions of neighbour bonds. They result in an additional exponential term: $\exp(-\ t/\Theta)$.

It is likely that in real chains several molecular processes belonging to each class are involved in the orientation change of a chain bond. Thus, the generalized orientation autocorrelation function should be given by:

$$\exp(-\ t/\overline{\Theta})\ \exp(t/\overline{\rho})\ \mathrm{erfc}(t/\overline{\rho})^{1/2}$$

or one of the equivalent relations [17, 18].

Now $\overline{\rho}$ and $\overline{\Theta}$ would be the mean correlation times corresponding to molecular processes belonging to each class and defined by:

$$\overline{\rho}^{-1} = \sum_i \rho_i^{-1}$$
$$\overline{\Theta}^{-1} = \sum_j \Theta_j^{-1}.$$

(c) Emission anisotropy

The above expression of $P_2(t)$ (Equation (6.11)) leads to the following results for the luminescence polarization:

$$r(t) = r_o \exp(-\ t/\overline{\Theta})\ \exp(t/\overline{\rho})\quad \mathrm{erfc}(t/\overline{\rho})^{1/2} \tag{6.12}$$

$$(r)^{-1} = (r_o)^{-1}[1 + \tau/\overline{\Theta} + ((\tau/\overline{\rho})(1 + \tau/\overline{\Theta}))^{1/2}]. \tag{6.13}$$

6.5 LUMINESCENT PROBES AND LABELS

The use of luminescence polarization for studying mobility of polymers requires an extrinsic luminescent molecule. Indeed, many polymers do not exhibit any luminescence, owing to their chemical structure. Even when the polymer chain contains luminescent groups as does polystyrene or poly(vinylnaphthalenes), the energy-transfer processes occuring between these groups lead almost to a complete depolarization of the emitted light, denying any possibility of studying the chain mobility.

In the case of bulk polymers, the mobility can be studied through extrinsic probes incorporated in the polymer matrix. By contrast, for polymer solutions studies, the luminescence group must be covalently bound to the polymer chain; it will be referred to as a label. The most common luminescent probes and labels will be mentioned below.

In choosing a luminescent probe or label, several features have to be considered. First, it is important that the electronic transition between the ground state and the 1st excited state could be achieved without any excitation of another transition. This is required for an unambiguous definition of the transition moments involved. In particular, molecules in which $\pi - \pi$ and $n - \pi$ absorption bands overlap must be avoided, since these transitions correspond to absorption transition moments that are, in aromatic molecules for example, 'in-plane' and 'out-of-plane' polarized, respectively. Secondly, the fundamental anisotropy, r_0, must have a high value in order to yield easily measurable polarizations. In the same way, it is convenient to get a high quantum yield of luminescence that does not vary greatly over a wide range of changes in the properties of the medium (temperature, solvent nature, chemical structure of polymers, ...). Finally, the absorption and emission bands must occur in a spectral range suitable for experiments, typically above 300 nm.

Table 6.1. Some typical fluorescent probes with their absorption and emission wavelength range. Double arrows indicate the transition moments.

Probe	*Formula*	*Absorption range (nm)*	*Emission range (nm)*
Anthracene		300 - 380	380 - 500
Diphenylanthracene (DPA)	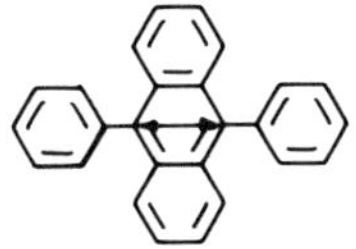	310 - 410	390 - 510
t. Stilbene		270 - 330	320 - 400
Diphenylbutadiene (DPBD)		290 - 360	360 - 460
Diphenylhexatriene (DPHT)		310 - 390	400 - 580
Diphenyloctatetraene (DPOT)		330 - 410	440 - 660

Table 6.2. Some typical phosphorescence probes with their emission wavelength range.

Probe	*Formula*	*Emission range (nm)*
Naphthalene		420 - 630
Fluorene		420 - 630
Biphenyl		400 - 560
Benzophenone	$C_6H_5-C(=O)-C_6H_5$	400 - 560
Acetophenone	$C_6H_5-C(=O)-CH_3$	370 - 560

6.5.1. *Probes*

Various probes can be used for fluorescence or phosphorescence polarization studies. Some of the most convenient ones are listed in Tables 6.1 and 6.2 with their optical characteristics.

In order to avoid any energy transfer between probe molecules, their concentration must be sufficiently low, usually less than 1 ppm. This concentration belongs to the range of concentration in which the luminescence intensity is linearly increasing.

6.5.2. *Labels*

Various types of labelling can be performed: at the chain end, along the main chain or on the side chain, depending on the chemical reaction used, and on the chemical structure of the polymer chain. For any labelling, one important feature concerns the position of the transition moments relative to the chain backbone. Indeed, it is necessary to bind the luminescent group covalently in such a way that rotations of its transition moments independently of the chain backbone should be avoided.

In the case of fluorescence, the most appropriate label is the anthracene group, for its optical properties are very convenient and many reactive derivatives are available for labelling. A review of the various ways of labelling a polymer chain with anthracene derivatives has been written by Krakoviak [19].

Only a few polymers have been labelled with phosphorescent groups, and they are reported in Rutherford and Soutar [20].

In this section, the main ways of labelling a polymer chain in various positions are described for the anthracene group. The various positions of anthracene are designated as follows:

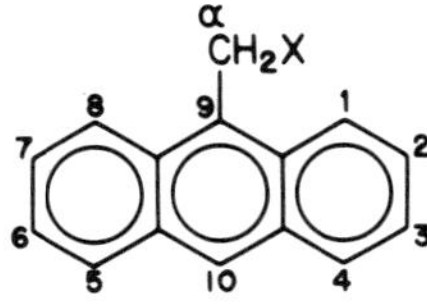

Most of them can be extended to other luminescent molecules.

(a) Side-chain labelling

With vinyl-type polymers, one easy way of putting an anthracene group in the side-chain is to perform a copolymerization with vinylanthryl monomers. Among them, 1- or 2-vinylanthracene can be introduced into polymers by free-radical or ionic polymerization, but this leads to labelled polymers (Schemes 1 and 2) in which the transition moment of the anthracene group can undergo rotations independently of chain motions. (Double arrows indicate the direction of the transition moment). The 9-vinylanthracene copolymerization gives rise to isomerization reactions that result in a mixture of side-chain and main-chain labelling (scheme 3), so that this derivative cannot be satisfactorily introduced. The best comonomer seems to be 9,10-styrilbenzylanthracene (scheme 4). Indeed, there are no side reactions during copolymerization and the direction of the transition moment is well defined relatively to the chain backbone, independently of rotational motions of the anthracene group around its linkage to the polymer chain.

Scheme 6.1:

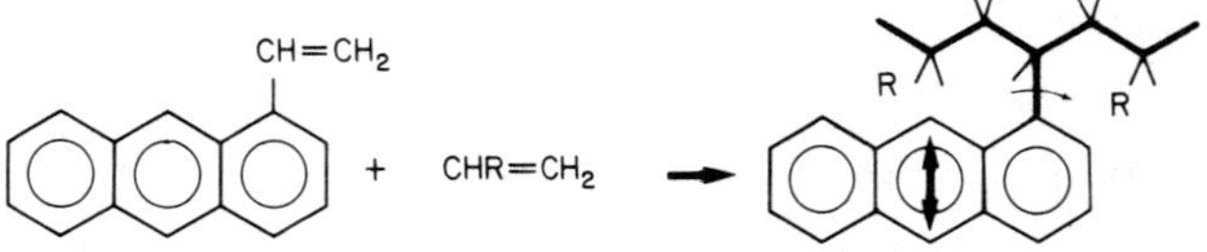

CH=CH$_2$ + CHR=CH$_2$ →

Scheme 6.2:

CH=CH$_2$ + CHR=CH$_2$ →

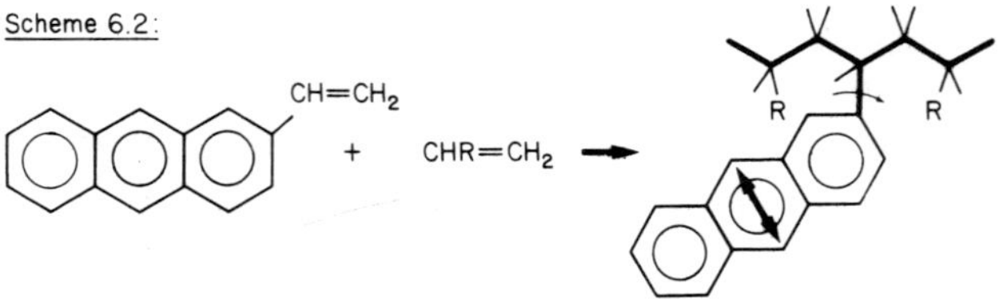

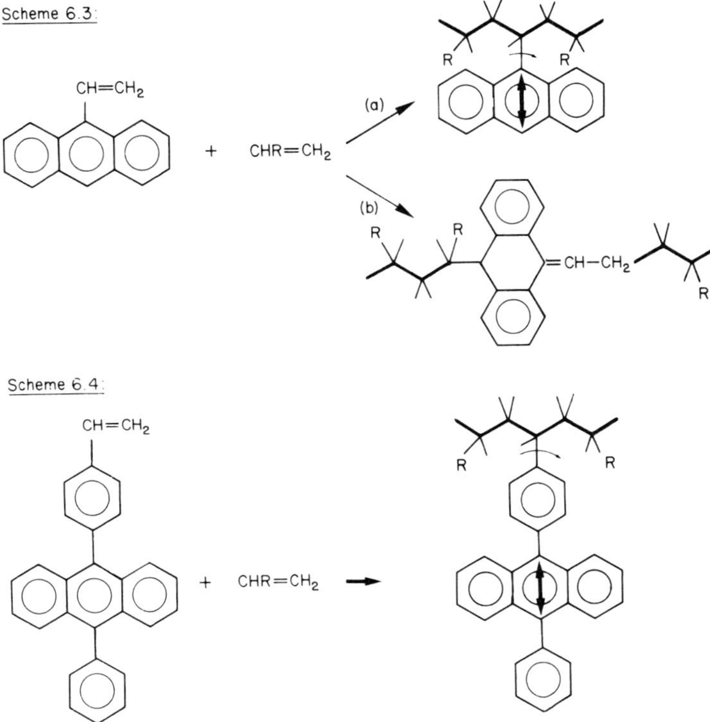

In the case of polymers bearing chemically reactive functions, such as COOH, OH, NH_2, anthracene derivatives such as 9-anthryl-diazomethane or 9-chloromethylanthracene can be used (schemes 5, 6), but they always imply a few bonds between the chain skeleton and the anthryl group.

One interesting case deals with the reaction of 9-anthrylcarbene with the C - H bond (scheme 7), which allows a labelling of polyethylene and polypropylene chains.

Scheme 6.5:

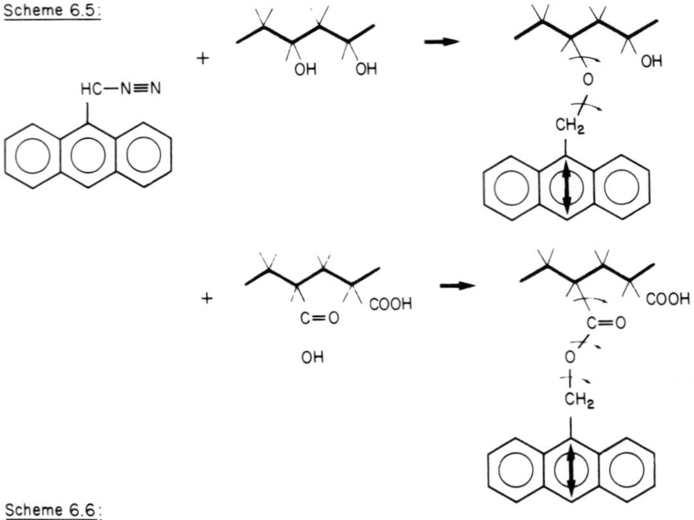

HC—N≡N

+ OH OH → OH O CH$_2$

+ C=O OH COOH → COOH C=O O CH$_2$

Scheme 6.6:

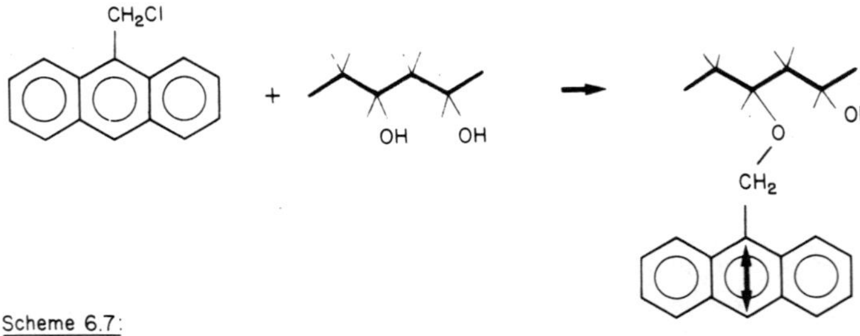

CH$_2$Cl

+ OH OH → OH O CH$_2$

Scheme 6.7:

:CH

H → CH$_2$

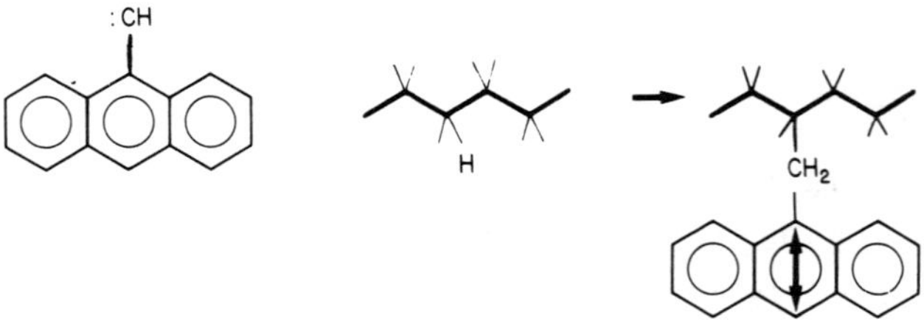

(b) Main-chain labelling

Though this labelling would be the most interesting for studying the intrinsic properties of the polymer chain, it is not very frequently used owing to experimental difficulties.

In the case of polymerizations, the best way, when possible, is to carry out an anionic polymerization and to deactivate the living chains with 9,10-(bisbromomethyl) anthracene (scheme 8). Polystyrene, polyisoprene and other polydienes have been labelled in this way.

When dealing with condensation polymers, such as polyesters, polyamides, some anthracene derivatives bearing ester, alcohol or amine groups in position 9,10 can be incorporated (schemes 9, 10, 11) by copolycondensation.

Scheme 6.8:

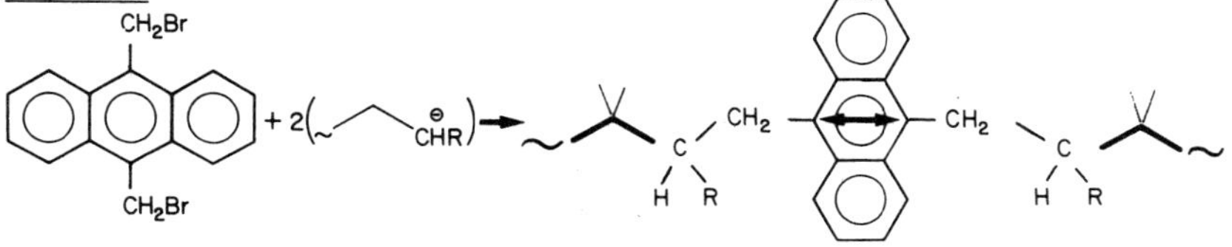

CH_2Br / CH_2Br + 2(~CHR⊖) → ~ C(H)(R) — CH_2 — anthracene — CH_2 — C(H)(R) ~

Scheme 6.9:

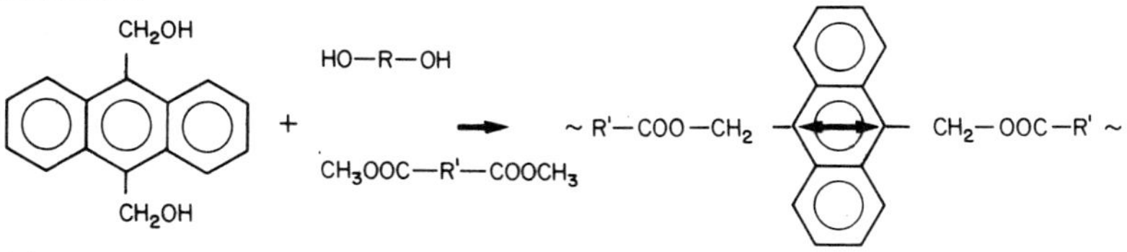

CH_2OH / CH_2OH + HO—R—OH, CH_3OOC—R'—$COOCH_3$ → ~ R'—COO—CH_2 — anthracene — CH_2—OOC—R' ~

Scheme 6.10:

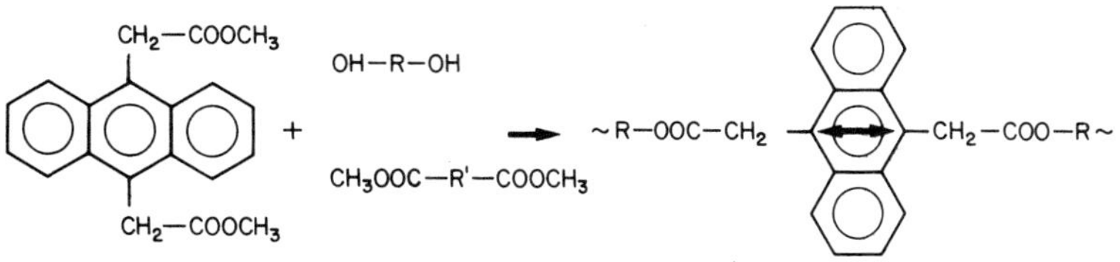

CH_2—$COOCH_3$ / CH_2—$COOCH_3$ + OH—R—OH, CH_3OOC—R'—$COOCH_3$ → ~R—OOC—CH_2 — anthracene — CH_2—COO—R~

Scheme 6.11:

CH_2NH_2 / CH_2NH_2 + NH_2—R—NH_2, HOOC—R'—COOH → ~R'—OC—NH—CH_2 — anthracene — CH_2—NH—CO—R' ~

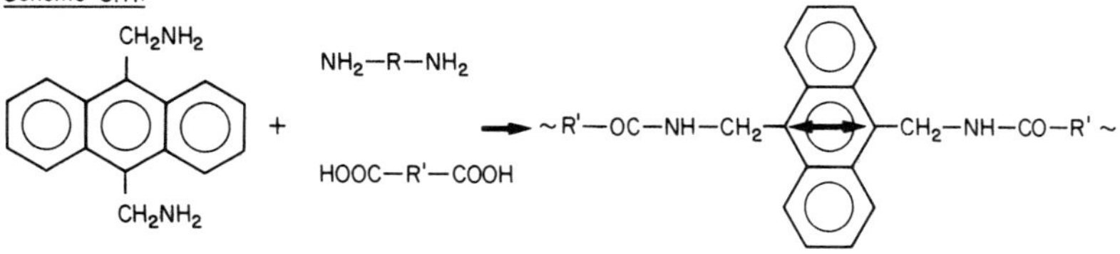

It is worth noting that in all these labellings, the transition moment of the anthracene group lies along the local axis of the chain backbone and cannot perform orientation changes independent of the main chain.

(c) End-chain labelling

With anionic polymerizations such a labelling can be achieved through deactivation with 9-bromomethylanthracene.

For polymers with functional end groups, such as polyethylene oxide, polyesters, polyamides, the labelling can be done by reacting with anthracene derivatives (scheme 12).

Scheme 6.12:

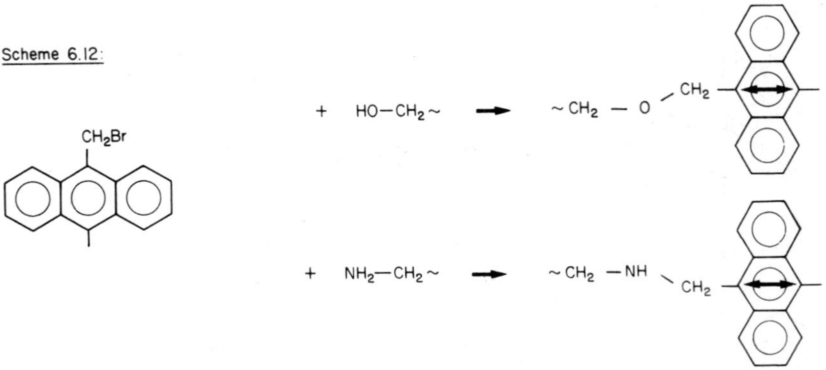

CH_2Br + HO—CH_2~ → ~CH_2 — O — CH_2—

+ NH_2—CH_2~ → ~CH_2 —NH — CH_2—

6.6 MOBILITY STUDIES OF POLYMER CHAINS IN SOLUTION

First, it has to be pointed out that only fluorescence polarization experiments can be performed in solution, since in such fluid conditions the triplet excited state is deactivated by radiationless processes, and is thus non-phosphorescent.

Second, the molecular mobility studied by fluorescence polarization deals with segmental motions of the polymer chain. Indeed, the polarization value does not depend on the polymer molecular weight over 20 000.

Owing to the different information that can be obtained from time-dependent or continuous-excitation measurements, the results corresponding to each technique will be separately presented.

6.6.1 *Time-dependent measurements*

Though this technique is the most powerful for analysing molecular mobility of labelled polymers, only a few papers have been published in this area [21-25]. Only the main results are presented below.

(a) Test of motional models

Polystyrene labelled in the middle of the chain by anthracene group with the following structure:

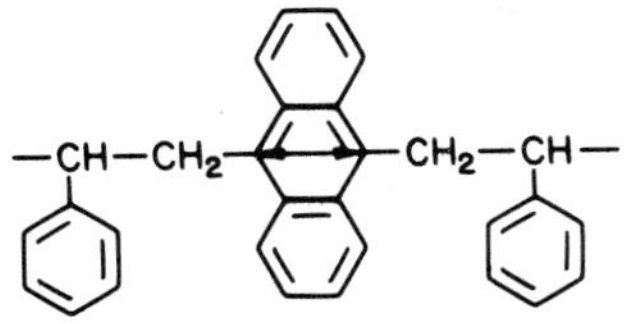

—CH—CH₂—(anthracene)—CH₂—CH— PS-A-PS

has been studied in solution in various mixtures of ethylacetate and tripropionin [23]. These solvents were chosen for they have very different viscosities but similar chemical nature in order to avoid any preferential solvation effect. As shown in Fig. 6.12 the motion is not isotropic ($r(t)$ would be monoexponential), but there is a good fitting of experimental data with the orientation autocorrelation function for flexible chains, leading to:

$$r(t) = r_0 \exp(-t/\Theta) \exp(t/\rho) \operatorname{erfc}(t/\rho)^{1/2}.$$

The validity of this autocorrelation function to account for dynamics of polymer chains in solution is also supported by other molecular relaxation techniques, such as ^1H and ^{13}C NMR dielectric relaxation.

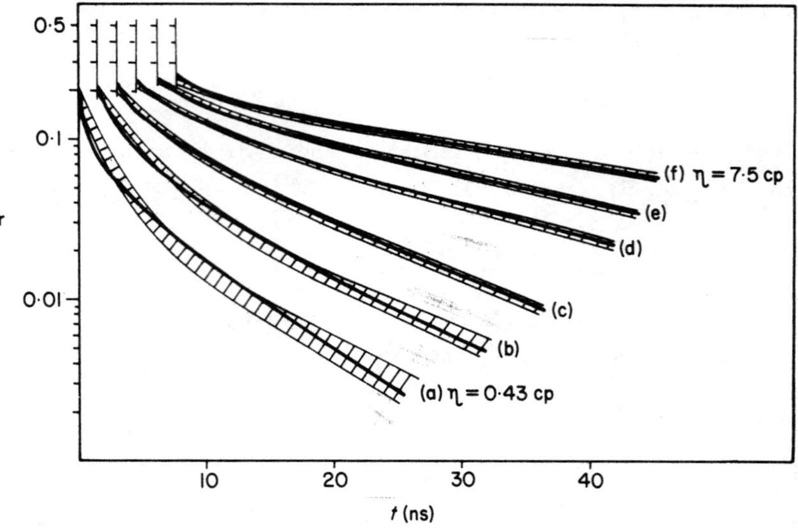

Fig. 6.12 Variation in anisotropy versus time for main-chain labelled polystyrene in various ethylacetate-tripropionin mixtures. The theoretical curves are drawn as full lines. For clarity, each curve has been shifted by 1.5 ns relatively to the preceding one. Viscosities (cp): (a) 0.43 (ethylacetate); (b) 0.89; (c) 1.50; (d) 2.88; (e) 4.35; (f) 7.47 (tripropionin.

(b) Effect of labelling position

In order to compare the mobility of anthracene group labelled to polystryrene chain at various positions, the following structures have been synthesized:

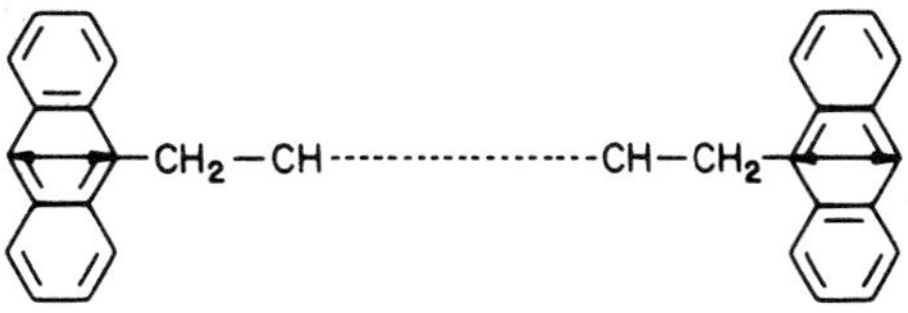

$-CH_2-CH$ .... $CH-CH_2-$ A-PS-A

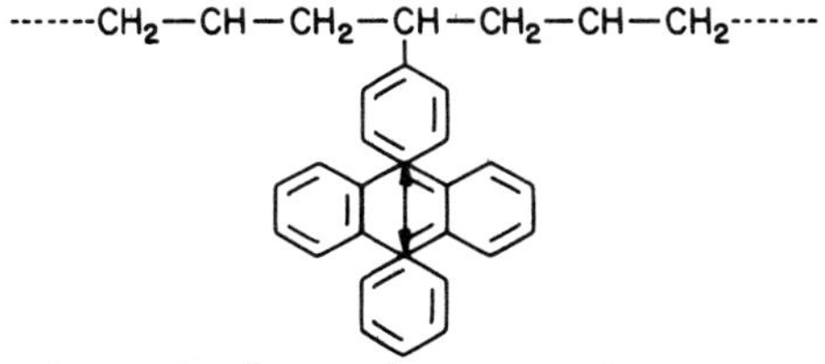

.......$CH_2-CH-CH_2-CH-CH_2-CH-CH_2$....... PS-DPA

For A-PS-A the emission anisotropy decay curves in ethylacetate-tripropionin mixtures can be fitted with the same autocorrelation function as for PS-A-PS, but with different Θ and ρ relaxation times [24].

By contrast, for PS-DPA, the fitting of $r(t)$ by this type of autocorrelation function is less satisfactory.

The Θ and ρ relaxation times for PS-A-PS and A-PS-A, as well as the mean correlation time $< \sigma >$ for the three polymers are reported in Table 6.3. As expected the end-chain motions are faster than the main-chain ones by a factor of 2 for Θ, 10 for ρ and 4 for $< \sigma >$. The PS-DPA behaviour is intermediate, indeed comparatively to PS-A-PS, the antracene group in the side position is sensitive to wagging motions of the main chain which are inefficient for changing the orientation of a transition moment of anthracene lying along the chain skeleton, as in PS-A-PS.

These data allow us to look at the influence of the solvent viscosity on the relaxation times. Thus, in Fig. 6.13 $< \sigma >$s corresponding to the various labelled polymers are plotted versus η. It appears that the relaxation time difference between PS-A-PS and PS-DPA slightly depends on η, whereas this difference strongly

Table 6.3 Correlation times Θ, ρ, $< \sigma >$ of polystyrenes labelled in main chain (PS-A-PS), at the ends (A-PS-A), in side-chain (PS-DPA) dissolved in various ethylacetate-tripropionin mixtures.

Tripropionion (%)	η_{cp}	PS - A - PS			A - PS - A			PS-DPA
		Θ	ρ	$< \sigma >$	Θ	ρ	$< \sigma >$	$< \sigma >$
0	0.43							
39.9	0.98	21.0	2.2	5.1	12.7	C.2	1.3	3.8
59.6	1.53	24.9	5.0	7.7	17.7	0.3	2.0	5.9
79.7	2.88	31.1	14.2	12.5	18.4	1.0	3.5	9.9
88.7	4.40	54.1	21.2	20.8	22.1	1.9	5.0	17.2
100	7.84	83.3	27.9	30.6	40.5	3.5	9.2	23.6

increases with η in the case of PS-A-PS and A-PS-A. In the latter comparison the main chain motions are four times more sensitive to the solvent viscosity than the chain end motions.

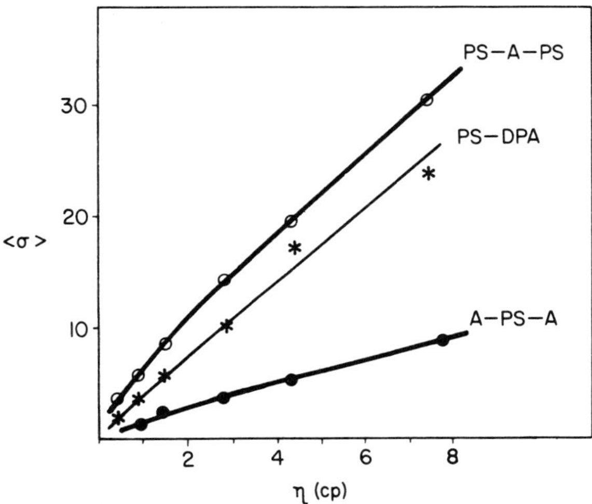

Fig. 6.13 Influence of viscosity at 25°C on the mean correlation times of polystyrenes labelled at different positions: main-chain (PS-A-PS), chain-ends (A-PS-A), side-chain (PS-DPA).

(c) Influence of solvent quality

The $r(t)$ curves have been determined for PS-A-PS in solution in a τ solvent (*trans*-decalin at 19°C) and in a good solvent (ethylacetate-tripropionin) of the same viscosity [25]. The relaxation times reported in Table 6.4 show a significant decrease of all of them in the good solvent solution in which the polystyrene chain is expanded.

Table 6.4. Correlation times of main-chain labelled polystyrene (PS-A-PS) in a Θ solvent (*trans*-decaline) and in a good solvent [ethylacetate (38%) tripropionin (62%) mixture]

Solvent	T(K)	η(cp)	< σ >(ns)	Θ(ns)	ρ(ns)
trans-decaline	291	2.25	14.5	44.7	10.6
	292	2.21	13.8	39.0	11.8
	294	2.11	12.9	35.2	12.1
	296	2.02	12.1	35.4	9.8
	298	1.95	11.2	31.1	10.1
	303	1.80	9.6	28.3	8.0
Ethylacetate - tripropionin	291	2.23	9.8	32.2	6.2
	292	2.19	9.7	28.3	7.6
	294	2.09	8.6	23.7	7.7
	296	2.05	8.1	23.9	6.2
	298	1.94	7.5	20.0	7.3
	303	1.87	6.8	21.3	4.7

6.6.2 *Continuous-excitation measurements*

Such measurements are the easiest to perform, but they lead to less information on polymer dynamics. In particular, only one mobility parameter can be determined. Nevertheless, by combining quenching and polarization experiments or by assuming a distribution of correlation times, it is possible to get information on the effect of various molecular characteristics of the chain structure.

(a) Combined quenching and polarization studies

The $\bar{r}$ value of the emission anisotropy upon continuous excitation is given by:

$$\bar{r} = (r_o/\tau) \int_o^\infty P_2(\cos \Theta(t)) \exp(- t/\tau)dt.$$

For isotropic motion, it leads to (see Section 6.4):

$$\bar{r}^{-1} = r_o^{-1}(1 + 3\ \tau/\tau_R)$$

and for flexible chains to:

$$(\bar{r})^{-1} = r_o^{-1}[1 + \tau/\Theta + ((\tau/\rho)(1 + \tau/\Theta))^{1/2}].$$

By adding a quencher to the polymer solution, τ can be changed in a continuous way and thus the shape of the autocorrelation function can be checked [26]. It only requires the use of a sufficiently efficient quencher in order to be able to change τ by more than one order of magnitude without appreciable change of the molecular environment.

Further, assuming a relaxation time distribution, it has been shown [27] that the mean relaxation time $< \sigma >$ is given by:

$$< \sigma > = \lim_{\tau \to \infty} \frac{\bar{r}\ \tau}{r_o} \tag{6.14}$$

Such a technique has been applied to polystyrene solutions by several authors [26 - 29]. Experimental results give unambiguous evidence that the segmental motion of a polymer chain is not isotropic, as proved by time-dependent measurements. A satisfactory fit has been obtained with the flexible chain autocorrelation function in the case of PS-A-PS [27] in 1,2-dichloroethane and chloroform solutions using carbon tetrachloride as a quencher, as shown in Fig. 6.14, in which $1/\bar{r}$ is plotted against the ratio of the fluorescence intensities I and I_o with and without quencher. The $< \sigma >$ values obtained from these experiments are 4.3 and 3.7 ns for 1,2-dichloroethane and chloroform solution, respectively, in excellent agreement with those obtained by time-dependent measurements, namely 4.2 and 3.8 ns, respectively.

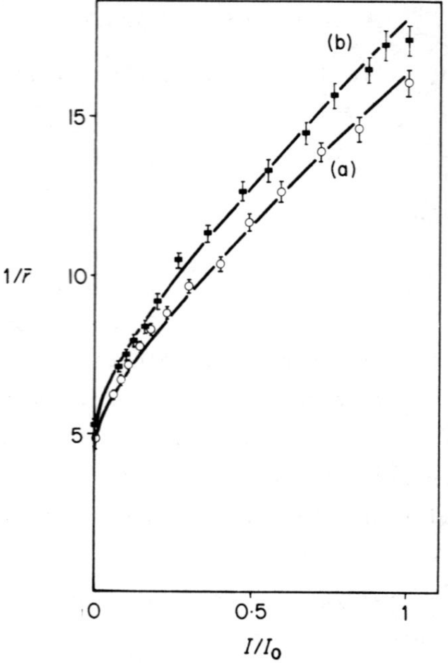

Fig. 6.14 Quenching of labelled polystyrene solutions by carbon tetrachloride. (a) 1,2-dichloroethane; (b) chloroform (after Valeur and Monnerie [27]).

(b) Combined changes of viscosity and polarization measurements

This procedure, first developed by Weber [2], has been extensively used by Anufrieva and Gotlib for studying the intramolecular mobility of various polymer systems. A review of these studies has been made [30], and hereafter only the basic principles and some typical results will be presented.

(i) Principles of measurements. The orientation autocorrelation function, $P_2(t)$ of a luminescent group covalently bound to a polymer chain is described by a spectrum of relaxation times $\tau_j/3$:

$$P_2(t) = \sum_j f_j \exp(-3\, t/\tau_j)$$

where f_j are relative weights of processes with times τ_j.

For continuous excitation with polarized light, this auto-correlation function yields the following expression for the emission anisotropy, $\bar{r}$:

$$\bar{r}^{-1} = r_o^{-1} \{\sum_j f_j/(1 + 3\,\tau/\tau_j)\}^{-1}$$

where τ refers to the life-time of the excited state. For simplicity, it is convenient to introduce a reduced measure of polarization changes:

$$Y = \frac{r_o}{\bar{r}} = \frac{1/p - 1/3}{1/p_o - 1/3} \qquad (6.15)$$

in which p is the polarization ratio.

In the case of polymer chains with internal rotation immersed in a viscous solvent with viscosity $\eta \sim 0.01\ P$ and undergoing segmental motions, relaxation processes with characteristic times $\tau_j > 10^{-9}$ s obey the condition of high friction [31] in such a way that the corresponding relaxation times are expressed by:

$$1/\tau_j = a_j(T/\eta)\ \exp(-\ U_j/kT)$$

where U_j are barriers to the internal motion in the chain or in side groups.

Consequently, Y becomes a function of (T/η). The general shape of $Y(T/\eta)$ is a curve convex toward the positive ordinate, as shown in Fig. 6.15 (a). If $U_j = 0$, then Y is a unique function of T/η and isotherms $Y(T/\eta)$ for different temperatures coincide when the solvent viscosity is varied. If $U_j \neq 0$, the dependences $Y(T/\eta)$ obtained when the temperature is varied will differ from those obtained by changing η at constant temperature. The latter case is usually found for polymer solutions.

Though the distribution of relaxation times could not be derived from the experimental curves $Y(T/\eta)$, some characteristic averaged parameters of the relaxation spectrum can nevertheless be obtained

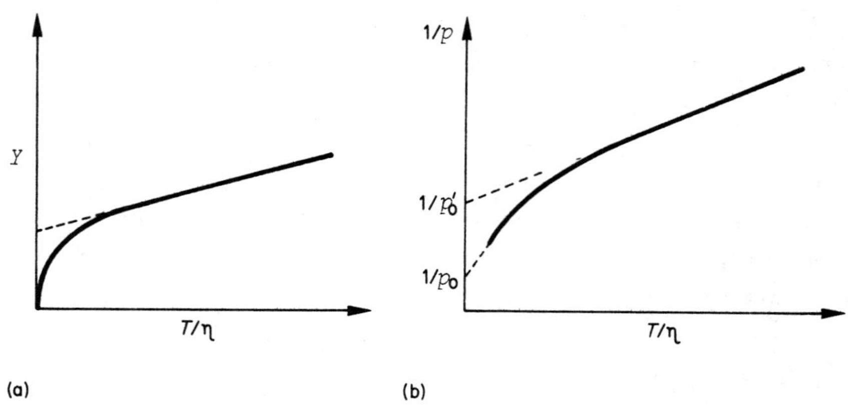

Fig. 6.15 Typical curves showing the influence of T/η on (a) the reduced polarization Y defined in the text and (b) the reciprocal polarization P^{-1}.

from the behaviour of $Y(T/\eta)$ over a limited range of T/η values [2]. In particular, it is possible to obtain the expression of the asymptotic dependence of Y for low viscosity solvents at $T/\eta \to \infty$, where the condition $\tau/\tau_j > 1$ is obeyed even for the longest τ_j:

$$Y \simeq <\tau^2>/<\tau>^2 + 3\,\tau/<\tau> - [<\tau><\tau^3> - <\tau^2>^2]/3\,\tau<\tau>^3 + \ldots$$

where
$$<\tau> = \sum_j f_j\,\tau_j$$
$$<\tau^2> = \sum_j f_j\,\tau_j^2$$
$$\tau_w = <\tau^2>/<\tau> \quad \text{(weight average relaxation time).}$$

Since the dependence of τ_j s is linear in T/η, the asymptotic behaviour of Y can be written:

$$Y_{T/\eta\to\infty} = Y'_o + C\,T/\eta \tag{6.16}$$

$$Y'_o = <\tau^2>/<\tau>^2 = \tau_w/<\tau>. \tag{6.17}$$

It is more convenient to plot the polarization data in a different way (Fig. 6.16 (b) Thus, $1/p$ is plotted against T/η and the linear asymptote at $T/\eta \to \infty$ intersects the ordinate at $1/\ _o'$. A new quantity Y' is defined by:

$$Y' = (Y - Y_o')/Y_o' = (1/p - 1/p_o')/(1/p_o' - 1/3) \qquad (6.18)$$

The linear asymptotic behaviour of which at $T/\eta \to 0$ yields:

$$Y' \simeq 3\ \tau/\tau_w. \qquad (6.19)$$

Consequently, the value of the weight average relaxation time τ_w can be derived from the slope of the asymptotic linear part of the curve Y' versus T/η.

It should be noticed that this result implies that $\tau/< \tau >> 1$ and $\tau/\tau_w > 1$. In practice the linear dependence $Y(T/\eta)$ at high values of (T/η) over a finite range of (T/η) is also observed at values of $< \tau >$ and τ_w comparable to τ. In this case, only a pseudo-asymptote is reached, which leads to a weight average $(\tau_w)_{eff}$ only over the slowest processes obeying the condition $\tau/\tau_j > 1$.

In practice, the corresponding procedure consists of measuring the polarization ratio at constant temperature for various viscosities then of plotting $1/p$ against (T/η) and determining $(\tau_w)_{eff}$ from the slope of the linear asymptote of Y' at large (T/η) values. Changes of viscosity are achieved by mixing solvents with similar chemical nature but different viscosities, for example: methyl or ethyl-acetate-triacetin; methanol-glycerol; water-sucrose; hexane, heptane, octane; toluene-decaline or cyclohexane.

(ii) Mobility of vinyl polymers. Values of τ_w for some polymers are listed in Table 6.5. It clearly appears that a methyl group in α position yields a significant decrease of chain mobility. Furthermore, in side-chains a phenyl ring increases the intramolecular hindrance more than an ester group.

The effect of the length of side chains has been studied for poly alkylmethacrylates of the following structure:

Table 6.5 Weight average relaxation time, τ_w, of various vinylpolymers in solution.

Polymer	*Solvent*	τ_w (ns)
Poly(methylmethacrylate)	Methyl acetate	3.9
Poly(methylacrylate)	Methyl acetate	< 1
Poly(methacrylic acid)	Methanol–water (60:40)	6.5
Poly(acrylic acid)	Methanol–water (60:40)	3.2
Poly(methyl styrene)	Toluene	8.6
Polystyrene	Toluene	5.1

$$
-CH_2-\underset{\underset{\displaystyle O-C_nH_{2n+1}}{|}}{\underset{\displaystyle CO}{\underset{|}{\overset{\overset{\displaystyle CH_3}{|}}{C}}}}-
$$

τ_w gradually increases from 2 ns (n = 1) to 10.5 ns (n = 22).

In the case of random copolymers a smooth mobility change with composition is observed for methylacrylate-methyl methacrylate copolymers, whereas the mobility goes through a maximum for styrene-α-methylstyrene copolymers (maximum occurs at 30% α-methylstyrene).

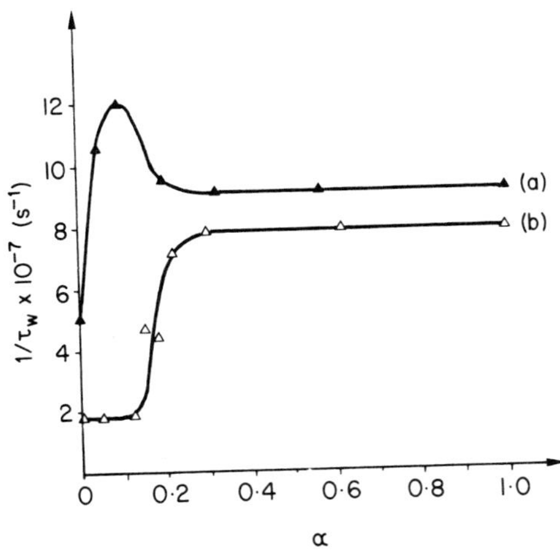

Fig. 6.16 Effect of the degree of ionization of carboxyl groups, α, on the weight average relaxation time, τ_w: (a) polyacrylic acid; (b) poly (methacrylic acid) (after Anufrieva *et al* [30]).

(iii) Mobility of polyelectrolytes. The fluorescence polarization technique is very sensitive to changes in the internal structure of polymers in solution. Any conformational change of polymer chain is reflected in the polarization ratio. Such sensitivity has been largely used in biology for studying helix-coil transitions of labelled proteins. In the field of synthetic polymers, an interesting example deals with the ionization effect on mobility of polyelectrolytes like polyacrylic acid and polymethacrylic acid. Fig. 6.16 shows the change of $1/\tau_w$ with the degree of ionization α for these two polymers. The sharp mobility change occuring at $\alpha = 0.2$ for PMA is in agreement with the observed increase in viscosity and corresponds to a transition from a globular conformation with hydrophobic groups inside (at low α) to a more expanded coiled conformation at higher α. For PAA at low α, intramolecular hydrogen bonding between carboxyl groups could exist and would be destroyed at higher α values, leading to a more flexible chain.

It has to be pointed out that many other studies have been performed by Anufrieva *et al* [30].

6.7 MOLECULAR MOTIONS IN BULK POLYMERS

That very few studies have been performed on mobility in bulk polymers may be due to the larger difficulties encountered in measurements on bulk samples. Indeed, luminescence polarization values can be affected by several causes as, for example, residual birefringence in the glassy state, development of voids or bubbles during heating or creep, leading to a deformation of the sample surface when measurements are performed far from the glass-rubber transition temperature for a too long time.

6.7.1 Fluorescence polarization

Owing to the short fluorescence lifetime (a few nanoseconds), only high-frequency motions ($10^8 - 10^{10}$ Hz) can be observed.

The first quantitative study was carried out on polyisoprene networks [32] using either probes free in the polymer matrix or labelled polymer chains. The main results are presented here.

Four probes of varying sizes were employed, their formulas and transtion moments being shown in Table 6.1. Cross-linked polyisoprene samples were swollen with solutions of the probes in benzene and then dried *in vacuo*. The final concentration in bulk samples was approximately 2 x 10^5 M.

Fluorescence polarization measurements were performed under continuous excitation and, for convenience, a mobility parameter, m, is introduced:

$$m = (r_o/\bar{r}(\tau, T)) - 1. \tag{6.20}$$

m increases from zero (molecular relaxation time much larger than the lifetime) to infinity (relaxation time much smaller than τ).

The plots of m against temperature for the probes quoted in Table 6.1 are shown in Fig. 6.17. For the DP probes, the onset of mobility occurs at about - 10°C, that is 50°C above the usual glass-rubber temperature T_g measured at 1 Hz. Such a shift agrees with what can be expected from the time-temperature superposition principle [33]: the higher the frequency of the experimental technique, the higher the temperature at which the transition is observed.

For the four probes, fluorescence anisotropy decay experiments at room temperature show that $M(t)$ is remarkably monoexponential and that the simple isotropic model (see Section 6.4.1) is valid. The characteristic correlation time, τ_{is}, can be calculated as a function of temperature from m and τ, through the relation:

$$\log \tau_{is} = \log \tau - \log m. \tag{6.21}$$

From the plots of log τ_{is} against $1/T$, shown in Fig. 6.18, it is possible to check more quantitatively if the molecular motions which are responsible for fluorescence depolarization belongs to the glass-rubber relaxation of polyisoprene. Indeed, the glass-rubber relaxation theory, based on the free-volume concept [33], predicts a temperature dependence of relaxation times which is given by the equation:

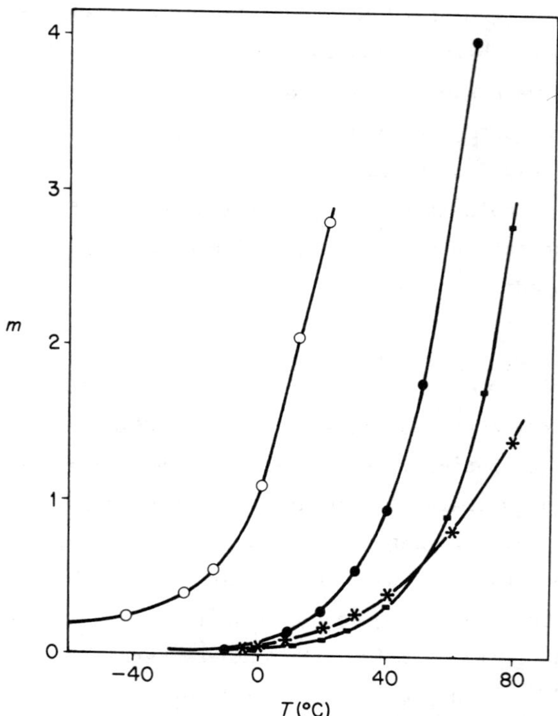

Fig. 6.17 Temperature dependence of the mobility parameter, *m*, for the various probes in polyisoprene: DPBD (✳), DPHT (●), DPOT (■), DMA (○) (after Jarry and Monnerie [32]).

$$\log \tau_{is} \text{ (or } \log \rho) = -\frac{A(T - T_g)}{B + T - T_g} + \text{constant} \qquad (6.22)$$

where $T_g = -60^{\circ}C$ is the glass-transition temperature of polyisoprene as measured by differential thermal analysis. Mechanical data [34] in the low frequency range for the same polymer lead to the values $A = 16$, $B = 57$. The corresponding curve is plotted as a dotted line in Figure 6.18. The slopes of $\log \tau_{is}$ for the three DP probes are in good agreement with that derived from Equation (6.22).

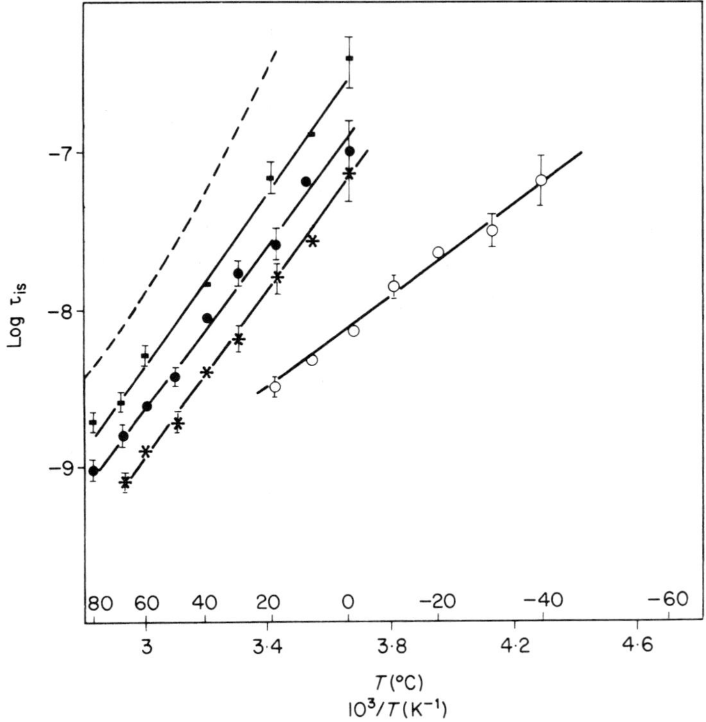

Fig. 6.18 Logarithmic plot of the isotropic correlation time, τ_{is}, versus $1/T$ for the four probes in polyisoprene (symbols identical to Fig. 6.17). The dotted line represents log τ_{is} + 3, calculated from Equation (6.22) (after Jarry and Monnerie [32]).

Such an agreement means that these large probes reflect molecular motions of the polymer that are involved in the glass rubber relaxation.

From Fig. 6.17 and 6.18 it is apparent that the DMA behaviour is quite different from those of the DP probes. It seems that, because of its smaller size, the mobility of DMA is related to the secondary transition of polyisoprene occuring around - 160°C at 1 Hz. A quite similar probe-size effect has been reported in ESR studies of the rotational mobility of nitroxide probes.

In order to get direct information on the chain mobility, fluorescent molecules covalently bound to the polymer main chain have been used. Anionic polyisoprene chains have been deactivated by 9,10-bis(bromomethyl)anthracene in such a way that the transition moment of the dimethylanthracene fluorescent group lies along the local chain axis (see scheme 8). Thus the fluorescence polarization only reflects the motions of the backbone. The labelled polymer (1%) and purified polyisoprene (99%) were mixed in solution before being cross-linked by dicumylperoxide. The final concentration of DMA in bulk samples was 10^{-5} M.

For the DMA label, the temperature dependence of the mobility parameter m is very similar to that of the DP probes. On the other hand, $r(t)$ is not a single exponential, in such a way that the relaxation times have been derived by using the flexible chain model with diffusive propagation of the segmental orientations along the chain backbone (see Section 6.4.3). The emission anisotropy is expressed as:

$$r(t) = \exp(t/\rho)\ \mathrm{erfc}(t/\rho)^{1/2} \tag{6.23}$$

where ρ is the characteristic correlation time, derived from continuous illumination measurements by the relation:

$$\log \rho = \log \tau - 2 \log m. \tag{6.24}$$

The plot of log ρ against $1/T$ is shown in Fig. 6.19 with the dotted line corresponding to Equation (6.22). A rather good agreement in the slopes is obtained between + 20 and + 80°C. However, in the low mobility range lying between - 10 and + 20°C, the slope of log ρ is smaller than that predicted. This disagreement could be due to the occurence of a secondary transition that could prevail over the primary glass transition.

Further information can be obtained by comparing label and probe mobilities. Indeed, since the autocorrelation $M(t)$ is most

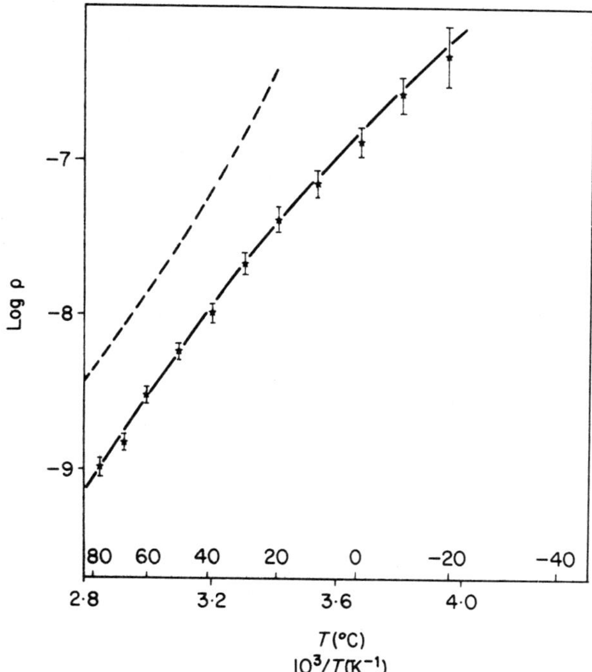

Fig. 6.19 Logarithmic plot of the diffusion correlation time, ρ, versus $1/T$ for labelled polyisoprene. The dotted line represents log ρ + 3 calculated from Equation (6.22) (after Jarry and Monnerie [32]).

sensitive to the smallest relaxation times, the characteristic time ρ of the DMA label principally describes the smallest segment motions. Besides, on the basis of the free-volume theory, the length a_s of the smallest segments involved in the glass-rubber relaxation should be comparable to that of a probe, whose relaxation time equals ρ. By comparing with the DP probes, we obtained $a_s \simeq 14$Å, which represents about five monomers in average conformation.

Very recently similar results have been observed in a series of elastomers: polybutadiene, butadiene-styrene random copolymers with various styrene contents and different butadiene micro structure.

All the results show that in the case of elastomers, the free-volume theory is in quantitative agreement with the temperature dependence of the correlation times of either sufficiently long probes or labels, which means that this theory can be applied to the high frequency region (10^8-10^{10} Hz) with the same parameters as those determined from mechanical measurements at low frequency (1-10^2 Hz).

Note that such a quantitative test requires fluorescence polarization measurements at temperatures higher than (T_g + 70°). For many polymers this corresponds to a temperature range in which fluorescence intensity is very weak.

6.7.2 Phosphorescence polarization

Before going to polarization studies, it is interesting to point out that, contrary to what happens with fluorescence, the phosphoresence intensity, involving the triplet excited state, is very sensitive to changes in the surrounding medium as well as to small amounts of oxygen quencher. Such effects have been used for observing secondary transitions of various polymers in which phosphorescent probes or labels are present [20, 35-37]. There is good agreement between observed transitions and those obtained by mechanical or dielectric measurements.

Although the phosphorescence polarization method was proposed many years ago [38] for studying molecular motions in a frequency range closer to mechanical experiments (0.1 - 10^2 Hz), such measurements have been achieved only very recently [20, 36, 37]. Indeed difficulties arise firstly, from very weak phosphorescence intensities available at moderate and high temperatures, secondly from the fact that in aromatic phosphorescent molecules the emission transition moment has a 'perpendicular' orientation relative to that of absorption, which results in negative and smaller fundamental polarization (Section 6.2.1). Furthermore, when mobility appears, the polarization ratio becomes closer and closer to zero, leading to very difficult measurements.

Experiments have been performed on both poly(methylacrylate) [36] and poly(methyl methacrylate) (PMMA) [20]. Only the latter results are presented here. Probes were naphthalene (N) and acenaphthene (A) and labelling was achieved by copolymerization with 1 and 2-vinyl-naphthalene (P1VN and P2VN), with 1 and 2-naphthylmethacrylates (P1NMA and P2NMA) and with acenaphthylene (PACE). The various modes of attachment of phosphorescent labels are shown on Fig. 6.20.

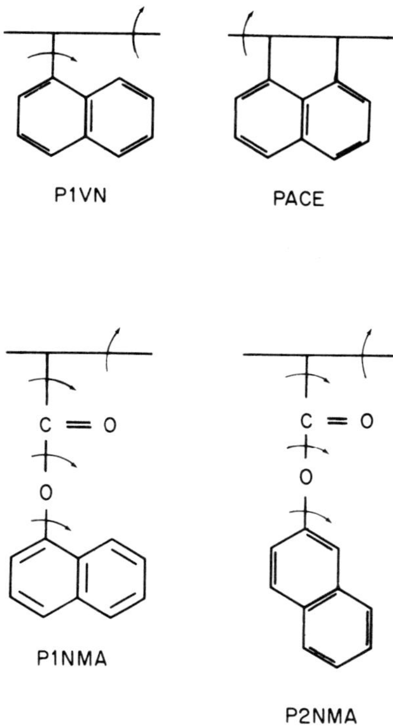

Fig. 6.20 Different types of phosphorescent labelling.

Graphs of phosphorescence intensities against temperature provide evidence of transitions that do not depend on whether the molecule is a dispersed probe or a label. The various transitions have been identified to the α, β, γ transitions observed in mechanical relaxation.

Polarization ratio is plotted against temperature for the various systems in Fig. 6.21. It has to be noticed that in the case of

P1NMA the emission is completely depolarized at all temperatures, probably owing to the angle between absorption and emission transition moments (at 55° complete depolarization is expected). The negative P_o values observed are in agreement with the fact that the $T_1 \rightarrow S_o$ transition is out-of-plane polarized. Variations in P_o between labels has been ascribed in part to the effects of motions that could occur during the phosphorescent lifetime.

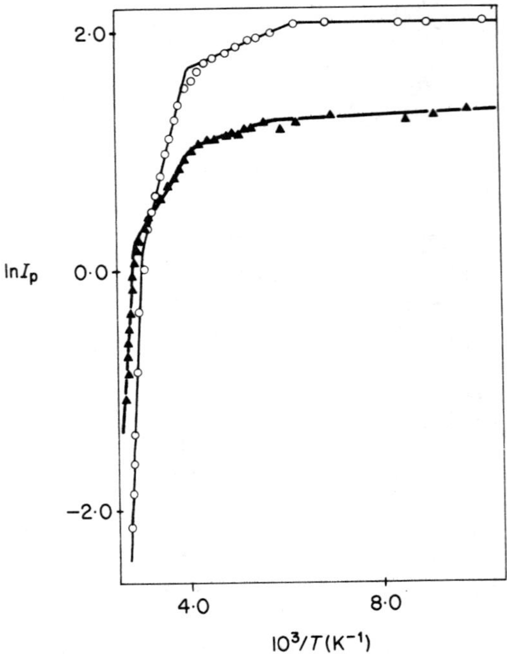

Fig. 6.21 Temperature dependence of the phosphorescence intensity for labelled PMMA (, P1NMA; PACE) (after Rutherford and Soutar [20]).

Assuming an isotropic motion for probes or labels, the correlation time τ_{is}, can be derived from polarization and life-time measurements. The corresponding results are shown in Fig. 6.22 and the transition temperatures as well as the activation energies are reported on Table 6.6. It is apparent from Fig. 6.22 that the temperature at which rotational motion occurs varies from one probe to another. Nevertheless rotational motions of labels and probes are absent at temperatures below that of the β relaxation in PMMA.

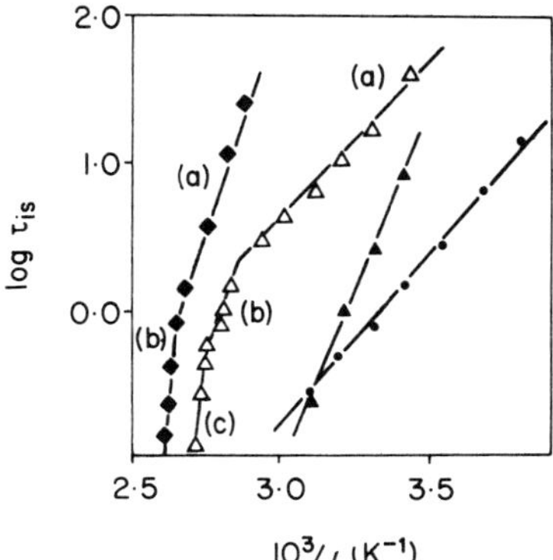

Fig. 6.22 Logarithmic plot of the isotropic correlation time, τ_{is}, versus $1/T$ for PMMA with phosphorescent probe (◆ acenaphtene) or phosphorescent labels (△, P1VN; ▲, PACE; ●, P2NMA) (after Rutherford and Soutar [20]).

For the phosphorescent probes, the activation energy is lower than that obtained by mechanical and dielectric relaxation for the β relaxation (70 - 90 kJ mole^{-1}), though the onset of rotational motion occurs within the range of the β transition.

The activation energy of P1VN indicates a relatively unrestricted rotation of the naphthyl label about its bond of attachment to the chain backbone. On the contrary, the naphthyl species in P2NMA appears to be cooperative with that of the ester group as a whole; it is quite well established that the β transition in PMMA is assigned to ester group motion.

Table 6.6 Transition temperatures and activation energies obtained by luminescence polarization for PMMA with phosphorescent probe (acenaphthene) or labels (P1VN, P2NMA, PACE) (after Rutherford and Soutar [20]).

		T(K)	E(kJ mole^{-1})
Acenaphthene		253	50
P1VN	(a)	265	44
	(b)	343	107
	(c)	367	478
P2NMA		265	94
PACE	(a)	343	115
	(b)	378	460

The case of PACE is of particular interest for there is a strong coupling of label motion with chain backbone motion. In Fig. 6.21 two regions of thermal dependence can be separated. The transition temperature (378K) is in the region of the α-relaxation extimated at 1 Hz and the activation energy in the higher temperature range is in excellent agreement with that of the α relaxation estimated by conventional techniques. A question remains about the interpretation of mobility observed at lower temperatures. Indeed, the activation energy is too high for a β-relaxation process and the onset temperature of depolarization (343K) is higher than that of the β-relaxation, but could correspond to the temperature region generally ascribed to the α' transition. Due to the 'out-of-plane' polarization of the $T_1 \rightarrow S_0$ transition, the motion observed between 343K and 378K could be tentatively assigned to rotation about the local chain axis, whereas the α transition would reflect rotation motions of this local chain axis. The appearance of such a motion at a temperature higher than that of the β-relaxation would suggest that there is not a direct coupling of ester group motion with the chain backbone. More likely it could be related to the earliest stage of the largest chain motions that occur at the glass-rubber transition.

Finally, two comments can be made about phosphorescence polarization measurements. The first one deals with an improvement in sensitivity that can be achieved by using the photon-counting technique in such a way that phosphorescence experiments could be performed in a larger temperature range. The second one concerns the angle between absorption and emission transition moments. Indeed, if the $T_1 \rightarrow S_0$ of aromatic compounds is 'out-of-plane' polarized whereas the absorption transition $S_0 \rightarrow S_1$ is 'in-plane' polarized, leading to perpendicular transition moments, in the case of the phosphorescence of ketones, which is usually (n, π^*), both transition moments are parallel leading to a positive fundamental polarization.

REFERENCES

1. PERRIN, F. (1929), *Ann. Phys.* 12, 169.

2. WEBER, G. (1948). Trans. Farad. Soc. 44, 185; (1952), *Biochem. J.* 51, 145 and 155; (1953), *Adv. Protein Chem.* 8, 415.

3. BECKER, R.S. (1969), *Theory and Interpretation of Fluorescence and Phosphorescence*, Wiley, New York.

4. SOLEILLET, P. (1929), *Ann. Phys.* 12, 23.

5. PARKER, C.A. (1968), *Photoluminescence of Solutions*, Elsevier, Amsterdam.

6. PINAUD, F., JARRY, J.P. and MONNERIE, L. (1982), *Polymer* 23, 1575.

7. JARRY J.P., SERGOT, Ph., PAMBRUN, C. and MONNERIE, L. (1978), *J. Phys. E.* 11, 702.

8. WILD, U.P., HOLZWARTH, A.R. and GOOD, H.P. (1977), *Rev. Sci. Instrum.* 48, 1621.

9. SPEARS, K.G., CRAMER, L.E. and HOFFLAND, L.D. (1978), *Rev. Sci. Instrum.* 49, 255.

10. KOESTER, V.J. and DOWBEN, R.M. (1978), *Rev. Sci. Instrum.* 49, 1186.

11. VERDIER, P.H. and STOCKMAYER, W.H. (1962), *J. Chem. Phys.* 36, 227.

12. MONNERIE, L. and GÉNY, F. (1970), *J. Polym. Sci. Pt C* 30, 93; (1969) *J. Chim. Phys. (Fr.)* 66, 1691.

13. SCHATZKI, T.F. (1962), *J. Polym. Sci.* 57, 496.

14. PECHHOLD, W., BLASENBREY, S. and WOERNER, S. (1963), *Colloid Polym. Sci.* 189, 14.

15. HELFAND, E., WASSERMAN, Z.R. and WEBER, T.A. (1980), *Macromolecules* 13, 526.

16. VALEUR, B., JARRY, J.P., GÉNY, F. and MONNERIE, L. (1975), *J. Polym. Sci. Polym. Phys. Ed.* 13, 667, 675 and 2251.

17. JONES, A. and STOCKMAYER, W. (1977), *J. Polym. Sci. Polym. Phys. Ed.* 15, 847.

18. BENDLER, J. and YARRIS, R. (1978), *Macromolecules* 11, 650.

19. KRAKOVIAK, M.G. (1981), *Adv. Polym. Sci.* 40, 1.

20. RUTHERFORD, H. and SOUTAR, I. (1980), *J. Polym. Sci. Polym. Phys. Ed.* 18, 1021.

21. WAHL, Ph., MEYER, G. and PARROD, J. (1970), *Eur. Polym. J.* 6, 585.

22. NORTH, A.M. and SOUTAR, I. (1972), *J. Chem. Soc. Faraday Trans. I.* 68, 1101.

23. VALEUR, B. and MONNERIE, L. (1976), *J. Polym. Sci. Polym. Phys. Ed.* 14, 11.

24. KASPARYAN-TARDIVEAU, N., VALEUR, B., MONNERIE, L. and MITA, I. (1983), *Polymer* 24, 205.

25. KASPARYAN-TARDIVEAU, N. (1980). Thèse Docteur-Ingénieur, Paris.

26. WEILL, G. (1971), *Compt. Rend. Acad. Sci. Paris* 272B, 116.

27. VALEUR, B. and MONNERIE, L. (1976), *J. Polym. Sci. Polym. Phys. Ed.* 14, 29.

28. BROWN, K. and SOUTAR, I. (1974), *Eur. Polym. J.* 10, 433.

29. BENTZ, J.P., BEYL, J.P., BEINERT, G. and WEILL, G. (1975),

Eur. Polym. J. 11, 711.

30. ANUFRIEVA, E.V. and GOTLIB, Yu.Ya. (1981), *Adv. Polym. Sci.* 40, 1.

31. KRAMERS, A. (1940), *Physica* 7, 284.

32. JARRY, J.P. and MONNERIE, L. (1979), *Macromolecules* 12, 927.

33. FERRY, J.D. (1970), *Viscoelastic Properties of Polymers*, 2nd edn., Wiley, New York.

34. SHEN, S.P. and FERRY, J.D. (1968), *Macromolecules* 1, 270.

35. SOMERSALL, A.C., DAN, E. and GUILLET, J.E. (1974), *Macromolecules* 7, 233.

36. RUTHERFORD, H. and SOUTAR, I. (1977), *J. Polym. Sci. Polym. Phys. Ed.* 15, 2213.

37. RUTHERFORD, H. and SOUTAR, I. (1978), *J. Polym. Sci. Polym. Lett. Ed.* 16, 131.

38. MILLER, L.J. and NORTH, A.M. (1975), *J. Chem. Soc. Faraday II* 71, 1233.

CHAPTER SEVEN

Fluorescence studies of synthetic polyelectrolytes

7.1 INTRODUCTION

Macromolecules with ionizable side-groups may undergo a range of conformational changes as the pH of aqueous solutions of the polymer is altered. The conformational transitions occurring in globular proteins and other biopolymers are well known, but recently considerable attention has been paid to both experimental [1-3] and theoretical [4-7] studies of synthetic polyelectrolyte solutions. Of particular interest has been the observation that certain polymers, such as poly(methacrylic acid) (PMA) (Fig. 7.1), in aqueous solutions at low degrees of dissociation adopt a highly compact conformation with dimensions less than the unperturbed values in non-aqueous solvents. Thus, the K_θ value in the Mark Houwink equation for PMA in 0.001 M HCl at 30°C is 6.6×10^{-4} [8], whereas K_θ for PMA in methanol at 26°C is 24.2×10^{-4} [9]. The K_θ value for PMA in acid is lower than the value of 9×10^{-4} obtained for poly(acrylic acid) (PAA) (Fig. 7.1) at 30°C [10], whereas bond rotation considerations would predict that PMA should have greater coil dimensions. These data suggest that at low pH the PMA chain undergoes a form of hypercoiling [11,12] in order to minimize unfavourable methyl group-water contacts. Further evidence for the existence of hydrophobic-type structures within the polymer coil has come from ^1H NMR linewidth measurements [13] and the ability of the compact form of PMA to solubilize both aliphatic and aromatic hydrocarbons [14].

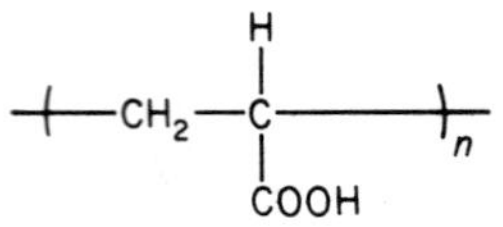

$$\left[CH_2 - \underset{\displaystyle COOH}{\overset{\displaystyle H}{C}} \right]_n$$

PAA

$$\left[CH_2 - \underset{\displaystyle COOH}{\overset{\displaystyle CH_3}{C}} \right]_n$$

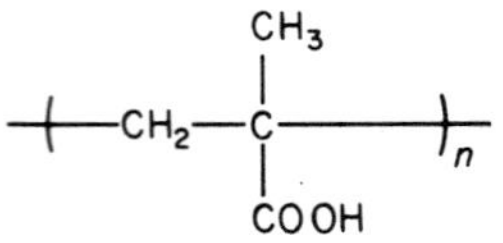

PMA

Figure 7.1 Structures for poly(methacrylic acid) (PMA) and poly(acrylic acid) (PAA).

Titration and viscosity data also illustrate the different conformational properties of PMA and PAA. When alkali is added to the polymer solutions, the carboxylic acid groups are ionized and acquire negative charge. For PAA this increase in coulombic repulsive forces results in a relatively smooth expansion of the polymer from a random coil to an extended conformation [15,16]. However, for PMA the expansion is non-uniform, and there is a sudden conformational transition from the hypercoiled form to a more extended random coil over a degree of neutralization (α) from 0.2 to 0.3 [16, 17].

Although the existence of anomalous conformational behaviour in PMA and other polyelectrolytes containing hydrophobic moieties [11,18] is well established, there remains considerable controversy about the nature of the polymer structure at different pH values and the intramolecular interactions involved in stabilizing the various conformations. The implication that similar behaviour may occur in more complex biopolymers has motivated further interest in these model synthetic polymer systems. The sensitivity of molecular luminescence to changes in the environment and motion of macromolecules makes fluorescence techniques particularly attractive for studying the conformational behaviour of polyelectrolyte systems. Indeed, an increasing number of such studies have been reported in the literature [1,19-25]. In this Chapter the application of steady-state and time-resolved fluorescence techniques and fluorescence anisotropy measurements for investigating the structure and conformational properties of synthetic polyelectrolytes will be discussed with particular emphasis on studies on PMA and PAA carried out in our own laboratories.

7.2 APPLICATION OF FLUORESCENCE TECHNIQUES

Details of fluorescence instrumentation and theory have been dealt with elsewhere in this book and in several recent reviews [26]. However, several aspects pertinent to studies of synthetic poly-electrolytes are worthy of emphasis here.

In most synthetic polyelectrolytes there is no useful fluorescent group present and it is necessary to introduce a fluorescent chromophore either covalently or non-covalently bound to the polymer chain. Changes in the spectral and temporal fluorescence properties of the probe molecule are then considered to reflect variations in the surrounding polymer environment. In the case of fluorescence polarization measurements, a further assumption often made is that motions of the fluorophore reflect those of the polymer. However, considerable care should be taken in the interpretation of fluorescence probe data. The photophysics of the probe should be well characterized and it should be ascertained whether the probe perturbs the polymer system. Evidence has been found, for instance, that the probes Auramine O and Crystal Violet stabilize the compact conformation in PMA and thus influence the conformational transition [27,28]. Perturbing effects when the dye Chrysophenine is bound to PMA in solution have also been reported [29].

In general, two approaches have been used for incorporating the probes in the polymers. Water-soluble dyes and water-insoluble hydrocarbons may be bound non-covalently to polyelectrolytes under certain conditions. Un-ionized PMA at a concentration of 0.1 M increases the solubility of anthracene in water from 4.47×10^{-7}M to 0.31×10^{-4}M [14] and similar observations have been reported for perylene [19] and pyrene [14]. Neutralization of the polymer results in a sudden decrease in solubility of the hydrocarbons and a corresponding decrease in fluorescence intensity of the probe in the region of the conformational transition in PMA. It has been suggested that the hydrocarbons may be considered to be solubilized within hydrophobic cores or structures in the compact form of the polymer [21]. Several water-soluble organic molecules also undergo marked changes in their fluorescence properties when incorporated in polyelectrolyte solutions. These molecules may be bound to the polymer by both electrostatic and non-electrostatic forces. 1,8-Anilinonaphthalene sulphonic acid (ANS) is virtually non-fluorescent in water but is strongly fluorescent when dissolved in PMA solutions at $\alpha = 0$ where the degree of binding is 19% [19]. The probe remains bound to an extent of 11% at $\alpha = 0.2$, but the fluorescence rapidly diminishes at higher degrees of neutralization. Although ANS appears to be bound entirely as a result of non-electrostatic forces, the binding of other dyes such as Rhodamine B certainly involves ionic and hydrogen bonding interactions [21].

In addition to the non-covalently bound probes referred to above, a second approach is to link the probe molecule to the polymer chain covalently. Techniques for incorporating covalently bound fluorescent probe molecules in polymers have been described by Anufrieva and Gotlib [1] and Monnerie [30]. In our own work, PMA and PAA containing 9,10-dimethylanthracene (9,10-DMA) end groups have been prepared by free radical polymerization of the monomer in the presence of 9,10-DMA as a transfer agent. The polymer may be separated from unreacted 9,10-DMA by gel permeation chromatography using methanol as solvent and Sephadex LH20. It is possible to incorporate fluorescent groups within the polymer backbone by polymerizing in the presence of dibromoanthracene, but our experience suggests it is difficult to eliminate multiple siting of the probe entirely. There are several advantages in studying covalently bound probe-polymer systems. In general, the location of the probe is known, allowing clearer interpretation of the data. In addition, for polyelectrolytes, the probe is firmly bound over the entire neutralization range, enabling detailed investigation of the conformational behaviour.

7.3 STUDIES WITH NON-COVALENTLY BOUND FLUORESCENT PROBES

The photophysics of anthracene derivatives have been extensively studied [31] and, like other hydrocarbons, their solubility in aqueous solutions is found to increase markedly in the presence of un-ionized PMA. Unfolding of the polymer structure by increasing the pH results in a sudden decrease in fluorescence intensity from the probe at a degree of neutralization corresponding to the conformational transition in PMA (see Fig. 7.2). This effect may be assigned to the collapse of hydrophobic domains within the hypercoiled structure that solubilize the hydrocarbon at low pH. We have used the fluorescence from various anthracene derivatives to investigate the structure and dynamics of the hypercoiled conformation of un-ionized PMA.

9-Methylanthracene (9-MeA) has a non-radiative excited-state relaxation process that is markedly dependent on solvent viscosity [31], whereas the fluorescence life-time of 9,10-Dimethylanthracene (9,10-DMA) is virtually independent of the surrounding environment. These probes were dissolved in atactic and syndiotactic PMA [20] determined by titration to be 0.1 M in the repeating units, to give a $< 10^{-5}$M solution. Under these conditions the ratio of probe molecule to PMA molecule is considerably less than one, thus making the presence of more than one probe per polymer chain unlikely. The fluorescence life-times measured by time-correlated single-photon counting [21] of 9-MeA and 9,10-DMA in several solvents and polymer solutions are given in Table 7.1.

The fluorescence decay of 9-MeA is single exponential in pure solvents but the life-time increases with increasing solvent viscosity However, the fluorescence decay of 9-MeA solubilized in PMA is biexponential and two life-times could be obtained by fitting the fluorescence decay data to a function of the form:

$$S(t) = S_0 \, [F \exp (-t/\tau_{F(1)}) + (1 - F) \exp (-t/\tau_{F(2)})] \qquad (7.1)$$

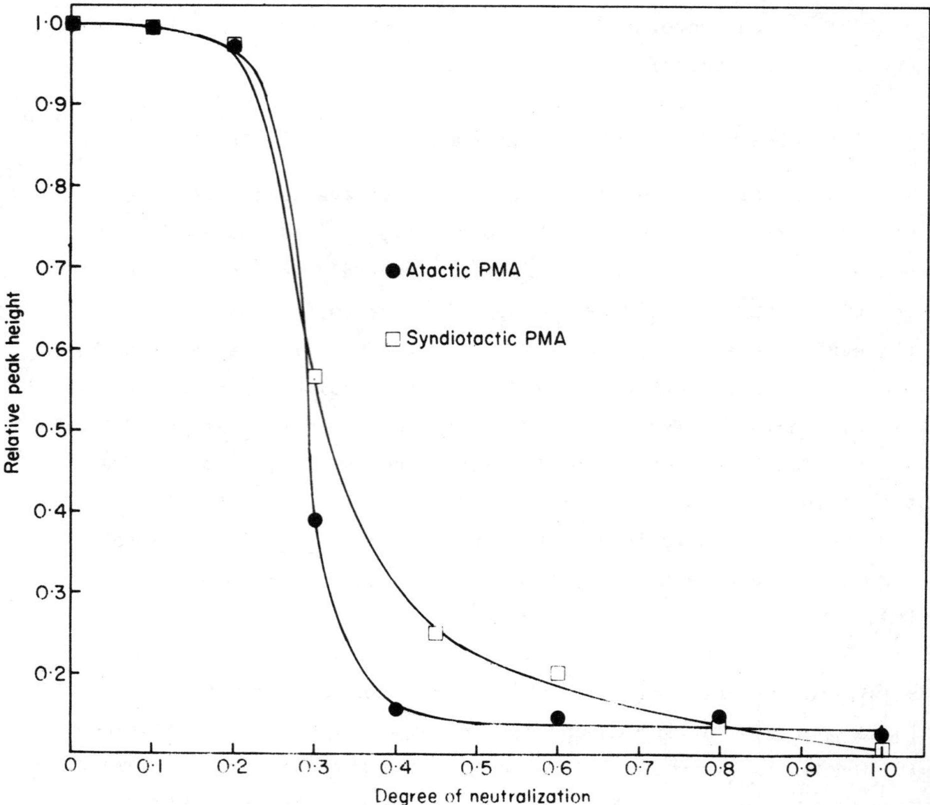

Fig. 7.2 Relative peak height of fluorescence for 9,10-dimethyl-anthracene (9,10-DMA) solubilized in atactic and syndiotactic PMA (0.1 M) at various degrees of neutralization of the polymer.

convolved with the instrument response function (see Fig. 7.3). The longer life-time component of about 13 ns approaches the maximum radiative life-time for 9-MeA of 15.4 ns [32], although the short life-time component is similar to the value found in low viscosity solvents. These results suggest that a high proportion of the solubilized 9-MeA is rigidly held in the polymer matrix whereas a smaller proportion is located in rather loose binding sites, perhaps partially accessible to the solvent. The decay of 9,10-DMA, whose fluorescence life-time is less sensitive to the environment, was well

Table 7.1 Fluorescence life-times of 9-methylanthracene (9-MeA) and 9,10-dimethylanthracene (9,10-DMA) in un-ionized aqueous PMA solutions and solvents

Solution/Solvent	*Probe*	*Fluorescence* $\tau_{F(1)}$ (ns)	*Life-time* $\tau_{F(2)}$ (ns)	*Initial fraction of (1)*
Hexane	9-MeA	4.1		
	9,10-DMA	12.1		
Methanol	9-MeA	5.0		
	9,10-DMA	15.6		
Paraffin Oil (Shell Ondina Oil 33)	9-MeA	7.7		
Propanediol	9-MeA	9.1		
Atactic PMA (MW. 216 000; α = 0)	9-MeA	12.9	5.0	0.75
	9,10-DMA	15.2		
Syndiotactic PMA (MW. 270 000; α = 0)	9-MeA	13.2	7.3	0.77
	9,10-DMA	15.9		

described by a single exponential process in each system.

Information concerning the size of the cluster surrounding the chromophore may be obtained from time-resolved fluorescence anisotropy measurements. In these experiments the decay of parallel ($I_{||}(t)$) and perpendicular ($I_{\perp}(t)$) components of fluorescence are recorded following excitation with polarized light pulses (Fig. 7.4). The anisotropy decay ($r(t)$) is given by the ratio of difference ($d(t)$) and sum ($s(t)$) decay functions as defined below [26,33]:

$$r(t) = d(t)/s(t) = (I_{||}(t) - I_{\perp}(t))/(I_{||}(t) + 2I_{\perp}(t)). \qquad (7.2)$$

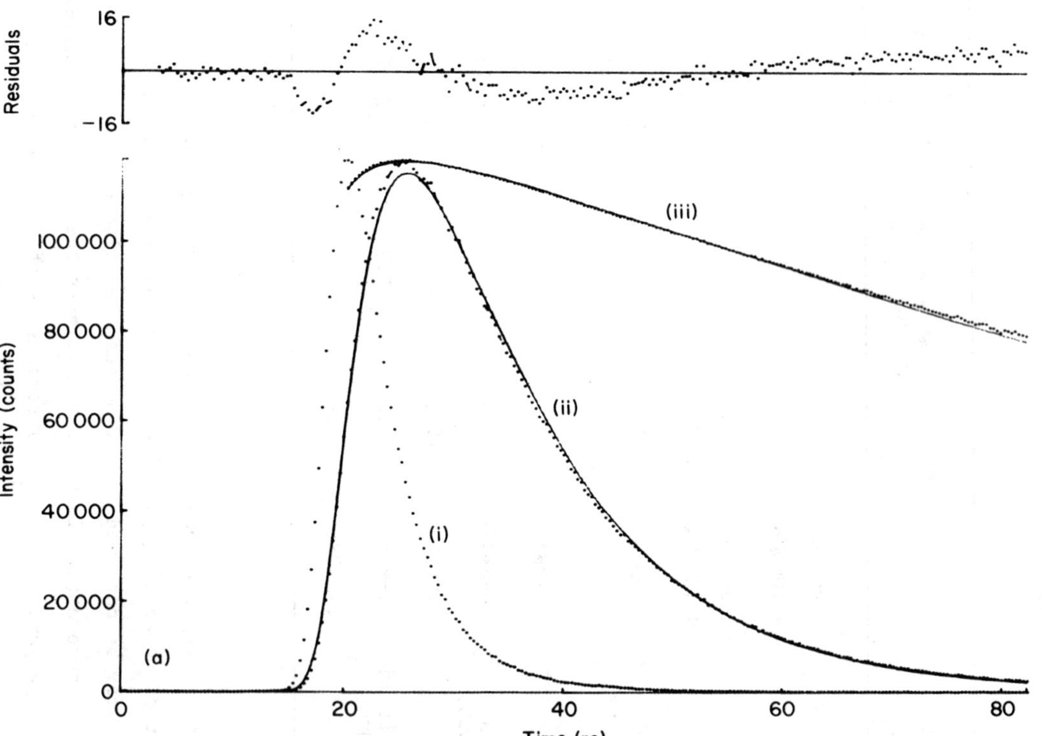

Fig. 7.3 Fluorescence decay of 9-methylanthracene (9-MeA) solubilized in atactic PMA at $\alpha = 0$. (a) Data fitted to a single exponential function with life-time 12.1 ns. Non-random distribution of residuals indicates poor fit of data [26].

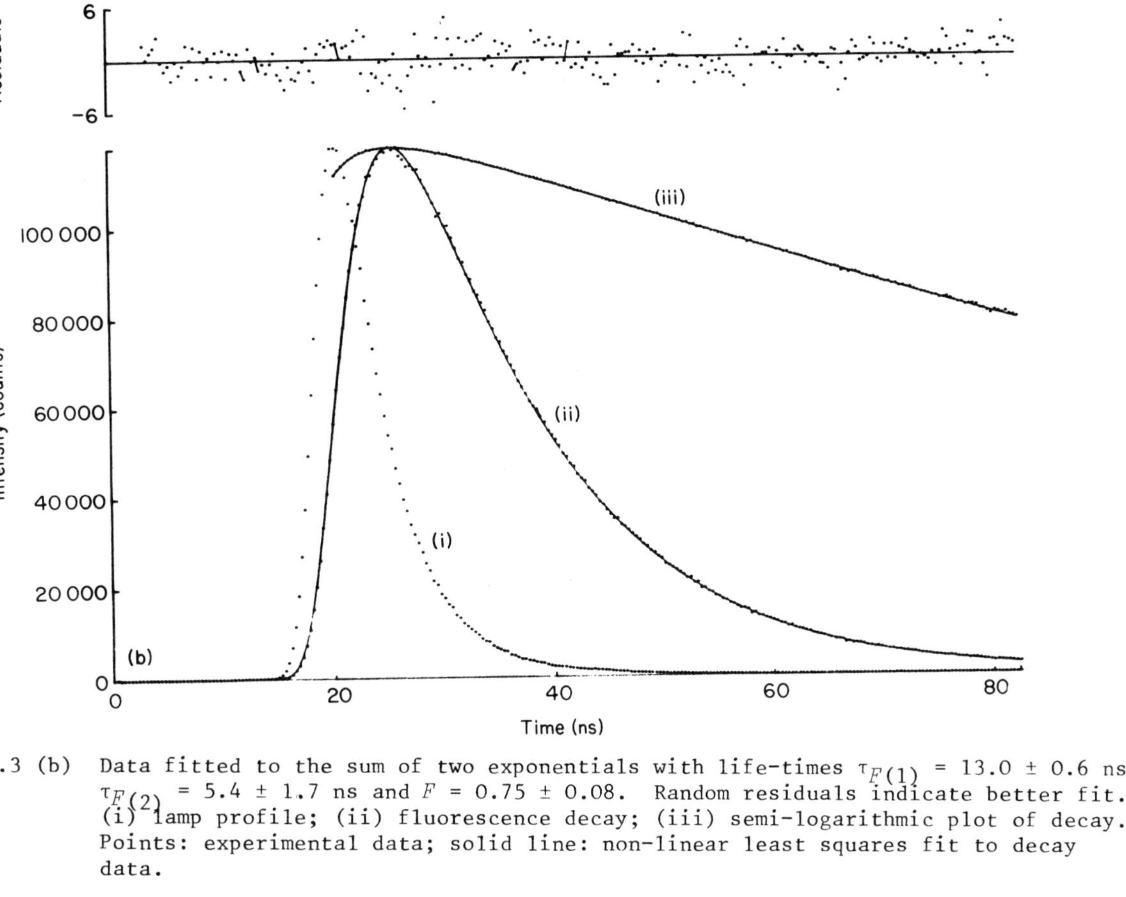

Fig. 7.3 (b) Data fitted to the sum of two exponentials with life-times $\tau_{F(1)} = 13.0 \pm 0.6$ ns, $\tau_{F(2)} = 5.4 \pm 1.7$ ns and $F = 0.75 \pm 0.08$. Random residuals indicate better fit. (i) lamp profile; (ii) fluorescence decay; (iii) semi-logarithmic plot of decay. Points: experimental data; solid line: non-linear least squares fit to decay data.

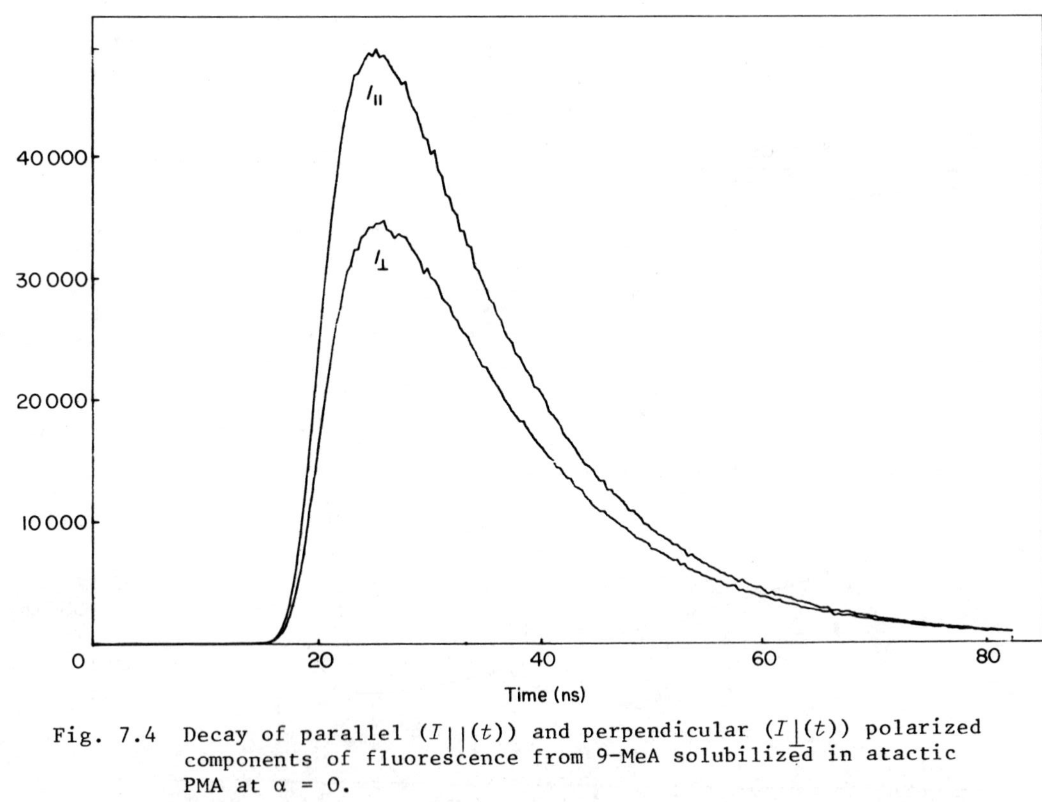

Fig. 7.4 Decay of parallel ($I_{||}(t)$) and perpendicular ($I_{\perp}(t)$) polarized components of fluorescence from 9-MeA solubilized in atactic PMA at $\alpha = 0$.

For the isotropic motion of a rigid sphere the anisotropy decay is single exponential and described by Equation (7.3):

$$r(t) = r_o \exp(-t/\tau_c) \tag{7.3}$$

where τ_c is the rotational correlation time and is related to the hydrodynamic volume V of the sphere and viscosity of the medium, η, through the Stokes-Einstein relationship (Equation (7.4)):

$$\tau_c = V\eta/RT \tag{7.4}$$

If the polymer is regarded as a close-packed sphere with radius r and molecular weight M, then τ_c can be related to the molecular weight of the rotating cluster through Equation (7.5):

$$[\eta] = (10\, N_A r^3)/3M = K_\theta M^{\frac{1}{2}} \tag{7.5}$$

where $[\eta]$ is the intrinsic viscosity of the polymer in solution and N_A is the Avogadro constant [19,20].

For the case of 9-MeA in PMA, if there are two fluorescent sites characterized by fluorescence life-times τ_1 and τ_2, the anisotropy decay will be described by Equation (7.6) [33]:

$$r(t) = d(t)/s(t) = \frac{[a_1\exp(-t/\tau_1)\; r_1(t) + a_2\exp(-t/\tau_2(t)]}{[a_1\exp(-t/\tau_1) + a_2\exp(-t/\tau_2)]} \tag{7.6}$$

In this case, the difference decay function, $d(t)$, will be double exponential if the rotating units incorporating the two sites are different and the rotational correlation times, τ_{c1} and τ_{c2}, may be extracted from the analysis of the difference decay function, viz:

$$d(t) = a_1\exp(-t/\tau_3) + a_2\exp(-t/\tau_4) \tag{7.7}$$

where $1/\tau_3 = (1/\tau_1) + (1/\tau_{c1})$ and $1/\tau_4 = (1/\tau_2) + (1/\tau_{c2})$. $d(t)$ was indeed found to be double exponential (Fig, 7,5) and, if the longer fluorescence life-time is assumed to be associated with the larger rotating unit, the rotational correlation times given in Table 7.2 may be obtained using Equation (7.7). For 9,10-DMA in PMA, for which only a single fluorescence life-time is observed, $d(t)$ and $r(t)$ could only be described by double exponential functions also consistent with the existence of two rotating units with the two correlation times reported in Table 7.2.

Table 7.2 Rotational correlation times for 9-MeA and 9,10-DMA solubilized in un-ionized aqueous PMA solutions (averaged values)

Polymer solution	*Probe*	τ_{c1}(±8)(ns)	τ_{c2}(±3)(ns)
Atactic PMA (MW 216 000;	9-MeA	33.0	11.2
α = 0)	DMA	42.1	5.0
Syndiotactic PMA (MW 270 000;	9-MeA	42.7	14.0
α = 0)	DMA	43.1	8.6

The longer correlation time of about 40 ns corresponds to the rotation of a spherical particle of radius 3.5 nm which may be assigned to a cluster of molecular weight 15 000 through Equation (7.5). Since the molecular weights of the polymers used are in excess of 200 000, there must be more than one such cluster within a single chain. It has been reported previously that perylene solubilized in PMA of molecular weight 10 000 undergoes rotational diffusion at a rate to be expected for a tight Einsteinian sphere of that molecular weight [19]. We conclude that when the molecular weight of PMA exceeds a value of 10 000, the hypercoiled form of the polymer does not exist as a tight single sphere but contains

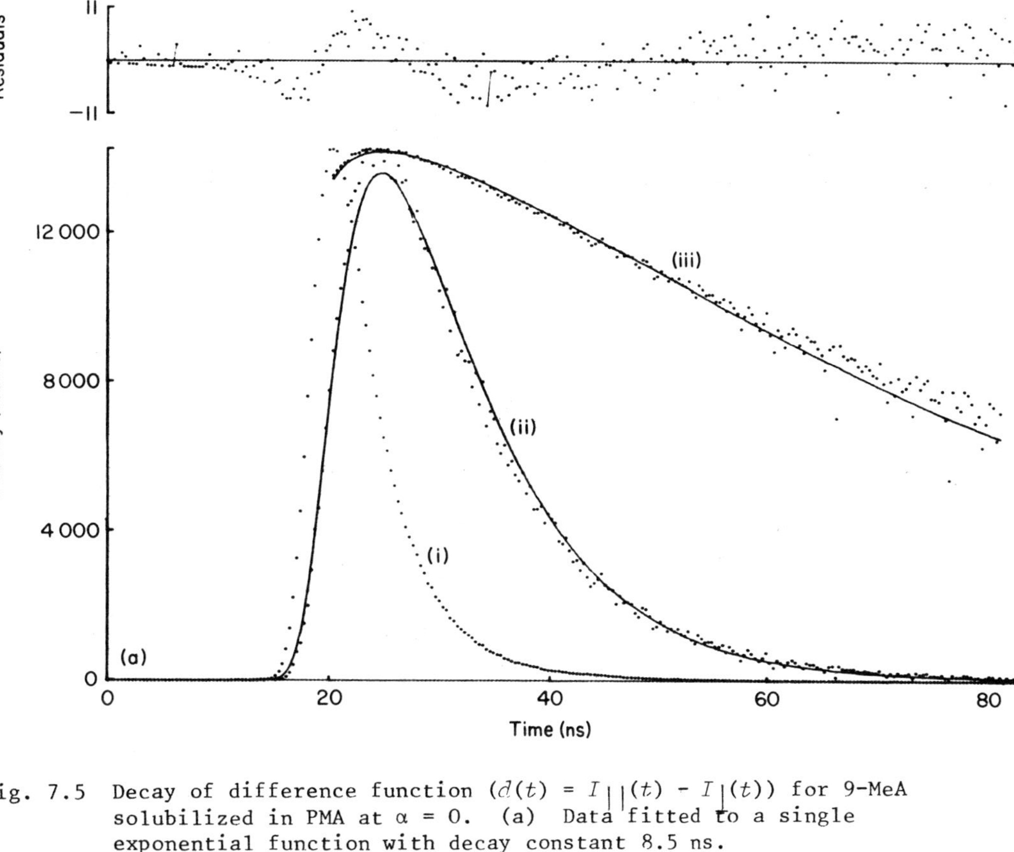

Fig. 7.5 Decay of difference function ($d(t) = I_{||}(t) - I_{\perp}(t)$) for 9-MeA solubilized in PMA at $\alpha = 0$. (a) Data fitted to a single exponential function with decay constant 8.5 ns.

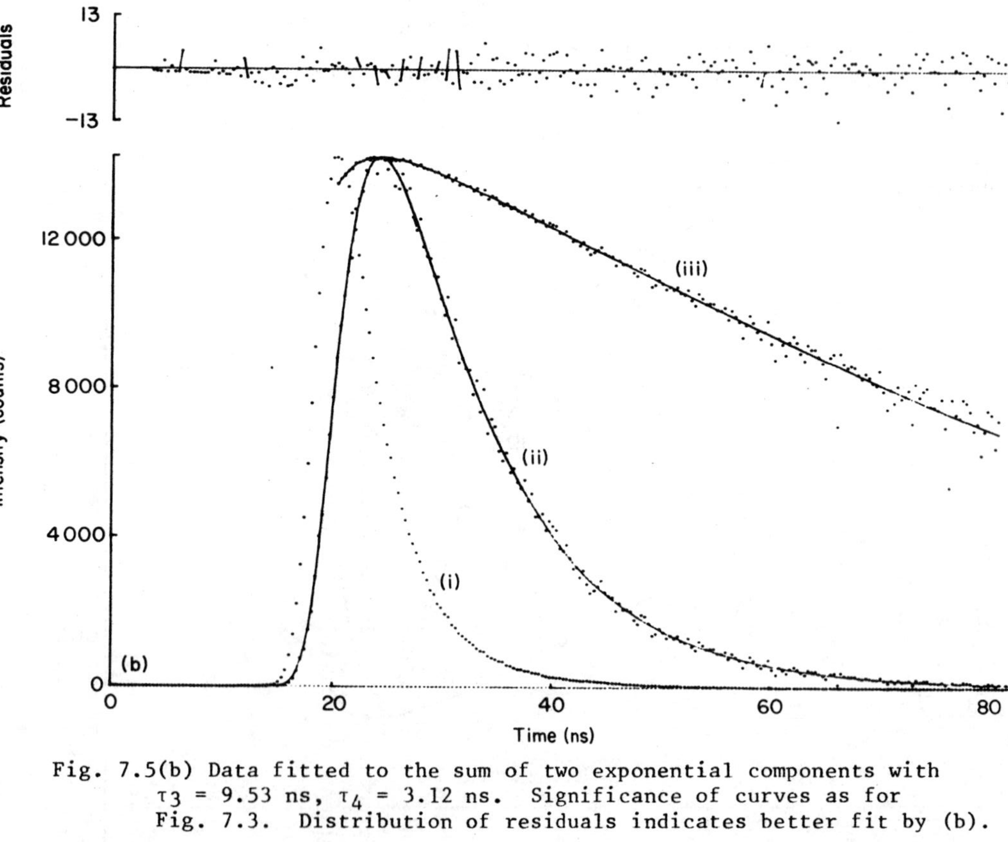

Fig. 7.5(b) Data fitted to the sum of two exponential components with τ_3 = 9.53 ns, τ_4 = 3.12 ns. Significance of curves as for Fig. 7.3. Distribution of residuals indicates better fit by (b).

several clusters separated by a sufficient section of the chain to permit isotropic rotation. Since the introduction of an aromatic hydrocarbon should strengthen rather than reduce a hydrophobic interaction, we might also conclude that the hypercoiled form of high molecular weight PMA in the absence of the hydrocarbon probe does not assume the form of a tight single sphere.

The question arises as to the origin of the shorter rotational correlation time, associated with the short fluorescence life-time for 9-MeA, in the polymers. The short correlation times correspond to the rotation of spherical particles with radii in the range 1.7 - 2.4 nm. A section of PMA chain of 20 monomer units constructed with space filling models was just sufficient to enclose a 9-MeA molecule and had a nearly spherical form with a radius of 1.5 nm. In the model, the methyl groups of the PMA are directed towards the 9-MeA, whereas the carboxylic acid groups face the aqueous exterior. One possibility is that such a structure could arise in response to unfavourable entropic interactions between the hydrocarbon and water molecules, resulting in the coiling of a short length of chain around the probe to remove it from its aqueous surroundings. This smaller cluster would not be expected to be part of the normal structure of PMA in the absence of aromatic hydrocarbons. Only minor, and possibly insignificant, differences are noted for the behaviour of the two probes in atactic and syndiotactic PMA. It is important to emphasize the usefulness of time-resolved fluorescence anisotropy measurements which have been able to detect dual siting and rotation of polymer-probe clusters in this system that would have been overlooked in steady-state emission experiments.

Since non-covalently bound aromatic hydrocarbons become insoluble once the hypercoiled conformation of the polymer disappears, they are of limited usefulness for investigating polyelectrolyte properties over a wide pH range. However, investigations using pyrene as a probe have shown that unfolding of the PMA polymer with increasing pH leads to an increase in the polarity of the pyrene environment, enhanced access of quenching solutes to the probe and the possibility of transfer of pyrene molecules from the polymer to surfactant micelles in solution [34].

The behaviour of the aromatic hydrocarbon probes in PMA may be compared to that of fluorescent water-soluble dyes that are also able to bind to the polymer. The interactions between PMA and the dyes ANS and Auramine O have been extensively investigated. For ANS, both the binding constant and the fluorescent properties of the probe alter with the change in PMA conformation [19]. The fluorescence behaviour of ANS and estimated rotational correlation times and fluorescence life-times from comparative steady state intensity measurements are reported in Table 7.3.

Table 7.3 Fluorescence emission maxima, life-times and rotational correlation times for ANS (3.2×10^{-5}M) bound to PMA (MW 10 000, 0.11 M). (From data of Treloar [19] at 20°C)

Degree of neutralization (α) of PMA	λ_F (nm)	τ_F (ns)	τ_c (ns)
0	476	15.3	21.8
0.027	480	11.6	15.8
0.045	484	9.4	11.7
0.09	488	5.0	5.5
0.18	500	3.0	2.5
No PMA	522	-	-

From this data, Treloar [19] proposed that the interior of PMA (where the ANS is bound) at $\alpha = 0$ contains no water, but the solvent immediately enters the polymer coil when neutralization commences, reducing the binding of the dye and enabling side chain rotation. The site polarity at $\alpha = 0$ was deduced to be similar to that of ANS in a 95% ethanol-water mixture and a similar correlation time to that of perylene in PMA was observed, indicating whole molecule rotation of the 10 000 molecular weight polymer. Like ANS, the cationic dye Auramine O is virtually non-fluorescent in aqueous solution but marked changes in the fluorescence properties are observed in solutions of un-ionized PMA, PAA, ionized PMA and alchohol-water mixtures of PMA [35]. At low degrees of ionization, the fraction of bound dye is distinctly higher for PMA than for PAA as a result of enhanced binding in hypercoiled sites. Although some dye is still bound in high pH solutions, it exhibits a low fluorescence yield presumably due to a reduction in the rigidity of the binding site [28].

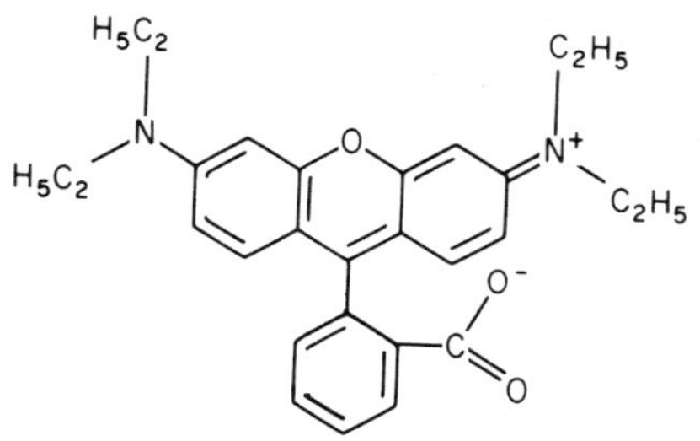

Fig. 7.6 Structure for Rhodamine B.

Recent experiments using Rhodamine B as a probe provide interesting data for comparison [21,36]. Rhodamine B (Fig. 7.6) exists as a fluorescent zwitterion over the entire neutralization range for PMA and the photophysics of the dye have been investigated in detail [37]. Non-radiative relaxation of the excited dye is influenced by solvent stabilization effects and torsional motion of the diethylamino groups of the molecule. The dye may be physically bound to the polymer by mixing dilute aqueous solutions of the dye and polymer at a pH of 7 where the polymer is partially expanded. The pH of the solution may then be adjusted back to $\alpha = 0$ and the excess salt and weakly bound dye may be removed by dialysis. The dye is very strongly bound to the polymer at low pH and resists separation by gel permeation chromatography on a Sephacryl S200 column, although efficient separation can be achieved at higher pH or by dialysis against methanol. The conformational behaviour of PMA, as assessed by titration curves, was unaffected by the presence of the small amount of dye ($< 10^{-6}$M). The fluorescence properties of Rhodamine B in several solvents and in polymer solutions are summarized in Table 7.4. The fluorescence life-times and rotational correlation times were measured using time-correlated single-photon counting and/or picosecond laser/streak camera techniques.

Table 7.4 Absorption maxima, emission maxima, fluorescence life-times and rotational correlation times for Rhodamine B zwitterion [36].

Solution	λ_{max}(*absorption*) (nm)	λ_{max}(*emission*) (nm)	τ_F (ns)	τ_c (ns)
H_2O	553	576	1.5	0.22
Glycerol	556	557	3.2	-
PMA (MW 290 000; .03 M; α=0)	563	579	4.2	23
PMA (α=1)	553	576	2.0	-
PAA (α=0)	560	581	3.0	10
PAA (α=1)	553	576	1.7	-

For the un-ionized polymer, the absorption maxima indicate that the binding site of the dye in PMA and PAA are of comparable local polarity, but the life-time data suggest a higher local viscosity surrounding the dye molecule in the PMA hypercoil (cf. Table 7.4). The fluorescence decay data for PMA depend on the degree of neutralization of the polymer. If the decays are fitted to a single exponential function characterized by an effective life-time, τ_F(eff), then this parameter is approximately constant for $0 \lesssim \alpha \lesssim 0.2$, but then decreases dramatically in the conformational region up to $\alpha = 0.55$ to a shorter constant τ_F(eff) value (see Fig. 7.7). However, although the decay can be characterized by a single fluorescence life-time for the extremes of the neutralization range at $\alpha = 0$ and $\alpha = 1$, at intermediate pH values, multiexponential decay is observed. Analysis of the decay data using a double exponential function (Equation (7.1)) suggests that two distinct

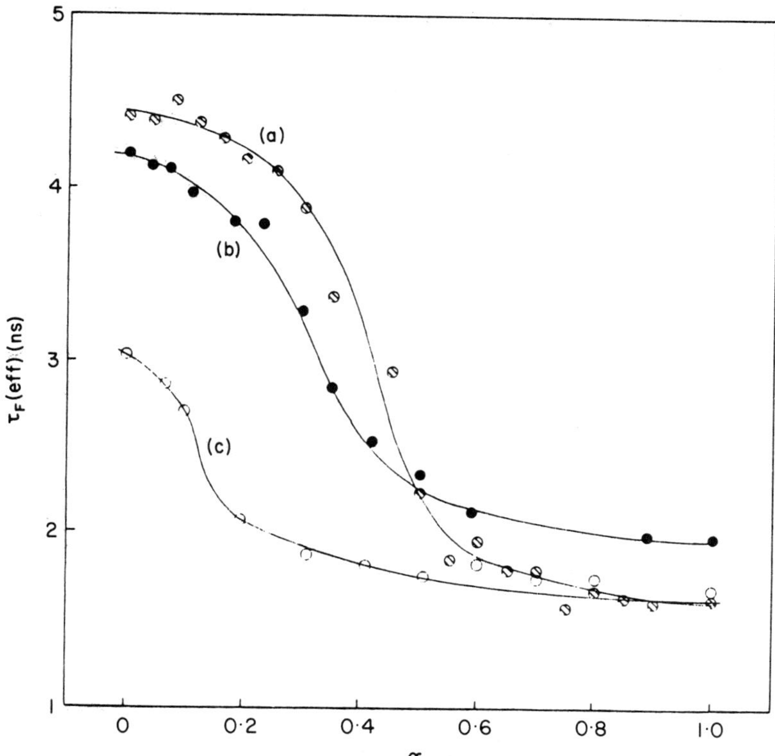

Fig. 7.7 Single exponential fluorescence decay time (τ_F(eff)) as a function of α for (a) Rhodamine 3B on PMA; (b) Rhodamine B on PMA; (c) Rhodamine B on PAA.

types of site are present in the intermediate neutralization region, characterized by decay times that resemble either solvated dye ($\tau_{F(1)} \simeq 1.7$ ns) or dye in a highly viscous region ($\tau_{F(2)} \simeq 4.5$ ns). The variation in the fraction F (cf. Equation (7.1)) of the components shows that the proportion of the two species in PMA alters slowly at first before a dramatic increase near $\alpha = 0.3$ until after $\alpha = 0.5$ only completely solvated dye molecule environments are present (Fig. 7.8). Thus, it appears that the binding sites in the PMA coil only become accessible to solvent after the conformational transition commences and it is the change in proportion of the viscous and solvated structures that causes the apparently smooth transition

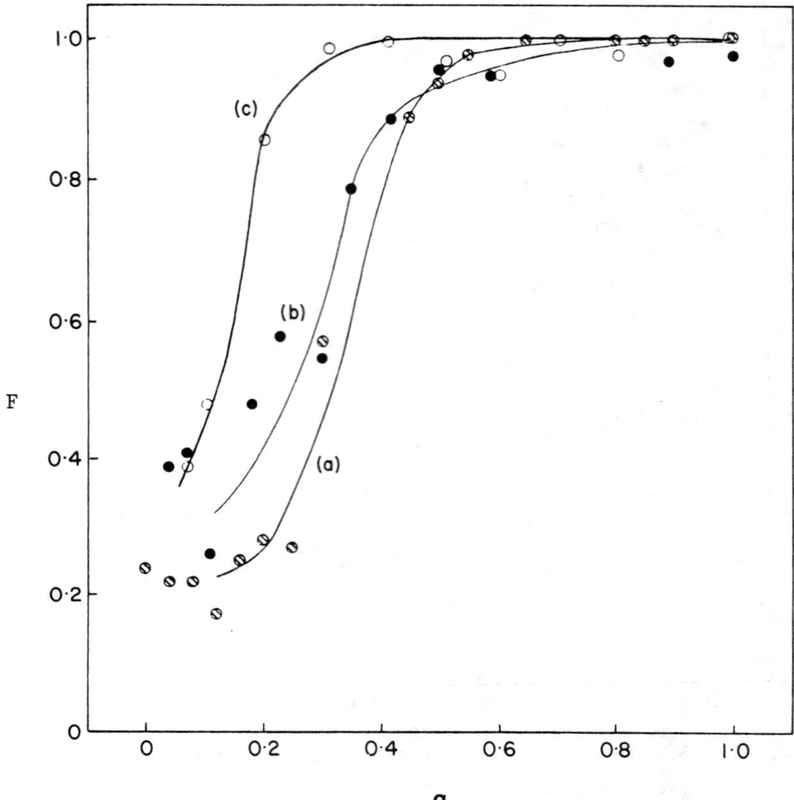

Fig. 7.8 Variation of fraction of short-lived component (F in Equation (7.1), with α for (a) Rhodamine 3B with PMA; (b) Rhodamine B with PMA; (c) Rhodamine B with PAA.

of averaged properties such as τ_F(eff). In contrast rapid solvation of the dye site occurs in PAA as the polymer neutralization commences, and the coil is either expanded for solvation or the dye released from the polymer by $\alpha = 0.2$. These results may be compared to the data of Erny and Müller [38] for Auramine O, who found that dye bound to PAA was accessible to solvent for all α, whereas for PMA the compact state shields the dye from access to water at low pH.

The results using Rhodamine B support a model consisting of two conformational structures occupied by the probe and that it is a shift in the proportion of the two limiting site types which causes an apparent shift in macroscopic properties during neutralization. This is in contrast to models that imply a gradual transition involving many configurations.

The size and molecular weights of the rotating units containing Rhodamine B at $\alpha = 0$ may be derived from fluorescence anisotropy measurements (Fig. 7.9). Both the difference decay, $d(t)$, and anisotropy decay, $r(t)$, may be described by single exponential functions to give rotational correlation times of 23 ns in PMA and 10 ns in PAA. From Equations (7.4) and (7.5) the rotating spherical unit in PMA may be assigned to a polymer cluster of 10 000 molecular weight. Since the total molecular weight of the polymer is 290 000 this result once again implies that the hypercoiled polymer is made up of many small clusters that may undergo isotropic rotation. Since the photophysical properties of the dye molecule indicate it is in a highly viscous binding site, independent rotation of the dye seems an improbable explanation for the short τ_c observed. The observation that both aromatic hydrocarbons and water-soluble dye probes detect similar cluster sizes in un-ionized PMA is strong evidence that such units are a normal feature of the hypercoil in the absence of the probes. As the chain is neutralized individual clusters may open to the solvent, resulting in an equilibrium between the hypercoil clusters and solvated segments of the chain.

Information concerning the rate of collapse of the hypercoiled structures has been obtained by Chen and Thomas [34]. These authors measured the rate at whcih solubilized pyrene is ejected from PMA solutions when the pH is raised above the transition by monitoring the accompanying decrease in fluorescence from the probe. Upon sudden neutralization of the polymer, the fluorescence from pyrene was observed to decay over several seconds and the decay kinetics implied that the expansion of the polymer coil depended on $(\text{time})^{1/3}$.

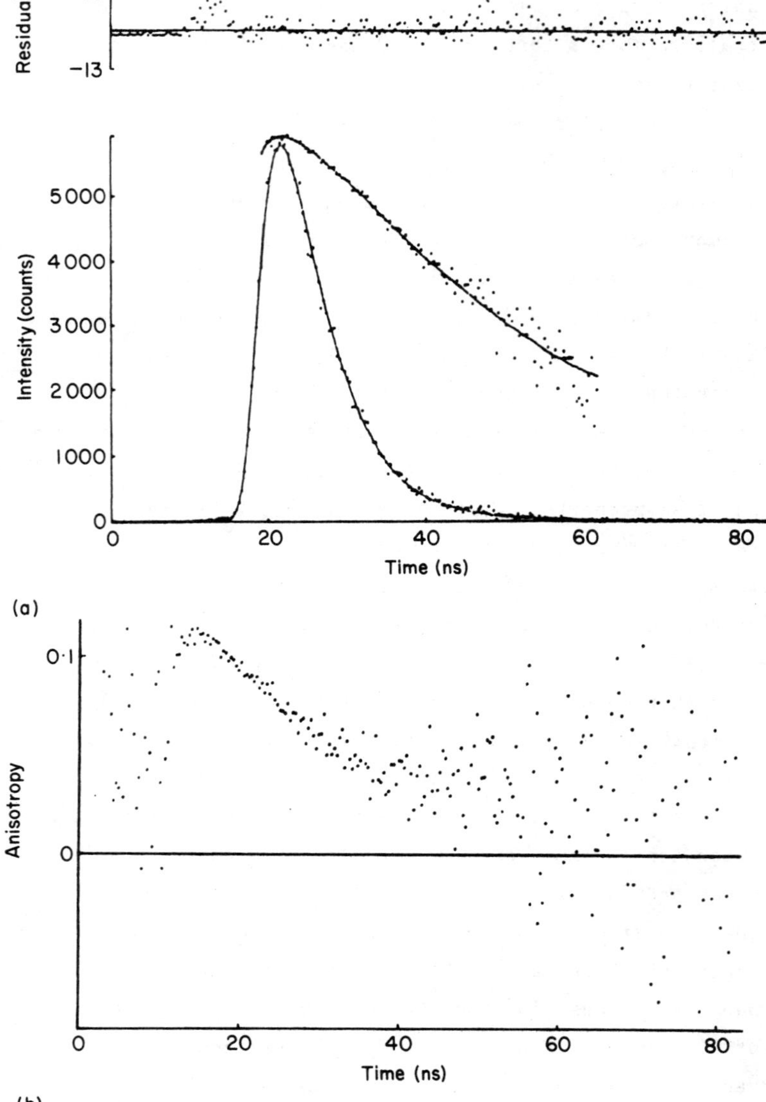

Fig. 7.9 Time-resolved fluorescence anisotropy data for Rhodamine B bound to PMA at $\alpha = 0$. (a) Difference decay function, $d(t)$, and semilogarithmic plot fitted by a single exponential function (solid line) with $\tau_D = 3.56 \pm 0.18$ ns. (b) Convolved anisotropy decay profile.

7.4 COVALENTLY BOUND FLUORESCENT LABELS

The introduction of covalently bound fluorescent probes enables the study of known regions of the polyelectrolyte macromolecules over the entire neutralization range. Fluorescent molecule may be incorporated as a chain end group, at the end of a side chain or within the main chain backbone during the polymerization process or through subsequent reactions involving functional groups of the polymer. A detailed review of the methods for incorporating anthracene derivatives in polymer chains, including polyelectrolytes, has recently appeared [1]. The anthracene chromophore has suitable photophysical properties for use as a fluorescent label and derivatives are available that enable covalent attachment of the molecule to various positions in the polymer chain. We have introduced 9,10-dimethylanthracene (9,10-DMA) as a covalently bound endgroup on PMA and PAA (PMA-DMA, PAA-DMA). The probe is incorporated by free radical polymerization of the monomer in the presence of 9,10-DMA used as a chain transfer agent (Fig. 7.10). Since the probe is attached to the chain end via reaction through the methyl substituent of the chromophore, little perturbation of 9,10-DMA photophysics is observed.

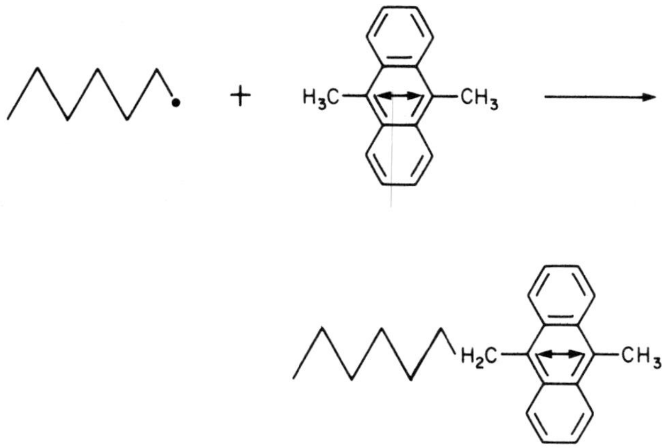

H_3C — CH_3 →

H_2C — CH_3

Fig. 7.10 Reaction scheme for incorporating 9,10-DMA as a chain end group.

The fluorescence quantum yield of 9,10-DMA is close to unity and is nearly solvent independent [39]. When attached to the polymer the fluorescence life-time remains virtually constant at 13.4 ± 0.1 ns over the range $0 \leqslant \alpha \leqslant 1$ and throughout the temperature range of $1^{o}C$ to $49^{o}C$ investigated. In contrast to the behaviour of PMA with solubilized 9,10-DMA (Fig. 7.2), there is no discontinuity in the change of fluorescence intensity of PMA-DMA with α confirming that the probe is covalently bound to the polymer. In the first instance, steady-state fluorescence polarization experiments provide useful information on the mobility of the end group as the carboxylic acid groups of the polymer are ionized. Measurements were made on an instrument based on the design of Bashford *et al* [40] where the perpendicular ($I_{\perp}$) and parallel ($I_{||}$) polarized components of fluorescence are measured simultaneously and the data processed electronically. The steady-state fluorescence polarization, P, is defined by Equation (7.8) and related to the fluorescence anisotropy, r, through Equation (7.9):

$$P = (I_{||} - I_{\perp})/(I_{||} + I_{\perp}) \tag{7.8}$$

$$r = 2P/(3 - P) \tag{7.9}$$

A plot of the variation of P with α for PMA-DMA and PAA-DMA is given in Fig. 7.11. It is apparent that at $\alpha = 0$ the probe in PMA has a much higher steady-state polarization than in PAA consistent with the chromophore experiencing slower rotation in the hypercoiled polymer. There is a rapid increase in mobility of the probe during the conformational transition of PMA to yield values of P similar to that observed for PAA-DMA. The slight increase in polarization values at high degrees of neutralization may be explained by the high ionic strength of the solution ($Na^{+} \simeq 0.01$ M) reducing coulombic repulsion between the acid groups thus increasing the size of the rotating cluster containing the end group.

The polarization values may be related to $\bar{\tau}_c$, the mean rotational correlation time assuming spherical geometry and isotropic rotation through Equation (7.10):

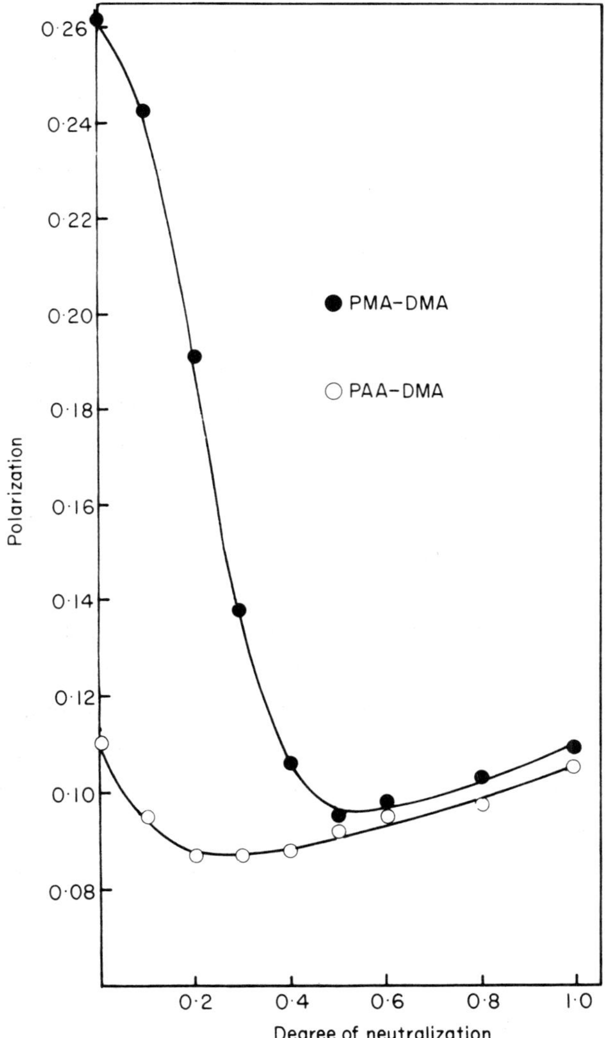

Fig. 7.11 Steady-state fluorescence polarization, P, as a function of degree of neutralization for PMA-DMA and PAA-DMA

$$[(1/P) - (1/3)] = [(1/P_0) - (1/3)][1 + (\tau_F/\bar{\tau}_c)] \qquad (7.10)$$

The limiting polarization P_0, was determined as 0.36 at 365 nm by extrapolating a plot of $1/P$ against T/η to infinite viscosity. The variations of $\bar{\tau}_c$ and the reduced viscosity, η_{sp}/c, with α for PMA-DMA are shown in Fig. 7.12. The rotational correlation time at $\alpha = 0$ again indicates a rotating unit of molecular weight 10 00 whereas, at higher degrees of neutralization, the short $\bar{\tau}_c$ of 3-4 ns corresponds to the rotation of an Einsteinian sphere of radius 1.4 - 1.6 nm. This smaller cluster may be associated with the rotation of a small section of the PMA chain that will just enclose the 9,10-DMA end group and shield it from entropically unfavourable hydrocarbon-water contacts. A similar effect was observed for the solubilized aromatic hydrocarbon probes (Section 7.3). The increase in reduced viscosity with increasing α demonstrates the overall expansion of the polymer chain dimensions as neutralization proceeds. Since $\bar{\tau}_c$ decreases with α it might be concluded that the original tight spherical units present in the hypercoiled polymer break down as the polymer expands, allowing smaller segments to undergo rotation independently.

However, steady-state polarization measurements will only provide a time-average picture of the molecular motions that occur during the fluorescence life-time of the probe. Further information can be obtained from time-resolved fluorescence anisotropy measurements. Rotational correlation times measured by the time-correlated single-photon counting technique are presented in Table 7.5. The fluorescence decay (or $s(t)$) was single exponential in each case, but for PMA-DMA, $d(t)$ and $r(t)$ could only be fitted adequately by a minimum of two exponentially decaying components over the entire neutralization range. The results are consistent with a model based on the existence of two rotating spherical clusters of different size that enclose the probe. The long and short rotational correlation times associated with the clusters for PMA-DMA at $\alpha = 0$ are slightly larger than those obtained with the solubilized probe molecules and may reflect additional constraints on rotation due to the covalent linkage and position of the probe. The larger clusters present in the hypercoiled PMA polymer compared to PAA are in accord with the existence of tight compact conformations in PMA. As neutralization proceeds, there is a marked decrease in the longer correlation time

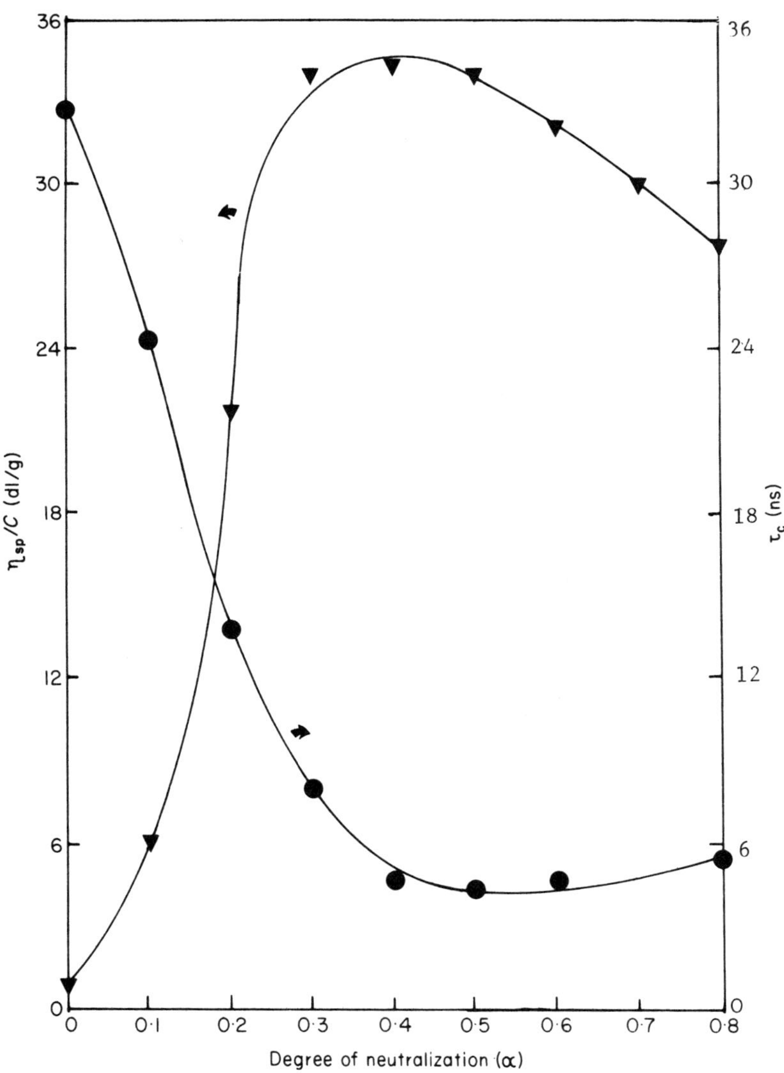

Fig. 7.12 Variation in reduced viscosity and mean rotational correlation time with α for PMA-DMA.

Table 7.5 Rotational correlation times at various degrees of neutralization for PMA (MW 120 000; [η] = 0.228 dl/g) and PAA with 9,10-DMA end groups

α	*Polymer*	τ_{c1}(ns)	τ_{c2}(ns)
0	PMA	58.8	7.4
	PAA	4.0	-
0.1	PMA	41.1	4.6
	PAA	5.4	2.0
0.2	PMA	13.6	4.0
	PAA	6.0	1.1
0.3	PMA	5.9	3.9
	PAA	11.2	2.1
1.0	PMA	8.7	0.8
	PAA	15.5	1.8

for PMA for $\alpha = 0 \rightarrow 0.3$ related to an increase in probe-polymer mobility and a dramatic decrease in the size of the rotating cluster through the conformational transition region. The shorter rotational correlation time, presumably associated with the rotation of a small segment of the polymer attached to the probe, also varies slightly as expansion of the polymer proceeds. In addition, the decay of anisotropy from PAA-DMA could only be adequately fitted by two exponential terms. This may indicate the presence of two small independently rotating units in the polymer. However non-exponential anisotropy decays can also arise if the rotating body is non-spherical (e.g. prolate ellipsoids), for complex segmental motions of polymers or if a combination of rotations influences the probe's mobility [26]. It is difficult to distinguish between these possibilities at this stage with the quality of fluorescence decay data available, and thus definite conclusions cannot be made.

Fluorescence techniques can also be used to investigate the accessibility of the probe to the solvent. Chen and Thomas [34] studied the quenching of excited pyrene in micelles and PMA by water-soluble molecules and found that the rate constants for quenching were considerably lower than expected for diffusion-controlled reactions. They attributed the low rate constants to restrictions these systems impose on the encounters of probes and quencher molecules. The dynamic quenching of fluorescence from a molecule in the presence of quencher, Q, may be described by the Stern-Volmer relationship (Equation (7.11)):

$$(I_o/I) - 1 = K_{SV}\ [Q]. \tag{7.11}$$

The quenching rate constant, k_q can be found from the Stern-Volmer constant, K_{SV}, if the fluorescence life-time in the absence of quencher, τ_o, is known:

$$K_{SV} = k_q\ \tau_o \tag{7.12}$$

The quenching of fluorescence from the PMA-DMA polymer by CH_3NO_2 was studied over a wide range of α. A typical Stern-Volmer plot is given in Fig. 7.13, whereas the calculated quenching rate constants at different degrees of neutralization are shown in Fig. 7.14. The increase in k_q with increasing ionization of the carboxylic acid groups for $\alpha = 0 \rightarrow 0.5$ can be interpreted as arising from increased access of CH_3NO_2 molecules to the DMA end group as the polymer coil expands. The further decrease in k_q for $\alpha = 0.5 \rightarrow 1$ may be associated with a slight recoiling of the polymer (and hence protection of the end group) as the ionic strength of the solution increases, reducing coulombic repulsion between ionized carboxylic acid groups. There is some correlation between decreasing cluster size, as measured by fluorescence polarization experiments, and increasing access to quencher and solvent with α. However, in the latter case the change in behaviour is less distinct in the region of the conformational transition, and optimum access to quencher occurs at $\alpha \simeq 0.5$.

The results of the fluorescence experiments on PMA point to a model for the expansion of the hypercoiled polymer shown in Fig. 7.15. Measurements using non-covalently bound aromatic hydrocarbons and water-soluble dyes, and covalently attached fluorescent labels indicate

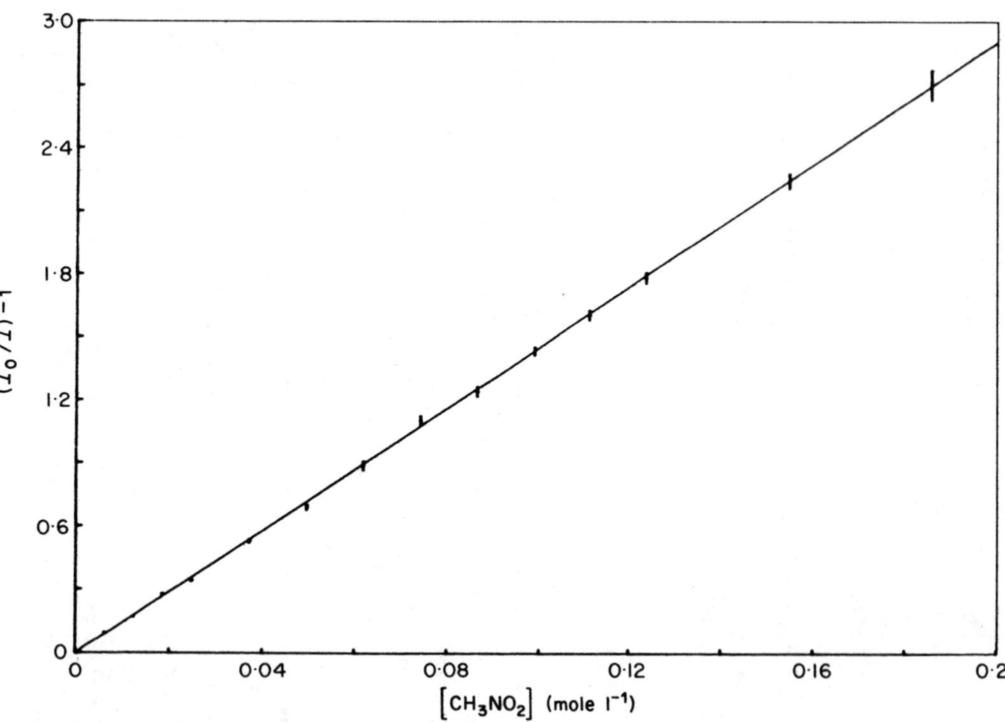

Fig. 7.13 Stern-Volmer plot for quenching of fluorescence from PMA-DMA by CH_3NO_2 at $\alpha = 0.2$: gradient = $14.5 \pm 0.06\ M^{-1}$.

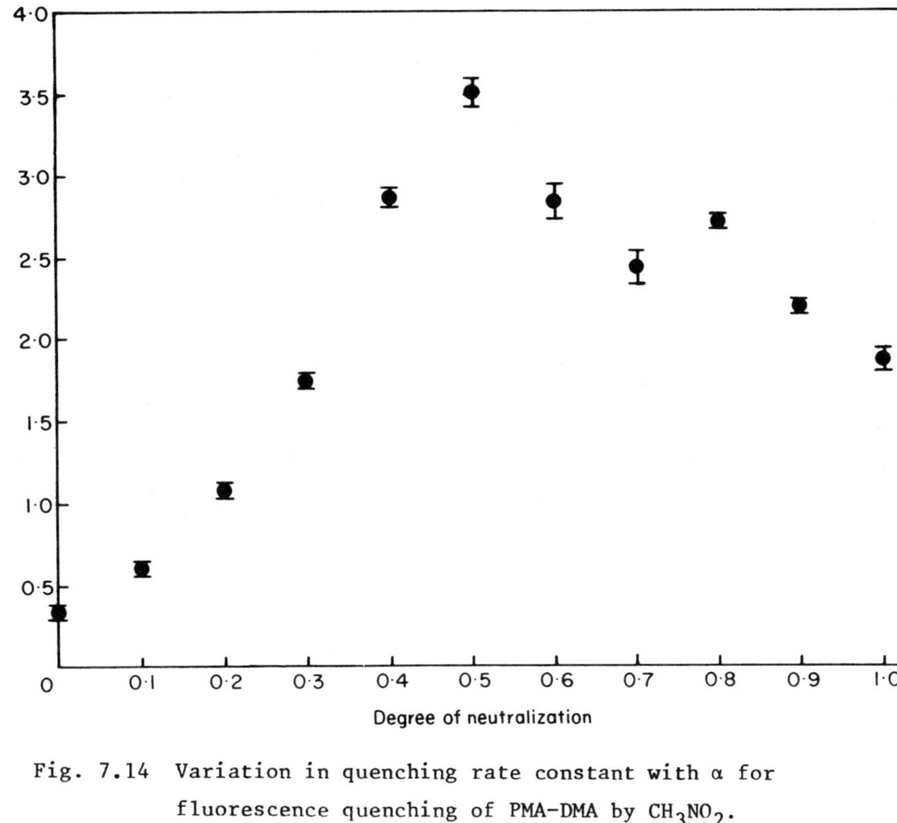

Fig. 7.14 Variation in quenching rate constant with α for fluorescence quenching of PMA-DMA by CH_3NO_2.

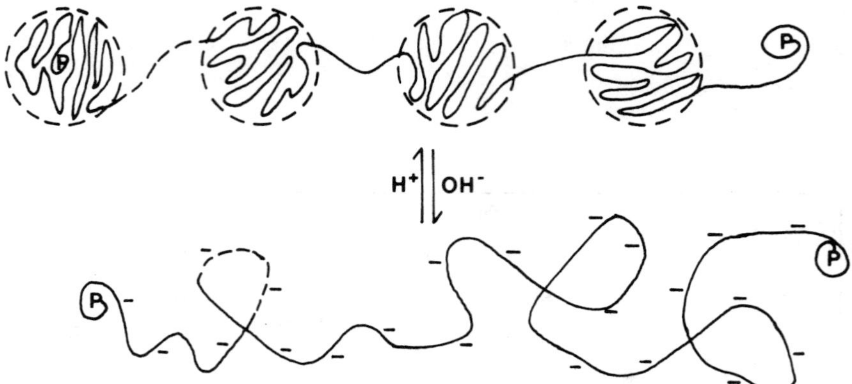

Fig. 7.15 Schematic diagram illustrating environment of fluorescent hydrocarbon probe, P, in hypercoiled PMA and neutralized polymer.

that the un-ionized form of PMA exists as several tightly packed clusters of molecular weight 10 000 - 20 000. Other experimental and theoretical studies are consistent with this model. Photoionization data from pyrene solubilized by pre-existing structures in PMA infer that the probe occupies a site between 1.6 nm and 4.0 nm from the aqueous phase [34]. These dimensions are of the same order of magnitude as the radii of PMA clusters of molecular weight 10 000 - 20 000.

In addition to experimental studies, considerable theoretical work has been aimed at describing the viscoelastic properties of polymer solutions. The approaches are centred on the development of mathematically tractable models that permit the derivation of the diffusion equation. One such approach is the 'bead and spring' model discussed by Bixon [41]. The beads of this model are not to be thought of as permanent structures but rather are statistical segments that can be treated as single units with respect to frictional properties. Theoretical treatments of polymer dynamics also partition the polymer chain into 'blobs', and this concept has recently been extended to theoretical studies of polyelectrolytes [4,7]. Khokhlov [7] discussed the coil-globule transition of a weakly charged polyelectrolyte in very dilute, salt-free solutions by representing the molecule as a sequence of blobs. Although this concept was introduced for mathematical simplification in the modelling of polymer collapse, the fluorescence probe experiments suggest that actual aggregates may exist in hypercoiled PMA, at least on the time-scale of fluorescence emission.

It should be emphasized that long-chain molecules are dynamic systems with a broad range of motions and dimensions over time. Thus measurements of the conformations and chain dimensions of polymers will depend on the sampling time of the experiment. Properties measured by viscosity, osmometry or light-scattering techniques may be quite different from those detected on a nanosecond time-scale by fluorescence. A discussion of polymer dynamics and further fluorescence polarization studies of polyelectrolytes have been undertaken by Anufrieva and Gotlib [1]. The orientation autocorrelation function of a luminescent group covalently bound to the polymer chain may be described by a distribution of relaxation times, but a weight-average relaxation time, τ_w, can be derived from variable viscosity and steady-state polarization experiments. Such measurements have been used to investigate polyelectrolyte intramolecular mobility in polymer-polymer complexes, cross-linked polymers and in the presence of surfactants and rare-earth ions [1].

In conclusion, fluorescence studies of synthetic polyelectrolytes may be used to investigate a wide range of polymer conformational behaviour. The information obtained complements data available from other techniques and can provide new perspectives on the dynamics of polyelectrolyte systems. Improvements in data collection and analysis procedures, particularly in time-resolved experiments, combined with developments in theoretical approaches should expand this area of investigation.

REFERENCES

1. ANUFRIEVA, E.V. and GOTLIB, Yu. Ya. (1981) *Adv. Polym. Sci.* 40, 1.
2. OOSAWA, F. (1971) *Polyelectrolytes*, Marcel Dekker, New York.
3. SELEGNY, E. (ed.) (1975) *Polyelectrolytes*, D. Reidel Publishing Co., Dordrecht, Holland.
4. KHOKHLOV, A.R. and KHACHATURIAN, K.A. (1982) *Polymer* 23, 1742.

5. NYSTROM, B. and ROOTS, J. (1982) *Prog. Polym. Sci.* 8, 333.
6. ODIJK, T. (1979) *Macromolecules* 12, 688.
7. KHOKHLOV, A.R. (1980) *J. Phys. A., Maths. Gen.* 13, 979.
8. KATCHALSKY, A. and EISENBERG, H. (1951) *J. Polym. Sci.* 6, 145.
9. WEIDERHORN, N.M. and BROWN, A.R. (1952) *J. Polym. Sci.* 8, 651.
10. NEWMAN, S., KRIGBAUM, W.R., LAUGIER, D. and FLORY, P.J. (1954) *J. Polym. Sci.* 14, 451.
11. DUBIN, P.L. and STRAUSS, U.P. (1975), Hypercoiling in hydrophobic polyacids, in *Polyelectrolytes* (ed. E. Selegny) D. Reidel Publishing Co., Dordrecht, Holland, pp. 3-13.
12. KODEM, O. and KATCHALSKY, A. (1955) *J. Polym. Sci.* 15, 321.
13. KAY, P.J., KELLY, D.P., MILGATE, G.I. and TRELOAR, F.E. (1976) *Makromol. Chem.* 173, 885.
14. BARONE, G., CRESCENZI, V., LIQUORI, A.M. and QUADRIFOGLIO, F. (1967) *J. Phys. Chem.* 71, 2341.
15. LIQUORI, A.M., BARONE, G., CRESCENZI, V., QUADRIFOGLIO, F. and VITAGLIANO, V. (1966) *J. Macromol. Chem.* 1, 291.
16. CRESCENZI, V. (1968) *Adv. Polym. Sci.* 5, 358.
17. LEYTE, J.C. and MANDEL, M. (1964) *J. Polym. Sci. A*1. 2, 1879.
18. MÜLLER, G., FENYO, J.C., BRAUD, C. and SELEGNY, E. (1975) Compact conformations of polyions stabilized by non-electrostatic short-range interactions, in *Polyelectrolytes* (ed. E. Selegny) D. Reidel Publishing Co., Dordrecht, Holland.
19. TRELOAR, F.E. (1976) *Chemica Scripta* 10, 219.
20. TAN, K.L. and TRELOAR, F.E. (1980) *Chem. Phys. Lett.* 73, 234.
21. SNARE, M.J., TAN, K.L. and TRELOAR, F.E. (1982) *J. Macromol. Sci. Chem.* A17(2), 189.
22. BRAUD, C. (1977) *Eur. Polym. J.* 13, 897.
23. VANDERELDE, M.C. and FENYO, J.C. (1979) *Eur. Polym. J.* 15, 431.
24. BRAUD, C., MÜLLER, G. and SELEGNY, E. (1978) *Eur. Polym. J.* 14, 479.
25. FENYO, J.C., MOGNOL, L., DELBEN, F., PAOLETTI, S. and CRESCENZI, V. (1979) *J. Polym. Sci. Polym. Chem. Ed.* 17, 4069.
26. GHIGGINO, K.P., ROBERTS, A.J. and PHILLIPS, D. (1981) *Adv. Polym. Sci.* 40, 69.
27. STARK, W.H.J., DE HASSETH, P.L., SCHIPPERS, W.B., KARMELING, C.M. and MANDEL, M. (1973) *J. Phys. Chem.* 77, 1772.
28. MANDEL, M. and STARK, W.H.J. (1974) *Biophys. Chem.* 2, 137.
29. LOUVRIEN, R. (1967) *Proc. Natl. Acad. Sci. U.S.A.* 57, 236.
30. MONNERIE, L. This Volume, Chapter 6.
31. BLATT, E., TRELOAR, F.E., GHIGGINO, K.P. and GILBERT, R. (1981) *J. Phys. Chem.* 85, 2810.
32. JUNG, Ch. and HECKNER, K.-H. (1977) *Chem. Phys.* 21, 227.
33. RIGLER, R. and EHRENBERG, M. (1973) *Quant. Rev. Viophys.* 6, 139.
34. CHEN, T.S. and THOMAS, J.K. (1979) *J. Polym. Sci. Polym. Chem. Ed.* 17, 1103.
35. ANUFRIEVA, E.V., BIRSHTEIN, T.M., NEKRASOVA, T.N., PTITSYN, O.B. and SHEVELEVA, T.V. (1968) *J. Polym. Sci. C.* 16, 3519.
36. SNARE, M.J. (1983) Ph.D. Thesis, University of Melbourne.
37. SNARE, M.J., TRELOAR, F.E., GHIGGINO, K.P. and THISTLETHWAITE, P.J. (1982) *J. Photochem.* 18, 335.

38. ERNY, B. and MÜLLER, G. (1979) *J. Polym. Sci. Polym. Chem. Ed.* 17, 4011.
39. RICE, J., McDONALD, D.B., NG, L.-K. and YANG, N.C. (1980) *J. Chem. Phys.* 73, 4144.
40. BASHFORD, C.L., MORGAN, C.G. and RADDA, G.K. (1976) *Biochim. Biophys. Acta* 426, 157.
41. BIXON, M. (1976) *Ann. Rev. Phys. Chem.* 27, 65.

CHAPTER EIGHT

Fluorescence emission properties of optically active vinyl polymers containing carbazole and related aromatic units

8.1 INTRODUCTION

Carbazole is a heteroaromatic molecule having 14 equivalent π electrons, including the unshared electron pair located on the nitrogen atom, distributed on a 13- atom planar skeleton. Typical of this kind of aromatic structure is a highly pronounced molecular polarizability with a relatively low ionization potential, and the presence of the nitrogen atom with a lone pair of electrons has the effect of lowering the electronic affinity relative to the iso-electronic molecule anthracene [1]. As a consequence, carbazole derivatives exhibit relatively low oxidation potentials in solution and readily participate in formation of electron donor-acceptor complexes with a wide range of acceptors. Charge-transfer absorption spectra arising from electronic transitions of ground-state complexes have been very well characterized for many carbazole derivatives and organic acceptor molecules [2-4]. It is equally well established that electronically excited carbazole moieties and their charge-transfer complexes can participate in charge-transport and electron-transfer processes [4].

From a materials science point of view, particular advantage could be expected by combining the rather unique electronic properties of carbazole with incorporation into a polymer matrix. Experimentally the most convenient way to incorporate carbazole into a polymer matrix is by homopolymerization of the readily available monomer, *N*-vinylcarbazole (NVC).

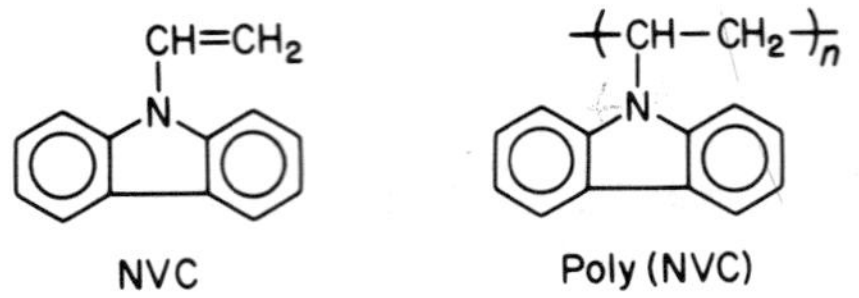

$CH=CH_2$ NVC

$-(CH-CH_2)_n-$ Poly (NVC)

Special interest in the properties of polymers containing carbazole units arises from the well established application of poly(NVC) to the development of composite layers highly efficient in charge-carrier generation and transport for application in electrophotographic copying equipments [5].

Homopolymerization of NVC is a well studied reaction that may be accomplished by either free radical or cationic processes [6], although there is still much uncertainty as to the presence or otherwise of any degree of configurational order in the polymer [7-15]. However, all the reports highlight an anomalous dependence of polymer microstructure on polymerization conditions that must be assumed to arise from a special combination of electronic and steric effects in the propagation process. Such effects appear to be important not only in affecting the polymerization processes of NVC, but also in controlling the properties of the polymeric products. The carbazole group is indeed sterically very bulky, and its accommodation in a vinyl homopolymer chain, via attachment at the nitrogen atom, can only be envisaged with a certain amount of interference or interaction between neighbouring chromophores. A most important manifestation of this interaction is evidenced by the ^1H NMR spectra of poly(NVC) where, independent of method of preparation, a significant component of the ring proton resonances appears at higher field than anticipated for any simple carbazole structure [16,17]. A similar interference of neighbouring carbazole units may be deduced from studies of ^{13}C NMR spectra of poly(NVC) [10,18].

The practical applications of poly(NVC) in electrophotography have generated substantial interest in studies of photoconductivity, photoluminescence, and cation radical (charge carrier) formation [4]. Arising from these studies it has been recognized for many years that, amongst vinyl polymers, the fluorescence of poly(NVC) is rather unique, consisting of two well resolved excimer emission bands [19, 20]. There is general agreement that the high- and low-energy excimer emission bands arise respectively from a built in partial carbazole-carbazole interaction, and a more usual rotationally populated fully eclipsed carbazole-carbazole sandwich-like structure. More recently, studies on low molecular weight model compounds reported independently by De Schryver *et al* [21] and Evers *et al* [22] suggest that the high- and low-energy excimers in poly(NVC) arise from syndiotactic and isotactic sequences as anticipated by Itaya *et al* [23].

It is well known that optical activity is very sensitive to conformational variations. Accordingly, chiroptical techniques (ORD and CD) have been widely used to investigate conformational equilibria of organic compounds, mainly in solution [24,25]. Theoretical interpretation of optical activity is very complex and satisfying, easily approachable, theories are not yet available [26]. In this connection, polymers, because of the linear repeating structure, offer some advantages and can be approached by semi-empirical methods that allow evaluation of the basic chiroptical properties for structurally ordered systems [27]. Natural polymers are generally optically active and can thus be studied directly with chiroptical techniques. This is not the case for most synthetic polymers, since circular dichroic absorption and optical rotation are observed only if optically active monomers or catalysts are used in synthesis [28]. The preparation of chiral monomers, particularly with high enantiomeric purity, is in general cumbersome and very expensive. The apparently convenient synthetic procedure based on the use of optically active catalysts and racemic or prochiral monomers gives good optical yields only in few cases [29-31]. Basic investigations on synthetic optically active polymers, especially when obtained by stereospecific processes, have shown that ordered secondary structures can exist in solution displaying often larger optical activity than low molecular weight structural models [32,33]. The extension of these investigations to copolymers of optically active vinyl monomers with prochiral aromatic monomers has evidenced both the value of chiroptical techniques in gaining a more definite insight into the conformational structure in solution of polymers containing both UV transparent and non-transparent monomeric units [34], and the convenience of the copolymerization procedure to achieve optically active polymers from prochiral monomers [25,28].

Application of chiroptical techniques to the study of suitably synthesized carbazole-containing polymers would have obvious value in interpreting inter-carbazole interactions in terms of configurational and conformational aspects. Accordingly for almost a decade we have studied, in a systematic way, the synthesis and properties of optically active polymers containing carbazole groups [35-44] in an attempt to evaluate further the correlation of structural and electronic properties. This review incorporates these studies together with related results for optically active polymers containing other polarizable aromatic groups. Fluorescence and chiroptical properties have been studied for copolymers of NVC, and other vinyl aromatic monomers, with (-)-menthyl acrylate (MtA), (-)-menthyl methacrylate (MtMA), and (-)-menthyl vinyl ether (MtVE).

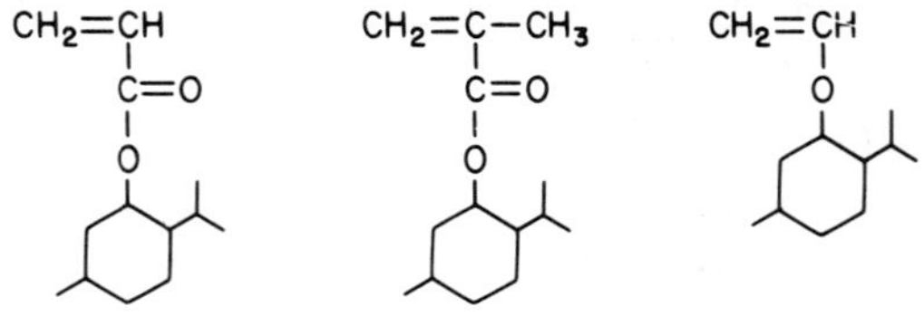

$CH_2{=}CH{-}C({=}O){-}O{-}$menthyl — MtA

$CH_2{=}C(CH_3){-}C({=}O){-}O{-}$menthyl — MtMA

$CH_2{=}CH{-}O{-}$menthyl — MtVE

Aromatic groups studied in this context were incorporated via the vinyl derivatives styrene (St), 1-vinylnaphthalene (1VN), 6-vinylchrysene (6VC), acenaphthylene (Ace), *trans*-4-vinylstilbene (4VS) and *trans*-4-acryloyloxystilbene (4AS).

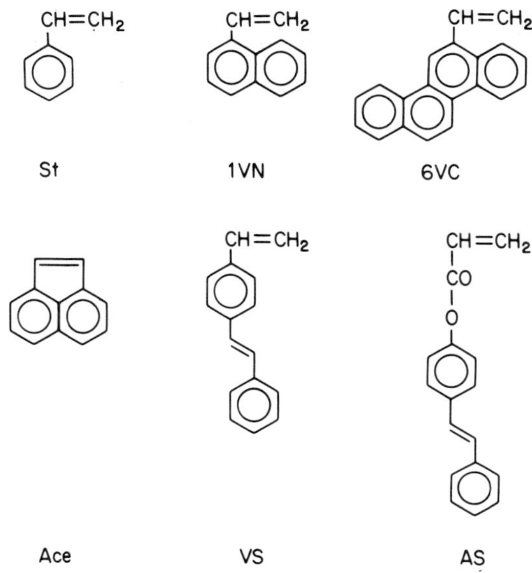

8.2 FLUORESCENCE EMISSION PROPERTIES

These are discussed in earlier chapters, but a brief resumé will be given here.

8.2.1 Carbazole-containing polymers (discussed also in Chapters 2 and 5) The dual excimer fluorescence behaviour of poly(NVC) is very well documented and discussed [19,20]. For all samples of poly(NVC) there is a high-energy structureless emission band at 370 nm and, depending on the method of preparation, a more or less significant second structureless emission band at 420 nm (Fig. 8.1). Fluorescence life-time determinations show clearly that both emission bands originate from excimer-like structures and are in complete contrast with the typical highly structured isolated carbazole fluorescence [20,45-47]. Several interpretations of the origins of the two excimer bands have been proposed and, based on the different effect of temperature on the two bands, it is widely supposed that the low-energy excimer arises from a normal eclipsed arrangement of adjacent carbazole units [20-23]. It is important to recognize that excimer formation in carbazole-containing polymers is observed only when the carbazole nucleus is attached directly to the polymer

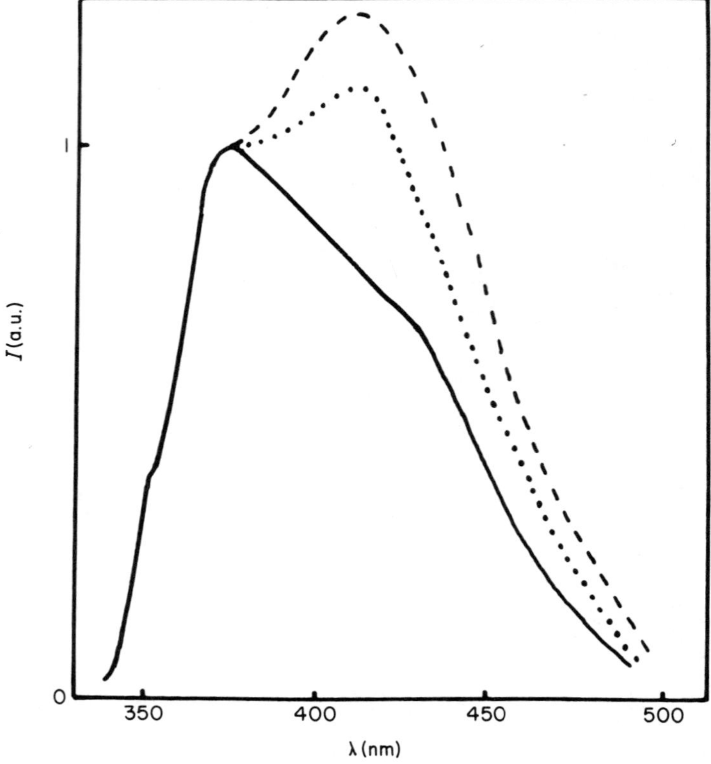

Fig. 8.1 Corrected fluorescence emission spectra normalized at 370 nm of poly(NVC) prepared by different catalytic systems: - - - $C_2H_5AlCl_2$; . . . $TiCl_3/Al(\underline{i}-C_4H_9)_3$; ——— AIBN.

backbone [41,44,48-50]. In this respect, carbazole behaves very differently from other aromatic molecules [51-54], suggesting that excimer emission in carbazole polymers arises primarily from substituent and backbone conformational constraints. Very recently it has been claimed that, as an infringing exception to the Hirayama's rule [55], configurationally and conformationally homogeneous poly{[N^{ε}-(9-carbazolyl)carbonyl]-L-lysine} with carbazole nuclei located distant from polymer main chain displays a definite and well resolved excimer emission in solution [56]. It should be noted however that the interacting groups in this case are carbazolylamido units rather than the usual alkyl carbazole groups. Amido groups may be expected to exhibit ground state dipolar effects.

For carbazole-containing polymers, the formation of two contemporary excimer sites appears to be restricted to poly(NVC). The ring substituted analogue poly[(*S*)-3-*sec*-butyl-9-vinylcarbazole] [poly(3R9VC)] gives homopolymers in which isolated carbazole emission is practically absent, as for poly(NVC), but with a relatively greater contribution from the corresponding high-energy excimer [39]. Such a result is to be expected on the basis of the proposed rotational eclipsing required for the formation of the low-energy excimer. It is also noteworthy that homopolymers of the ring vinylated carbazoles, poly[(*S*)-9-(2-methylbutyl)-2-vinylcarbazole] [poly(9R2VC)] and poly[(*S*)-9-(2-methylbutyl)-3-vinylcarbazole] [poly(9R3VC)], seem to exhibit only isolated carbazole and eclipsed (i.e. low-energy) excimer emission, as anticipated [57,58].

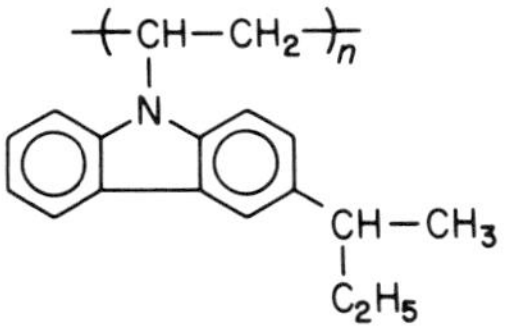

Poly (3R9VC) 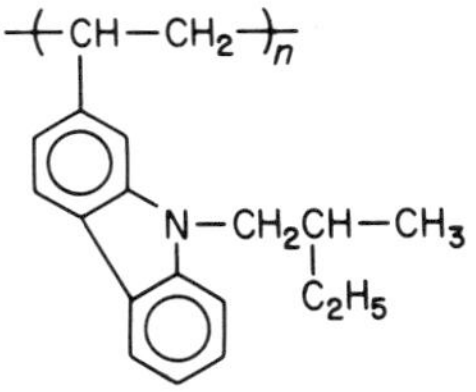 Poly (9R2VC) 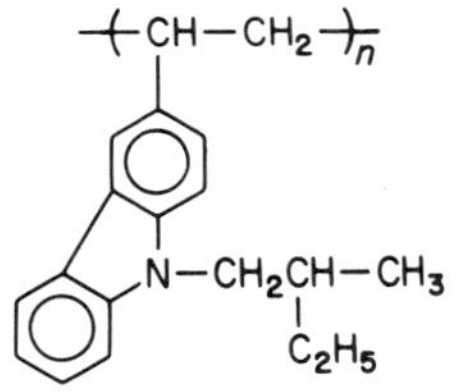 Poly (9R3VC)

An insight into the origins of the two excimer sites in poly(NVC) is provided by recent work of De Schryver *et al* [21] and Evers *et al* [22]. These workers have independently reported rather elegant studies on the diastereomeric isomers of the model compound 2,4-di(9-carbazolyl)pentane (D9CP). The *meso* isomer of D9CP corresponds to isotactic diads in poly(NVC) and significantly gives rise to emission of both isolated carbazole at 350 nm and eclipsed, low-

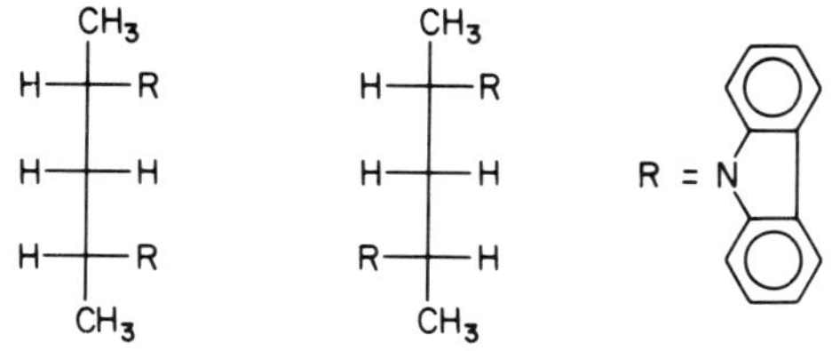

meso-D9CP *racemic*-D9CP

energy excimer at 420 nm. In contrast, the *racemic* isomer of D9CP, which corresponds to syndiotactic diads in poly(NVC), gives rise to a structureless emission band at 370 nm, similar in energy to that of isolated carbazole, and corresponding to the high energy excimer emission in poly(NVC) (Fig. 8.2). It is reasonable to conclude therefore that the origins of high- and low-energy excimer emissions in poly(NVC) arise mainly from syndiotactic and isotactic sequences respectively.

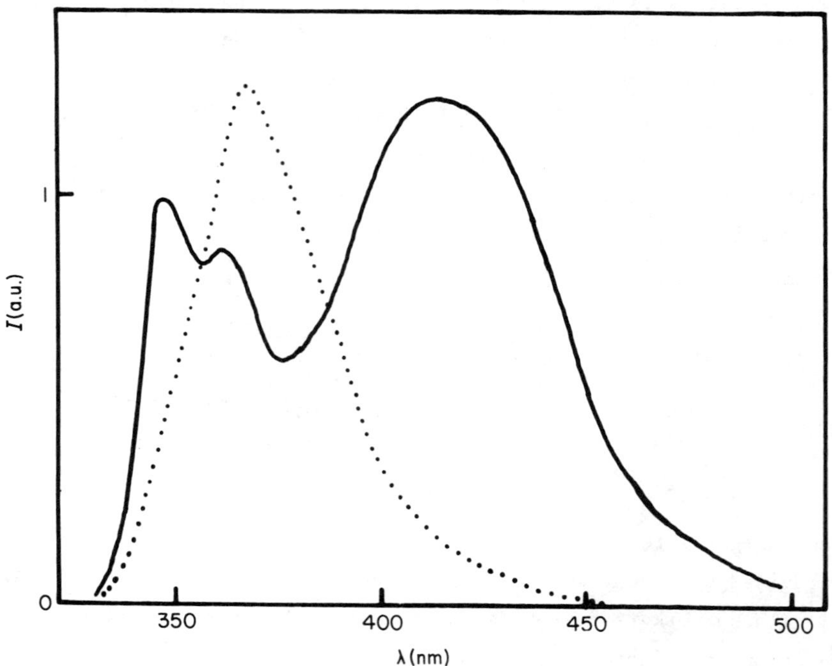

Fig. 8.2 Fluorescence emission spectra of diasteromeric 2,4-di-(9-carbazolyl)pentane (D9CP): ——— *meso* isomer; ··· *racemic* isomer (after De Schryver *et al* [21]).

From these results, and from the reported extrapolated values of 276°C (syndiotactic) and 126°C (isotactic) for the glass transition temperatures of poly(NVC) [15], it has been concluded that rotational eclipsing in isotactic NVC sequences is much more favourable than in corresponding syndiotactic sequences. This conclusion is strongly supported by examination of space-filling molecular models. Such considerations must be borne in mind when considering the optical and chiroptical properties of carbazole-containing polymers, and it has to be expected that copolymers of NVC will have fluorescence emission properties that can be related to both main chain tacticity and distribution of monomeric units.

Free radically polymerized mixtures of NVC and acrylic derivatives of (-)-menthol have fluorescence spectra very similar to that of the isolated carbazole chromophore (up to 40% NVC units) in accordance with the anticipated random distribution of NVC units [42]. Increasing NVC content in the region 40-70% gives rise to a small but significant excimer emission. In complete contrast, the cationic copolymerization of NVC with MtVE produces block-like structures in which excimer emission may be detected even at 0.1% incorporation of NVC units [39] (Fig. 8.3). Cationic polymerization of MtVE gives isotactic polymers [59] and, under the experimental conditions chosen, copolymerization with NVC would be expected to yield sequences of both monomers having predominantly isotactic character [9,60]. All polymeric samples do exhibit fluorescence spectra in which both types of excimer emission are recognizable. In particular, the low-energy excimer becomes progressively more important with increasing NVC content, in keeping with a rather high degree of isotacticity. Correlations of excimer emission with copolymer composition are represented in Fig. 8.4.

It is interesting to note that in apparently alternating copolymers of NVC and optically active fumarate esters, (-)-ethyl menthyl fumarate (EMtF) and (-)-dimenthyl fumarate (DMtF), the fluorescence emission spectra indicate some degree of excimer-like interactions [43,61]. In these cases, it is extremely unlikely that

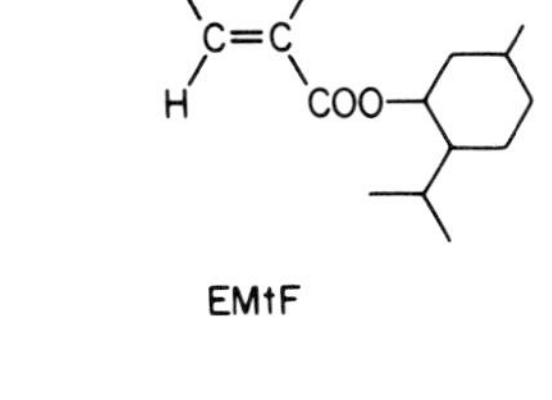

C_2H_5OOC–CH=CH–COO–(menthyl)

EMtF

(menthyl)–OOC–CH=CH–COO–(menthyl)

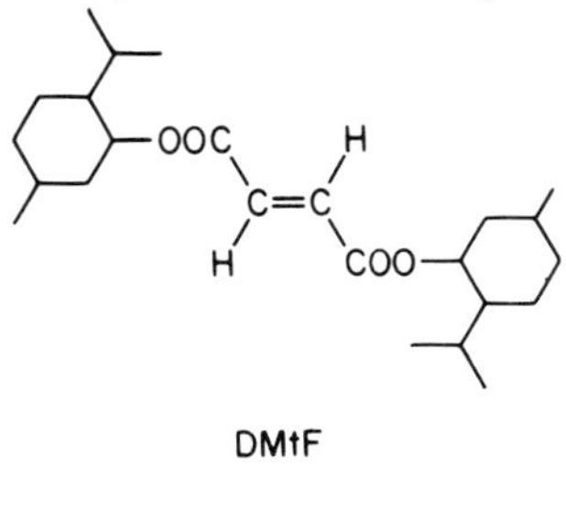

DMtF

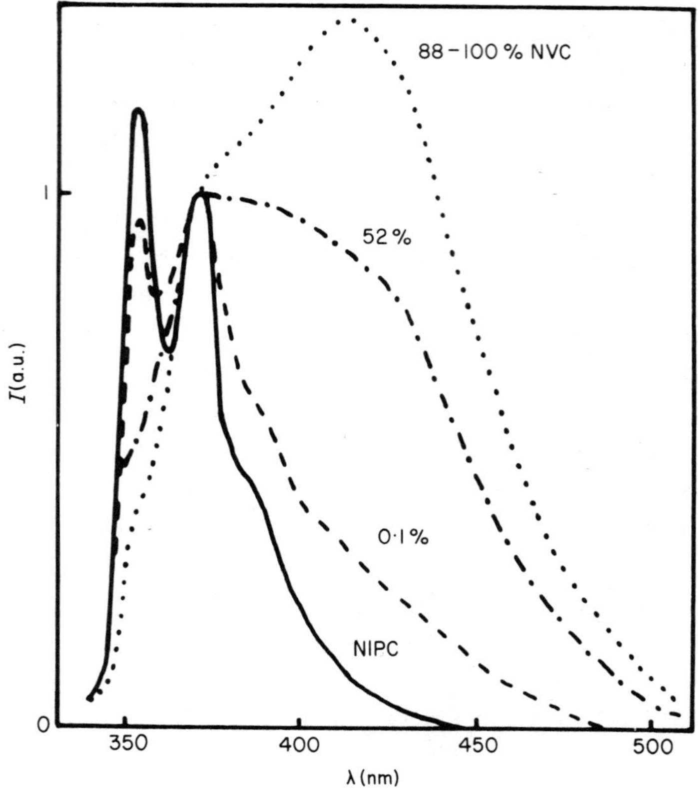

Fig. 8.3 Corrected fluorescence spectra normalized at 370 nm of copolymers of NVC with MtVE, and of 9-*iso*propyl-carbazole (NIPC).

excimer formation could arise from nearest-neighbour carbazole units, and according to suggestions put forward by Ghiggino and coworkers [62] long-range interactions may be responsible.

Comparing published reports [41,49] it is clear that removal of the carbazole moiety from the polymer backbone by any kind of spacer group eliminates the formation of excimer emission sites. This applies even to homopolymers of the rather rigid carbazole-substituted styrenes, *m*-(9-carbazolylmethyl)styrene (*m*9CSt) and *p*-(9-carbazolylmethyl)styrene (*p*9CSt) [44].

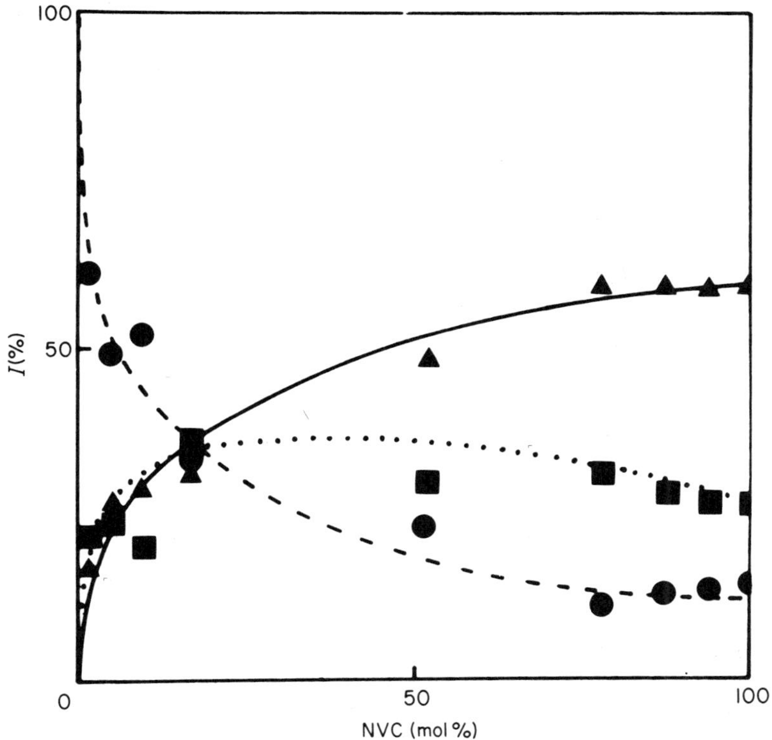

Fig. 8.4 Trends of relative intensity (I) of different components of fluorescence emission of copolymers of NVC with MtVE against NVC content: —·●—— monomer; ··■·· high energy excimer; —▲— low-energy excimer.

$CH=CH_2$ CH_2 N

m9CSt

$CH=CH_2$ CH_2 N

p9CSt

8.2.2 *Styrene-containing polymers* (discussed also in Chapters 2, 3, 5, 6 and 7).

The fluorescence spectrum of polystyrene shows emission bands at about 289 and 380 nm, attributed to monomer and excimer emission, respectively [63]. Relative intensities of the two peaks are significantly affected by polymer main chain tacticity [64] and, in St copolymers, by sequence distribution of aromatic units [65]. Such relationships were of special value in the characterization of copolymers of St with (*S*)-4-methyl-1-hexene (4MH) prepared in the presence of a Ziegler-Natta catalyst [66]. In fact, reactivity

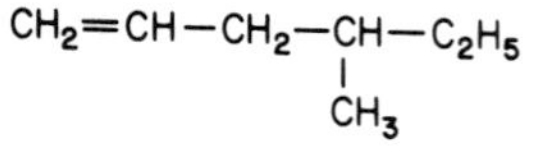

$$CH_2{=}CH{-}CH_2{-}\underset{\displaystyle CH_3}{\underset{|}{CH}}{-}C_2H_5$$

4MH

ratios suggest an almost random distribution of monomeric units (r_{St}=0.76 and r_{4MH}=1.65), whereas fractionation by solvents indicates that these copolymers are blocky in nature. Fluorescence spectra of unfractionated copolymers show the typical dependence of excimer emission on the content of aromatic units (Fig. 8.5). However spectra of copolymer fractions soluble in diethyl ether or in cyclohexane exhibit monomeric emission only, whereas the spectra of fractions soluble in chloroform show the pattern typical of polystyrene. This demonstrates that copolymer samples are constituted by a mixture of macromolecules containing almost isolated units of one comonomer flanked by long sequences of the other one, in agreement with solubility data. The validity of such interpretation has been independently confirmed by ^{13}C-NMR analysis [67]. Additionally, fluorescence emission properties of copolymer fractions containing long styrene sequences indicate the existence of a rather high degree of isotacticity.

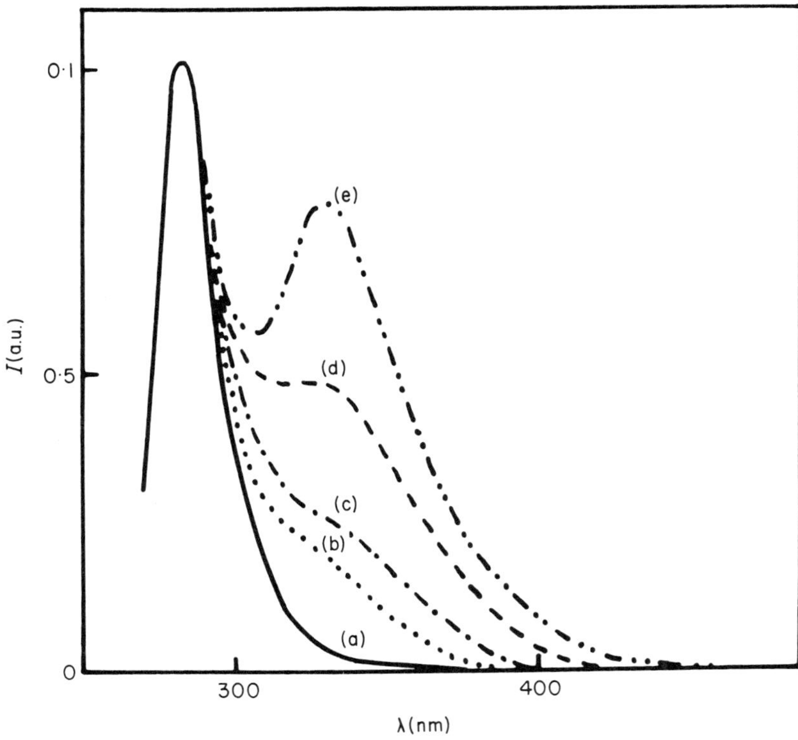

Fig. 8.5 Fluorescence spectra normalized at 283.5 nm of: (a) *iso*-propylbenzene; (b)-(d) unfractionated copolymers of styrene (St) with (*S*)-4-methyl-1-hexene containing 15.4, 68.7, and 87.2 mol-% of St units, respectively; (e) isotactic polystyrene.

8.2.3 *Naphthalene-containing polymers*

The fluorescence spectra of copolymers of 1VN with MtA and MtMA [68] consist of monomer and excimer emissions centred at about 328 and 415 nm, respectively, as in poly(1VN) and in other copolymer systems based on 1VN [69-72]. Some typical fluorescence spectra for the 1VN-*co*-MtMA series are shown in Fig. 8.6. The ratio of the excimer-to-monomer fluorescence intensity (I_E/I_M) increases with increasing content of vinyl aromatic units in all the radically prepared copolymers. This is consistent with previous observations on other types of copolymers [69-71, 73].

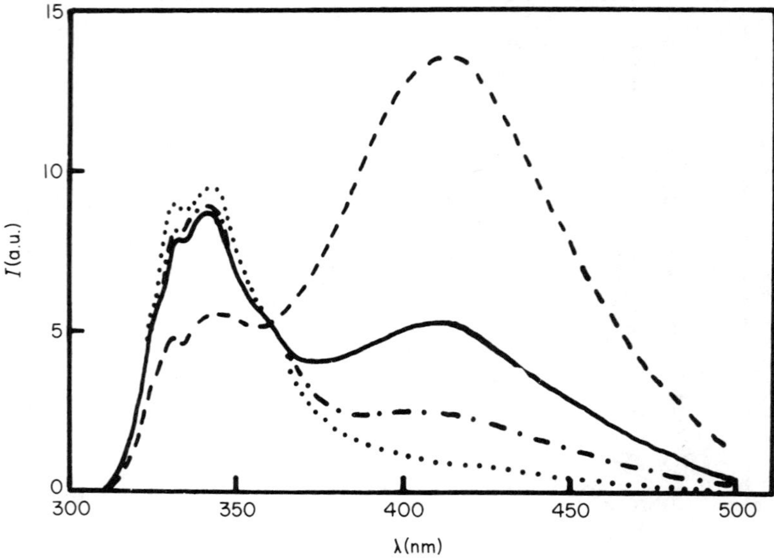

Fig. 8.6 Fluorescence emission spectra normalized at 360 nm of copolymers of 1VN with MtMA having different content of 1VN units: ---57.3 mol.%; —— 33.2 mol.%; ··· 8.5 mol.%; -·- poly(MtMA-*alt*-1VN).

It has been suggested that, in copolymers in which excimer formation occurs predominantly between nearest-neighbour chromophores, the ratio I_E/I_M should be proportional to the fraction of diads of fluorescent units, as observed for the system styrene-*co*-methyl methacrylate [65]. However, this linear trend was not found to occur over a substantial copolymer composition range for many copolymer systems [72], including copolymers of 1VN with optically active MtA and MtMA (Fig. 8.7) [68]. Other correlations have been proposed to describe the dependence of excimer emission intensity on polymer structure also in terms of energy migration [72]. However, recent work [74] has highlighted the very low efficiency of energy migration in naphthalene-containing polymers and, although formation of excimer emitting sites is well proven, it is no longer worthwhile to make correlations of polymer microstructure with excimer emission intensity.

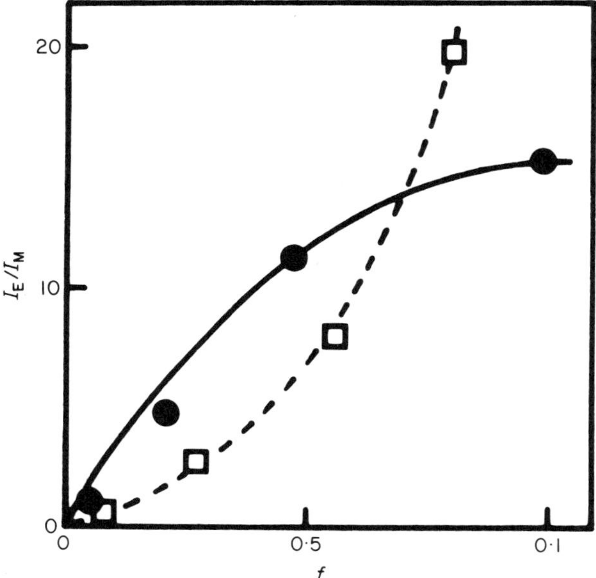

Fig. 8.7 Dependence of the ratio excimer-to-monomer fluorescence intensity (I_E/I_M) on the fractions of links between naphthalene units (*f*) for copolymers of 1VN with —●— MtA and —□— MtMA, having different 1VN content.

8.2.4 Chrysene-containing polymers

Systematic investigations of homopolymers and copolymers of 6VC have not yet been carried out, nevertheless fluorescence properties appear to be of great potential in the assessment of primary and secondary structure of these materials [75]. Indeed, homopolymer samples obtained according to different initiation processes display marked differences in their fluorescence spectra. In particular, cationically prepared polymers show only excimer emission at about 450 nm (Fig. 8.8), whereas samples prepared by Ziegler-Natta catalysts exhibit a rather intense monomeric component centred at 400 nm, intermediate behaviour being observed in samples obtained by free radical initiation. Such results must reflect different stereochemical features in various polymeric products and it may be envisaged that samples prepared by cationic initiators possess a higher degree of main-chain stereoregularity, and conceivably the typical anionic-coordinate mechanism is not operative in the polymerization of 6VC so that Ziegler-Natta catalysts behave very much as free radical initiators.

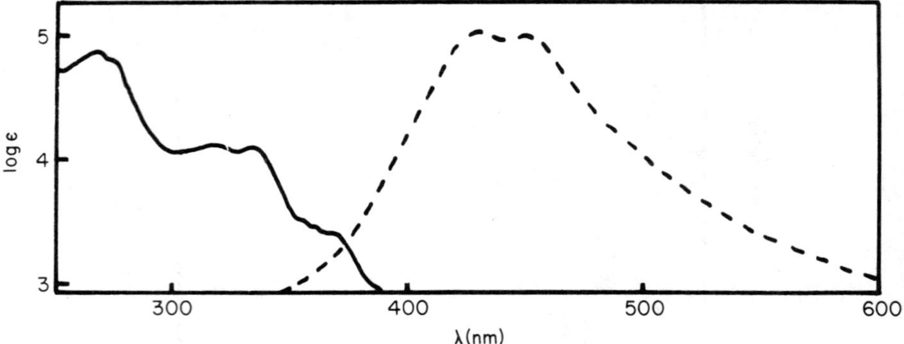

Fig. 8.8 UV (——) and fluorescence emission (---) spectra of poly(6-vinylchrysene) prepared by cationic polymerization.

It is worth stressing that cationically prepared poly(6VC) displays exclusively excimer emission, in analogy to poly(NVC) and poly(1-vinylpyrene) [76]. This suggests that in polymer systems with extended and highly polarizable pendant aromatic units there exists a high concentration of excimer forming sites that are easily populated by energy migration, with a consequent very efficient quenching of monomer type emission.

The fluorescence spectra of the copolymerization products of 6VC with optically active acrylic comonomers displayed solely monomeric emission in free radically prepared samples containing up to 45% mol of aromatic units, whereas a combination of monomer and excimer components was detected in the anionic-coordinate copolymer with 4MH having the same (45%) molar content of chrysene moieties (Fig. 8.9). Such apparent discrepancy of emission properties may be explained by assuming a random distribution of monomeric units in the former polymer systems, whereas a more blocky structure must occur in the copolymer obtained by Ziegler-Natta catalysis.

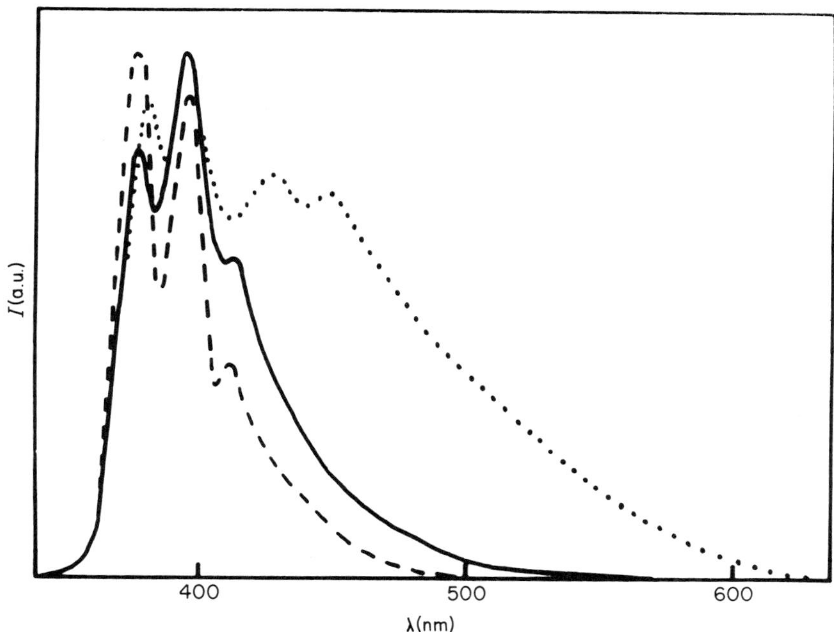

Fig. 8.9 Fluorescence emission spectra of copolymers of 6-vinylchrysene (6VC) with (---) MtMA, (——) MtA and (···) (*S*)-4-methyl-1-hexene, containing about 45 mol% of 6VC units.

8.2.5 *Acenaphthylene-containing polymers* (discussed also in Chapter 5)

Acenaphthylene homopolymers [77] and copolymers [78-82] have attracted much interest, due to the peculiar way in which the monomer unit is incorporated in the polymer backbone, prohibiting the typical sandwich-like conformation between nearest-neighbouring chromophores.

It was first reported by Wang and Morawetz [79] that copolymers of Ace with monomers capable of giving a quaternary carbon atom in the polymer main chain (e.g. methyl methacrylate) could, under appropriate conditions, exhibit excimer formation in excess of that of the homopolymer of Ace. In contrast, similar copolymers of Ace with methyl acrylate showed a much reduced tendency to form excimers. The apparent difference in ease of excimer formation between copolymers of Ace with methyl acrylate and methyl methacrylate having similar compositions has prompted more detailed studies by Reid and Soutar [80] and Anderson *et al* [81] and by David *et al* [82]. In these later works the effects noted by Wang and Morawetz have been confirmed, and the question of energy migration in both series of copolymers has been investigated by means of fluorescence depolarization experiments.

The peculiarities observed in the fluorescence of Ace copolymers can be explained only on the basis of distributions of monomeric units and local configurational and conformational order. It seemed therefore profitable to investigate homologous series of copolymers of Ace with MtA and MtMA [83,84], whose fluorescence and chiroptical properties could be related to primary and secondary structure in solution.

Fluorescence emission spectra of Ace with MtA and MtMA are characterized by a structured band located at about 340 nm, whose position is almost independent of the chemical composition, and a broad, structureless band centred at about 400 nm, whose intensity and profile strongly depend on copolymer composition. In particular, the intensity of the latter emission increases in a monotonic fashion in poly(Ace-*co*-MtA) samples, whereas it reaches a value in excess of that of poly(Ace) in the poly(Ace-*co*-MtMA) series (Fig. 8.10). The two bands can be attributed to a monomer and excimer emission, respectively, but it should be noted that, owing to the structural peculiarity of the aromatic unit in Ace homopolymer and copolymers, emitting excimer species cannot be formed through the interaction of two nearest-neighbour chromophores [77]. Excimer-to-monomer emission ratios exhibit a clear maximum for the methacrylate series at approximately 80 mol. % Ace content, in both CH_2Cl_2 and 2-methyltetrahydrofuran solvents (Fig. 8.11). However, an overall reduction of the capability of excimer formation takes place in the latter solvent, probably due to differences in chain conformations and/or changes in quenching efficiences. On the contrary, the I_E/I_M ratio decreases monotonously with decreasing Ace content for the acrylate series (Fig. 8.12). This trend is similar to that previously reported in analogous copolymer systems of Ace containing achiral comonomers [80,81]. The different behaviour of the two series of copolymers can be interpreted assuming that in the methacrylate copolymers there is a structural effect of the comonomer that serves to enforce the alignment of two Ace units required for excimer formation, whereas this effect is apparently not present in the much more flexible acrylate copolymers.

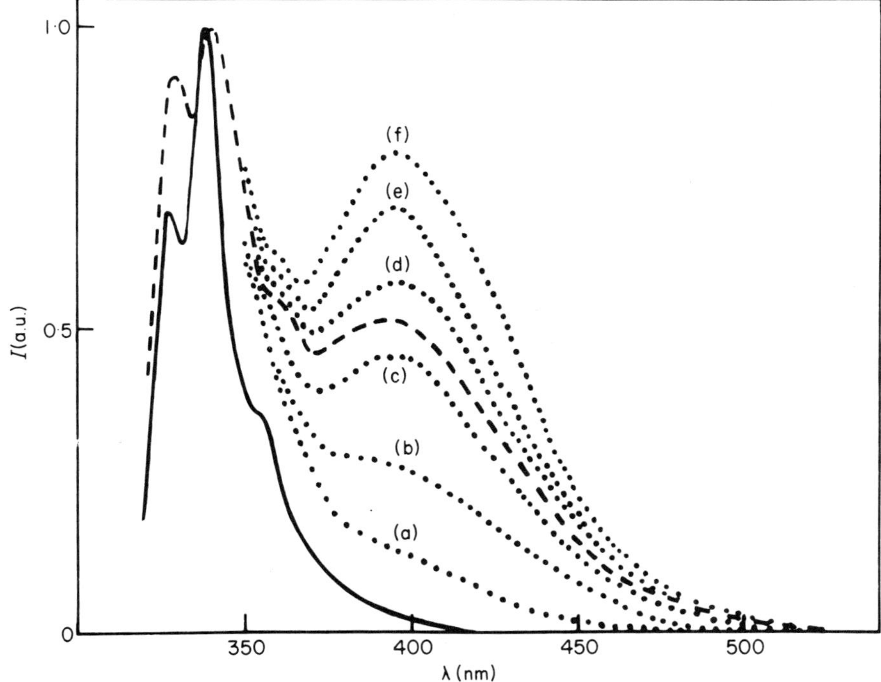

Fig. 8.10 Fluorescence emission spectra normalized at the maximum intensity of (——) acenaphthene, (---) poly-(acenaphthylene), and (a)-(f) copolymers of acenaphthylene (Ace) with MtMA containing 14.7, 33.9, 50.1, 79.0, 88.8, and 96.8 mol% of Ace units, respectively.

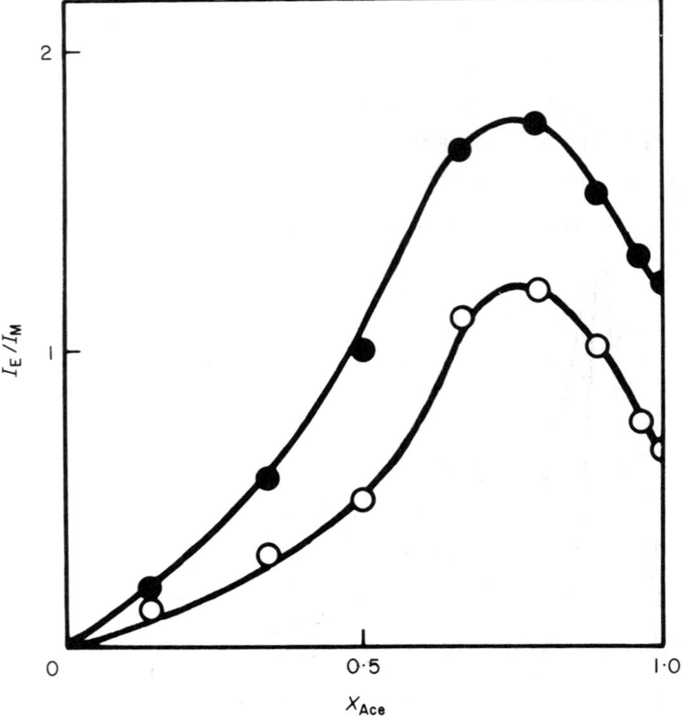

Fig. 8.11 Variation of the excimer to monomer emission intensity ratio (I_E/I_M) with Ace molar fraction in copolymers of Ace with MtMA in —●— dichloromethane and in —○— 2-methyltetrahydrofuran.

It is generally assumed that polymerization of Ace involves overall trans addition to the double bond [85], but any preference for cis addition would yield an obvious configurational difference, which might be reflected in enhanced excimer formation. Alternatively it is also possible that enhanced excimer formation in the methacrylate copolymers is caused by conformational effects arising from the presence of quaternary carbon atoms in the polymer backbone, as originally proposed by Wang and Morawetz [79]. These would have the obvious effect of reducing segmental mobility in the poly(Ace-co - MtMA) series.

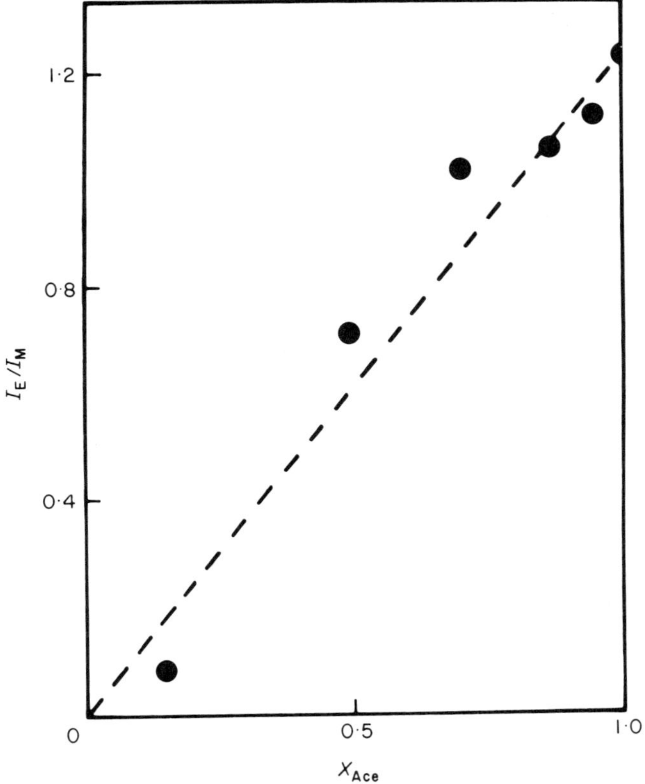

Fig. 8.12 Variation of the excimer to monomer emission intensity ratio (I_E/I_M) with Ace molar fraction in copolymers of Ace with MtA.

Reid and Soutar [80] and Anderson *et al* [81] have proposed a kinetic scheme to accommodate the peculiar fluorescence properties in acenaphthylene-containing copolymers and have found that excimer formation is adequately quantified over all the composition range by the equation:

$$I_E/I_M = k\Sigma_2 f_a$$

where Σ_2 is a factor descriptive of concentration of excimer forming sites on the basis of probability of different triads, and f_a is the molar fraction of aromatic units correlated with the concentration dependence of energy migration, as evaluated by measurements of fluorescence depolarization. However, recent studies have questioned whether this statistical treatment can be applied to Ace copolymers because extensive depolarization of fluorescence was observed not only for all the copolymers with MtA and MtMA($p<0.022$), but also for the low molecular weight model compound acenaphthene ($p=0.045$). Depolarization in acenaphthene could arise in part as a result of the mixed polarization of the vibrational levels in the long wavelength absorption band of the chromophore [82]. For all the copolymers as noted above, there is extensive depolarization and differences in residual polarization are very small. It does not seem reasonable therefore to make quantitative correlations of low degrees of residual polarization (especially in view of the likely experimental errors) with copolymer structure. For the moment, the extent and mechanism of energy transfer in polymers containing acenaphthylene units remain to be better established.

8.2.6 Stilbene-containing polymers

Polymers containing side chain *trans*-stilbene chromophores have attracted particular attention in view of the possible conformational variations in the macromolecular backbone induced by side chain *trans*→*cis* photoisomerization [52,86,87].

Homopolymers of the acrylic ester of *trans*-4-hydroxystilbene (4AS) exhibit both monomer and excimer emissions, and in the case of copolymers with MtA the relative intensity of the excimeric component can be at least qualitatively related to monomer sequence distribution (Fig. 8.13).

Copolymers of *trans*-4-vinylstilbene (4VS) with MtMA show a somewhat different behaviour, in fact on increasing the 4VS content from 15 to 100% the fluorescence emission maximum shifts from 362 to 415 nm (Fig. 8.14). This finding is obviously related to the mean sequence length of 4VS units. However, it is not completely understood whether the effect arises from increased contributions by a not well resolved, long wavelength excimeric component or, at least partially, to the distortion of the geometry of the stilbene

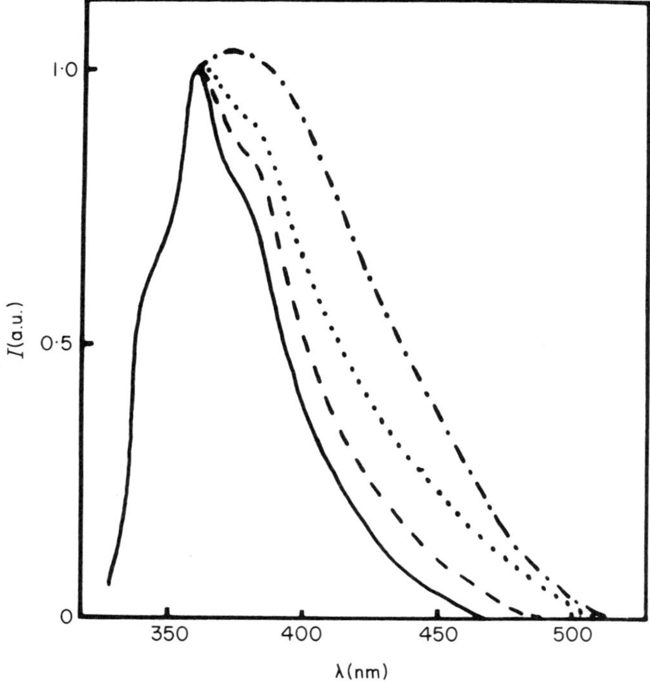

Fig. 8.13 Corrected fluorescence emission spectra normalized at 360 nm of (___) 4SP and of copolymers of 4AS with MtA containing (---) 25.0, (···) 68.6, and (-·-) 100 mol% of 4AS units.

chromophore as indicated by the UV spectra [86].

On irradiation in the first $\pi\rightarrow\pi^*$ electronic transition, the stilbene chromophore undergoes *trans→cis* isomerization. Fluorescence quantum yields for the *cis*-stilbene chromophore are negligible and accordingly in copolymer samples of 4AS and 4VS, on increasing the *cis* content, the emission intensity decreases, whereas the excimeric component progressively disappears. This has been attributed to the reduced mean sequence length of *trans*-stilbene units, affecting both energy migration and concentration of excimer forming sites, and supported also by the non-linear dependence of emission intensity on extent of isomerization [52,89].

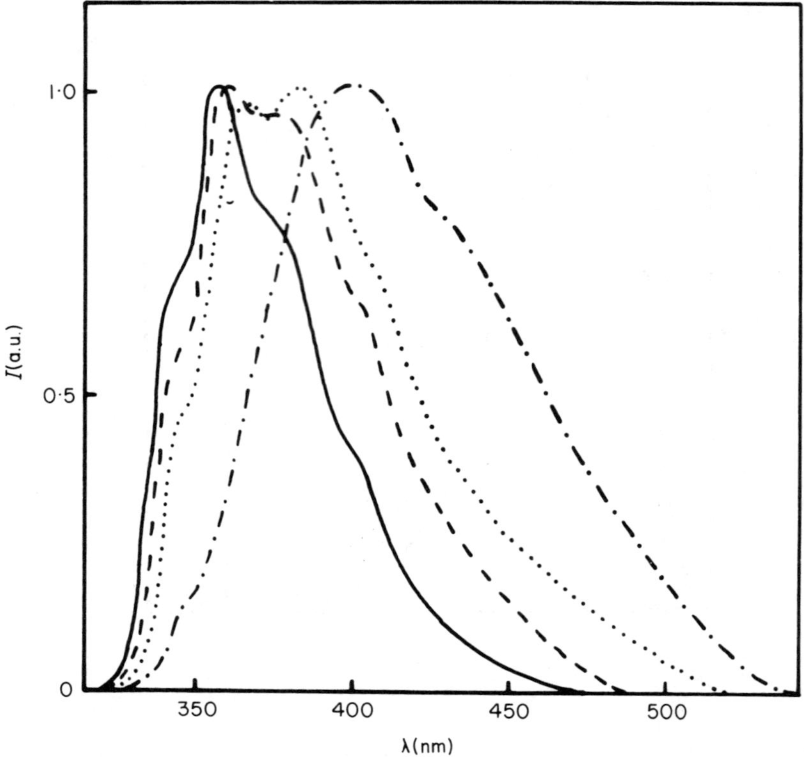

Fig. 8.14 Corrected fluorescence spectra normalized at the maximum intensity of (——) *trans*-4-methylstilbene and of copolymers of *trans*-4-vinylstilbene (4VS) with MtA containing (---) 14.9, (···) 41.0, and (-·-) 100 mol% of 4VS units.

The singlet state of *trans*-stilbene is very short lived and therefore measurements of fluorescence depolarization can be carried out in solution at room temperature. In copolymers containing mainly isolated *trans*-stilbene moieties the degree of fluorescence polarization is larger than that observed for the corresponding low molecular weight model compounds, *trans*-4-hydroxy-stilbene-2-methylpropionate (4SP) and *trans*-4-methylstilbene (4MS), thus indicating a restricted mobility of the stilbene group when attached to a polymer backbone.

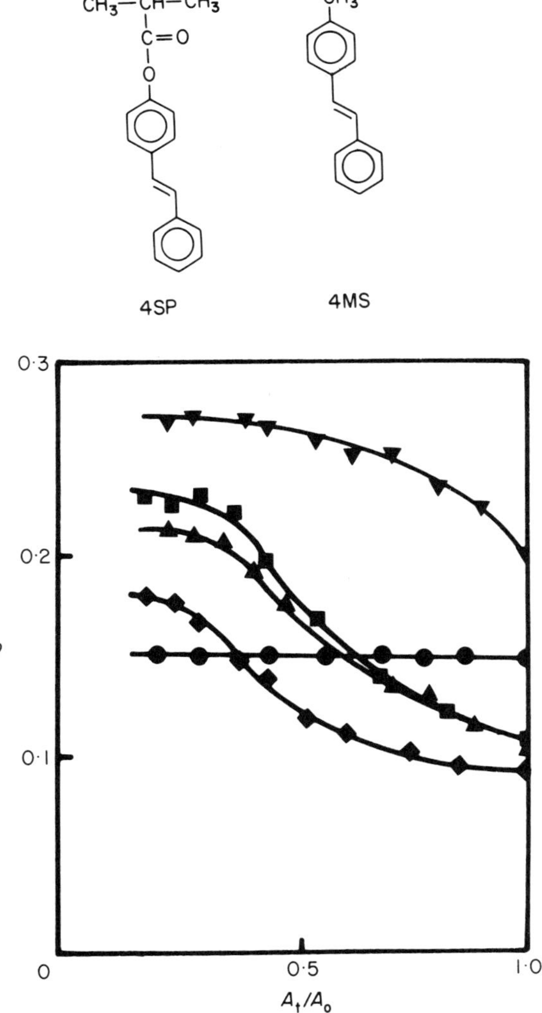

Fig. 8.15 Variation of the degree of fluorescence polarization (p) with the relative absorbance (A_t/A_o) on irradiation at 330 nm of (-●-) 4SP and of copolymers of 4AS with MtA containing (▼) 12.6, (■) 25.0, (▲) 87.0 and (◆) 100 mol. % of 4AS units.

In contrast, a degree of polarization lower than that of the models is observed in polymers containing sequences of *trans*-AS or *trans*-VS units, due to energy migration. On UV irradiation, the stilbene-containing polymers progressively undergo *trans*→*cis* photoisomerization of the stilbene chromophores and correspondingly the degree of polarization measured for polymers with varying *cis*/*trans* ratios increases up to the value of isolated *trans*-stilbene groups [52,89] (Fig. 8.15). This is because the chromophores in the *cis* configuration do not participate in energy migration.

8.3 CHIROPTICAL PROPERTIES

Generally optical rotatory measurements at 589 nm ($[\alpha]_D^{25}$), and even at shorter wavelengths, in copolymers of optically active monomers and prochiral comonomers can seldom provide clear indications of any preferentially induced asymmetry for the prochiral comonomer units [34,90,91]. A substantial improvement in analysis of macromolecular conformation in solution by chiroptical techniques can be gained in particular by circular dichroism (CD) measurements on copolymers of optically active 'transparent' monomers with comonomers containing an aromatic chromophore, which acts as a spectroscopic probe [34,92].

8.3.1 Styrene-, naphthalene-, and chrysene-containing polymers

In copolymers of styrene with α-olefins the only absorbing moiety in accessible spectral regions is the benzene chromophore, the electronic transitions related to its π electron system being located at wavelengths greater than 180 nm. In the coisotactic copolymers with optically active 3,7-dimethyl-1-octene, in which the asymmetric carbon atom is in the α position to the main chain, the benzene chromophore is optically active as shown by the presence of a dichroic band around 265 nm where its lowest energy $\pi\rightarrow\pi^*$ electronic transition ($^1L_b\leftarrow{}^1A$) is located. The ellipticity of this dichroic band is considerably larger than in low molecular weight structural models [90]. Successive examinations down to 185 nm showed [93] that the $^1L_a\leftarrow{}^1A$ and $^1B\leftarrow{}^1A$ electronic transitions also gave rise to CD bands,

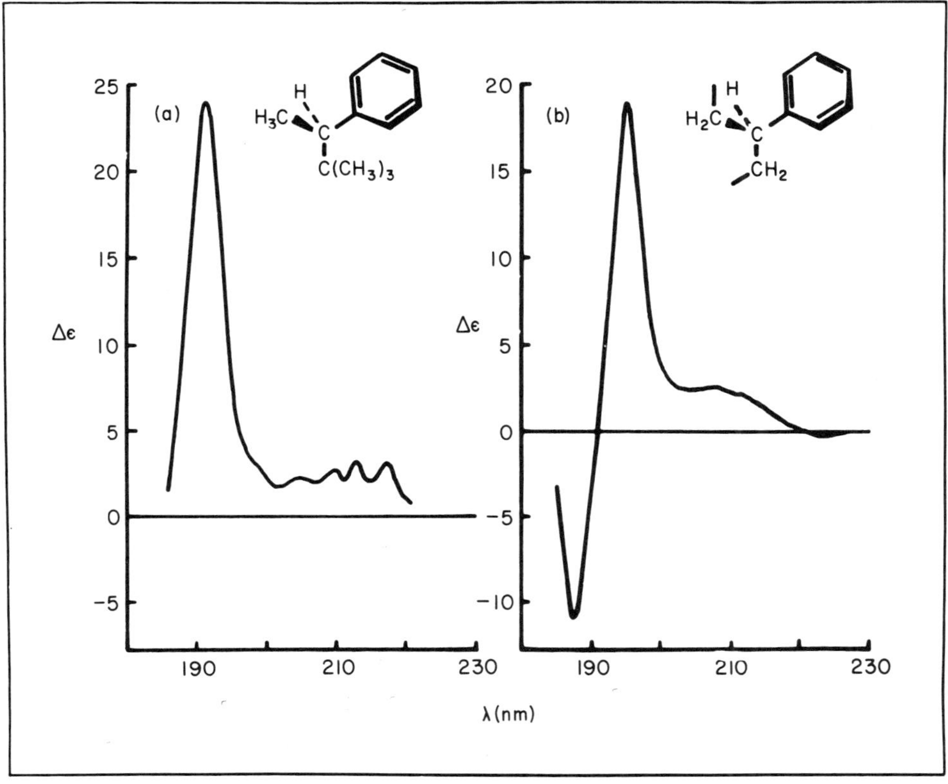

Fig. 8.16 CD spectra of (a) (*S*)-2,2-dimethyl-3-phenylbutane, and (b) copolymer of styrene (St) with (*S*)-3,7-dimethyl-1-octene containing 20 mol% of St units.

the last one assuming the typical exciton splitting into two dichroic bands with similar ellipticity and opposite sign (Fig. 8.16). The conformational low molecular weight model (*S*)-2,2-dimethyl-3-phenylbutane, for which the conformational analysis predicts a similar situation as in the macromolecules, displays comparable optical rotation and ellipticity as the isotactic copolymer [94]. However, this conformational model of styrene copolymers, even if showing dichroic bands between 270 and 185 nm of comparable intensity to the polymer, does not show the above splitting, only one dichroic band being observed in the region of the $^1B \leftarrow {}^1A$

electronic transition (Fig. 8.16). Thus, although conformational homogeneity gives ellipticity increase in all $\pi \rightarrow \pi^*$ electronic transitions, the occurrence of the exciton splitting of the ^1B band must be related to electrostatic dipole-dipole interactions between aromatic chromophores disposed in a mutual dissymmetric arrangement in the same molecule. This situation can be achieved in helical sections of the copolymer macromolecules having a single screw sense.

Confirmation of this interpretation is provided by semiempirical calculations of CD spectra by the De Voe theory [95]. The agreement between experimental and calculated CD spectra was excellent, particularly in the region of the $^1B \leftarrow {}^1A$ transition where a couplet comparable to that observed is predicted [27]. More detailed structural information can be obtained by looking at the influence of composition on the chiroptical properties and the different correspondence between calculated and experimental spectra for the various dichroic bands. The stronger dependence of ellipticity on aromatic unit content, observed in the calculations performed assuming statistical distributions of the co-units, can be explained by taking into account the substantially block distributions experimentally detected [96]. Extension of these concepts to analogous copolymers of 1VN [97] and 2VN [98] with the same optically active α-olefin led to similar conclusions, even if the situation appears more complicated because of the larger complexity of naphthalene chromophores with respect to benzene. In fact, even if exciton splitting occurs also in these cases in the $^1B_b \leftarrow {}^1A$ transition range, the sign of the couplet is opposite for the two systems. Moreover in the latter copolymers a second splitting seems to be present at slightly shorter wavelengths (Fig. 8.17).

Isotactic copolymers of styrene with 4MH that at least in principle are structurally similar to the above reported copolymers, were expected to behave similarly. Nevertheless their CD spectra in the 300-200 nm region do not show the typical exciton splitting pattern. In styrene/α-olefin copolymers, CD spectra below 240 nm can be recorded only on cyclohexane-soluble fractions, that in the case

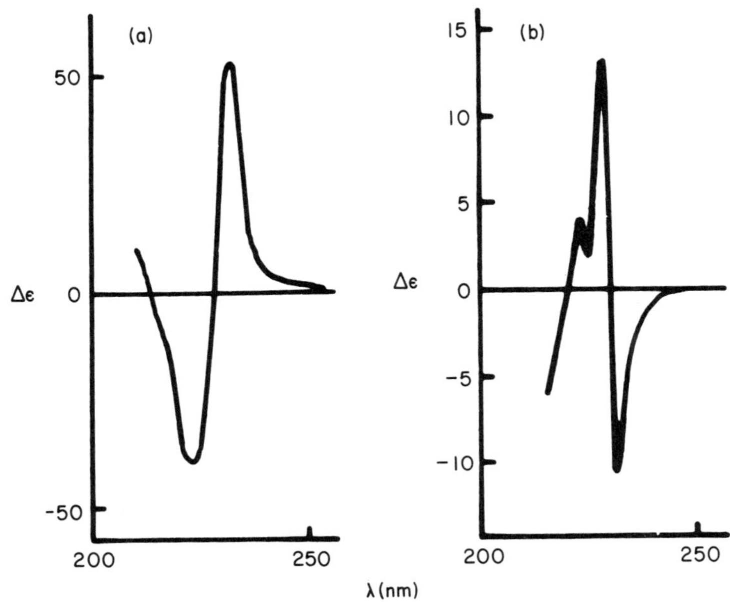

Fig. 8.17 CD spectra of coisotactic copolymers of (*S*)-3,7-dimethyl-1-octene with (a) 1VN and (b) 2VN, containing 90 mol% of optically active units [97,98].

of 4MH are mainly constituted by isolated aromatic units flanked by long sequences of aliphatic chiral units, as independently demonstrated by fluorescence [66] and ^{13}C NMR [67] techniques. Exciton splitting, as already mentioned, arises from suitable short-range interactions between aromatic chromophores and therefore must be rather sensitive to the distribution of monomeric units along the macromolecular backbone. Consequently it is reasonable that in the case of copolymers St-*co*-4MH the large separation between aromatic units prevents the coupling responsible for exciton splitting.

Styrene [99,100], 1VN [68] and 6VC [75] have been copolymerized in the presence of free radical initiators with MtA and MtMA, both absorbing below 230 - 220 nm, whereas the longest wavelength absorption band of the aromatic chromophores is located at a wavelength greater than 250 nm.

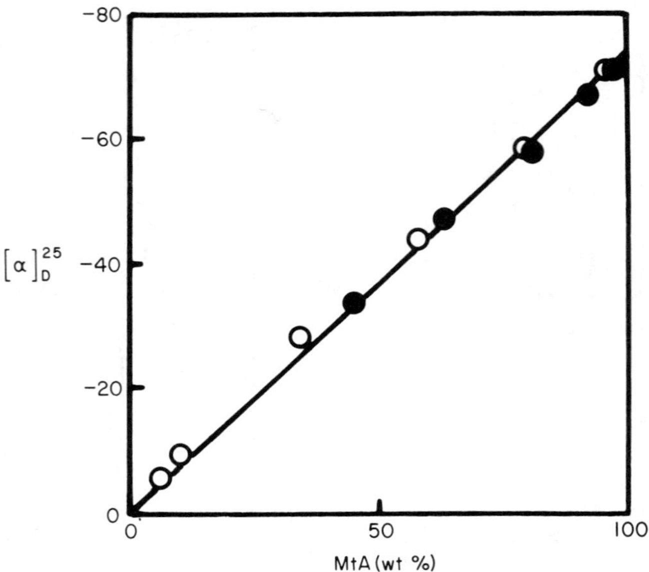

Fig. 8.18 Dependence on composition of the specific rotatory power of copolymers of MtA with —●— St or —○— 1VN [68,100].

Optical rotation of all these copolymers depends linearly on composition, thus apparently excluding any dissymmetric effect of (-)-menthyl groups on the aromatic units (Fig. 8.18). However, CD spectra show appreciable dichroic bands in the region where the various aromatic chromophores have absorption bands (Figs. 8.19 and 8.20).

The copolymers of St with MtA have been investigated in detail as far as the lowest energy absorption band is concerned. This last is located between 250 and 280 nm (Fig. 8.19), well out of the region of the ester chromophore absorption, which is below 225 nm [101] and can act as a convenient probe for investigating the chiral perturbation of the phenyl ring in these copolymers. The ellipticity of the ^1L_b band, and then the dissymmetric perturbation of St units, increases with increasing content, and mean sequence length ($\overline{l}_{MtA}$) of the chiral counits, reaching an asymptotic value for a molar content of MtA units larger than 70% ($\overline{l}_{MtA}$=4-5) (Fig.8.21) [100]. This result is consistent with that observed in the case of

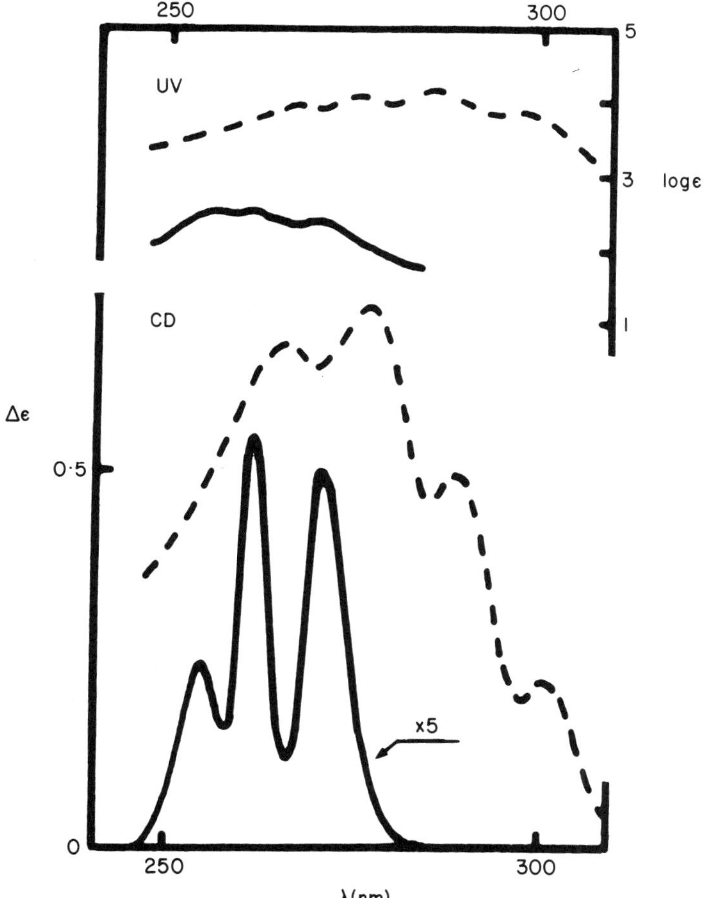

Fig. 8.19 UV and CD spectra of copolymers of MtA with (——) styrene [99,100] and (---) 1VN containing about 5 mol% of aromatic units [68].

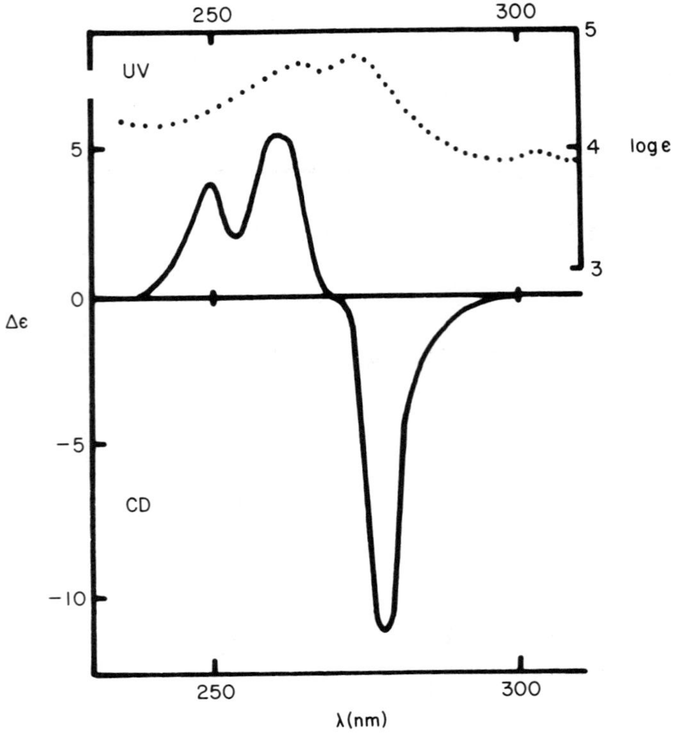

Fig. 8.20 UV (···) and CD (——) spectra of a copolymer of MtA with 6VC containing 44.5 mol% of 6VC units.

isotactic copolymers of optically active α-olefins and St [91] and is probably associated with a cooperative conformational effect. The value of the differential dichroic absorption at saturation is comparable to that of the Ziegler-Natta cataly ed copolymer of St with (*S*)-5-methyl-1-heptene where the side-chain asymmetric carbon atom is also in γ position with respect to the main chain [91]. Moreover the ellipticity in poly(MtA-*co*-St)s is higher at any composition than in the low molecular weight structural model compounds (−)-menthyl 4-phenylbutyrate (MtPB, and (−)-dimenthyl 4-phenylheptanedioate (DMtPH). This last has Δε = −0.018, whereas the corresponding poly(MtA-*co*-St) containing 34.7 mol. % St (two menthyl groups per phenyl ring) exhibits a Δε value of + 0.062 per St unit [100].

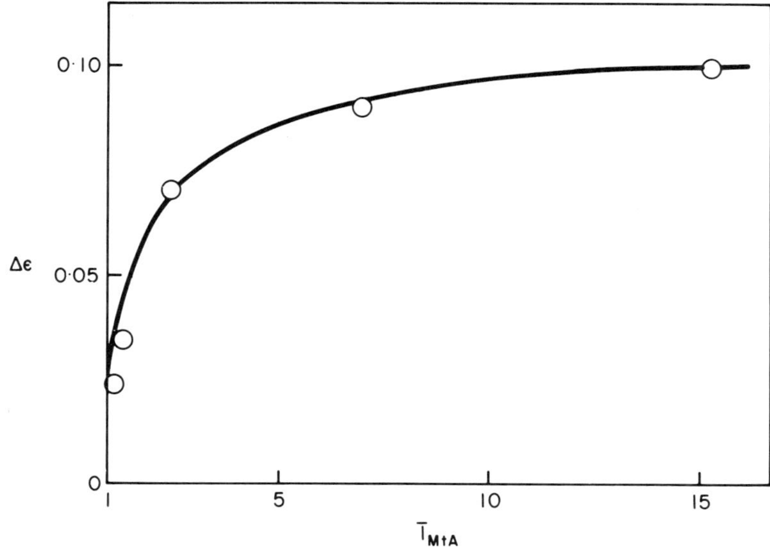

Fig. 8.21 Variation of molar dichroic absorption coefficient ($\Delta\varepsilon$) with the mean sequence length of MtA units ($\bar{l}_{MtA}$) in copolymers of MtA with styrene [100].

MtPB DMtPH

Further confirmation of the conformational cooperative effect derives from the lower ellipticity of the alternating copolymer with respect to that of the random copolymer with low content of St units. Local surroundings of the phenyl chromophore are the same in both copolymers with the St units substantially isolated. Additional evidence that the dichroic absorption band of the $^1L_b \leftarrow {}^1A$ electronic transition of the phenyl chromophores is very sensitive to structural effects is provided by the opposite sign of ellipticity in the alternating copolymers with MtA and MtMA [100].

Copolymers of MtA with 1VN prepared by radical initiation show dichroic bands in the region of the $^1L_a \leftarrow ^1A$ electronic transition of the naphthalene chromophore, the ellipticity again increasing with an increase of MtA units [68]. The maximum dichroic absorption was observed when practically all 1VN units were isolated between relatively long blocks of MtA units, the alternating copolymer displaying a lower $\Delta\varepsilon$, as observed for the corresponding St copolymers. Interestingly, in these copolymer systems even the sign and shape of the dichroic absorption were found to be highly dependent on composition and sequence length of monomeric units (Fig. 8.22) [68] Experimental results were explained by assuming that, when the content of 1VN units is lower than approximately 25 mol. %, the predominant contribution of positive sign to the ellipticity of the ^1L_a band derives from some prevailing chiral conformation of the main chain induced by long sequences of chiral co-units. However, when the content of 1VN units is larger than 50%, the prevailing chiral arrangement of the main chain should decrease [102]. Thus, the local site asymmetry of naphthalene chromophore could give a reverse sign contribution to CD maxima at about 277 and/or 299 nm. In addition, particularly in the case of poly(MtA-*alt*-1VN), and of poly(MtA-*co*-1VN)s containing a rather small amount of 1VN units, negative or positive contributions to induced optical activity can arise from the interactions between optically active ester groups and naphthalene chromophores, as observed in some biopolymers [103].

Calculations on stereoregular, optically active, hydrocarbon copolymers based on 1VN have also clearly shown that CD of the ^1L_a band of the naphthalene chromophore is dependent both on local site asymmetry and dipole-dipole interactions [98].

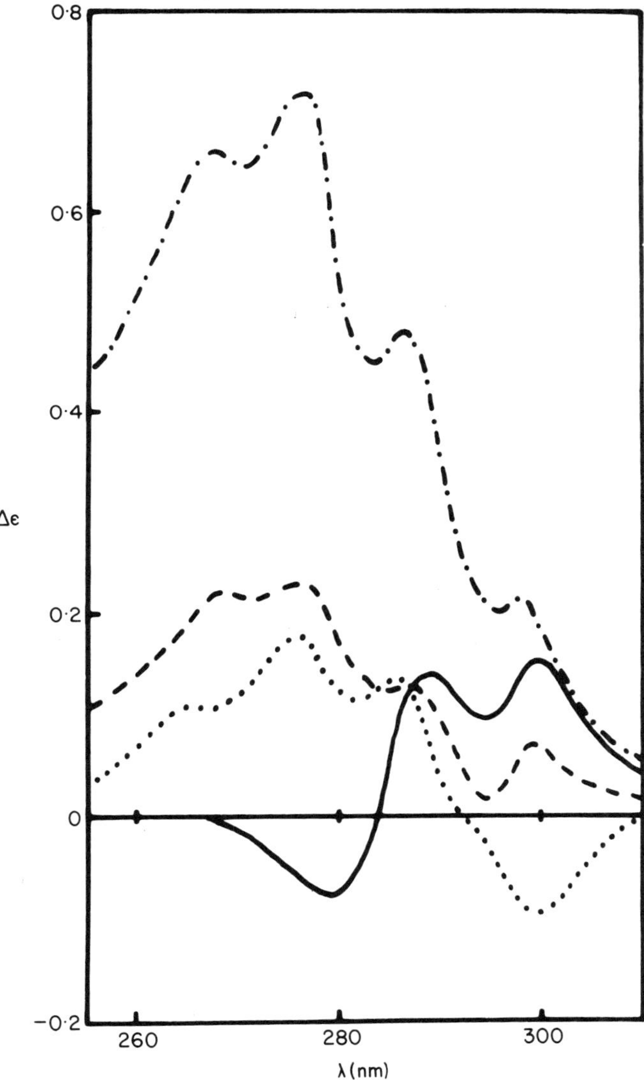

Fig. 8.22 CD spectra of copolymers of MtA with 1VN containing (−·−) 4.3, (---) 24.0 and (——) 50.6 mol% of 1VN units, and of poly(MtA-*alt*-1VN) [68].

8.3.2 *Acenaphthylene-containing polymers*

The copolymers of Ace with MtMA exhibit in the CD spectrum in dichloromethane solution a negative band at 235 nm and a structured positive band centred at about 300 nm, in close correspondence, respectively, with the $^1B_b \leftarrow {}^1A$ transition and the overlapping $^1L_a \leftarrow {}^1A$ and $^1L_b \leftarrow {}^1A$ transitions of the aromatic chromophore (Fig. 8.23) [84].

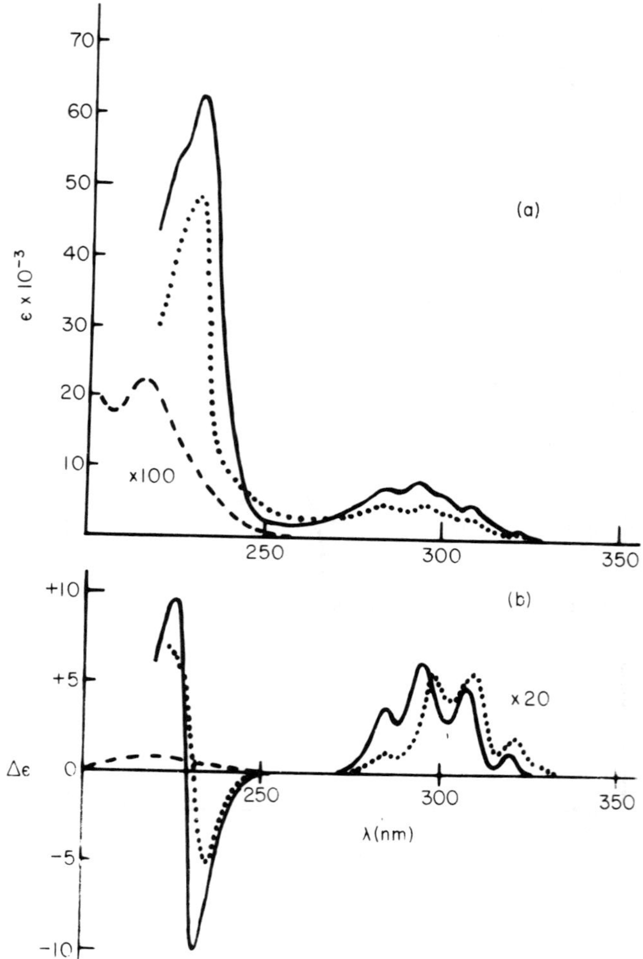

Fig. 8.23 (a) UV and (b) CD spectra of a copolymer of Ace with MtMA containing 14.7 mol% of Ace units in (——) *n*-heptane and in (...) dichloromethane, and of (---) poly-(MtMA) in *n*-heptane.

In the far end of the CD spectrum (λ <230 nm) a positive band starts to appear, which cannot be fully investigated owing to the strong absorption of the halogenated solvent in that spectral region. However, the copolymer with the lowest Ace content (14.7 mol. %) is soluble also in transparent hydrocarbon solvents, and the spectrum in n-heptane (Fig. 8.23) shows two strong dichroic bands of opposite sign having maxima at 232 and 225 nm, respectively, attributable to exciton splitting. For all the CD bands, whose intensities incidentally are of the same order of magnitude as those observed for coisotactic copolymers of vinyl aromatic monomers with α-olefins [25,34] and alkyl vinyl ethers [38], the molar differential extinction coefficient increases up to a maximum value at approximately 50 mol. % of Ace in the copolymer and then decreases with further increase in the content of the aromatic comonomer (Fig. 8.24). This anomalous trend suggests that the induced dissymmetric perturbation of aromatic units is maximized not for the isolated chromophore, but rather for appropriate sequences of units derived from the relatively rigid achiral comonomer adjacent to sequences of intrinsically chiral counits. The sequence distribution of monomeric units, as evaluated on the basis of the reactivity ratios, indicates that the optimization in the induced CD originates from the interposition of relatively short aromatic unit sequences (2-4 units) in relatively long sequences of the inherently chiral comonomer (3 or more units).

To account for this ability to transmit the chiral perturbation over some distance from the optically active groups, we have to assume that neighbouring aromatic chromophores can strongly interact, as clearly evidenced by the exciton splitting observed in the shorter wavelength region of the CD spectrum and by the excimer emission noted earlier.

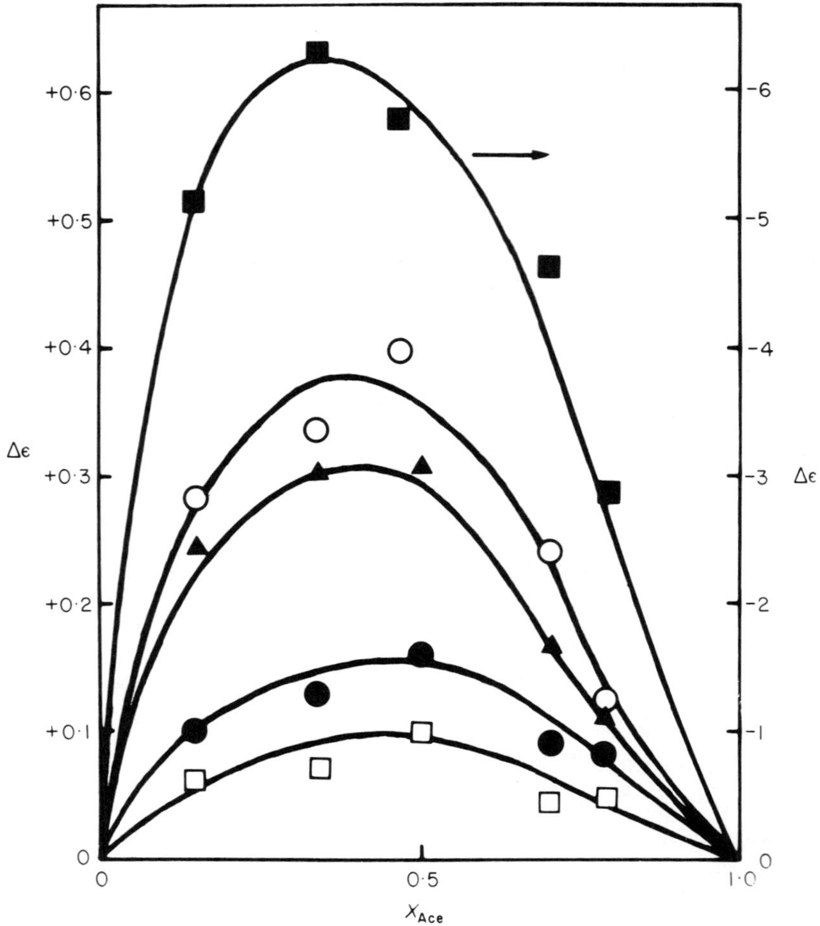

Fig. 8.24 Variation of the dichroic molar absorption coefficient (Δε) with Ace molar fraction (X_{Ace}) in copolymers of Ace with MtMA at different wavelengths: —●—, 321; —▲—, 310; —○—, 298; —□—, 285; and —■—, 235 nm.

The copolymers of Ace with MtA do not exhibit any appreciable induced circular dichroism in the Ace chromophore, despite the fact that MtA is known to produce significant induced CD effects in a wide variety of vinyl aromatic units [42,68,100]. At the present, no conclusive explanation of the different behaviour between acrylate and methacrylate copolymers has been put forward. However, both different distribution of monomeric units in the two copolymer series and lower flexibility of the main chain in the methacrylate copolymers must play a significant role in determining the asymmetry of the local environment in which the Ace units are embedded.

8.3.3 Carbazole-containing polymers

Optically active carbazole-containing polymers have been prepared by two alternative routes:

(1) Polymerization of inherently chiral heteroaromatic monomers.

(2) Copolymerization of a prochiral carbazole-containing monomer with a suitable chiral comonomer.

According to the former procedure, optically active vinyl polymers containing carbazole moieties have been obtained by polymerization of optically active monomers, such as (*S*)-3-*sec*-butyl-9-vinylcarbazole (3R9VC), (*S*)-9-(2-methylbutyl)-2-vinly-carbazole (9R2VC), and (*S*)-9-(2-methylbutyl)-3-vinylcarbazole (9R3VC) [36,37]. Polymer samples obtained in the presence of different catalytic systems show rather significant differences in molar optical rotation that might be accounted for, in part, on the basis of different molecular weights. In the case of poly(3R9VC) the change in optical rotation in going from the monomer or low molecular weight model compound (*S*)-3-*sec*-butyl-9-*iso*-propylcarbazole to the polymer, and the differences observed among samples prepared under different conditions can be interpreted in terms of a substantial difference in the degree of stereoregularity of the backbone tertiary carbon atom, the average degree of polymerization of the various samples being rather comparable ($\overline{DP}_n$ = 70-90) [37]. These data accord in fact with NMR tacticity studies on poly(NVC) [9-11] and indicate a comparatively large degree of isotacticity in samples obtained with conventional cationic initiators. However, an appreciable induced circular dichroism is observed only for poly(9R2VC) (Fig. 8.25). In this context, it may be of interest to recall that also in poly-(*S*)-*sec*-butylstyrene the asymmetric carbon atom in the *para* position

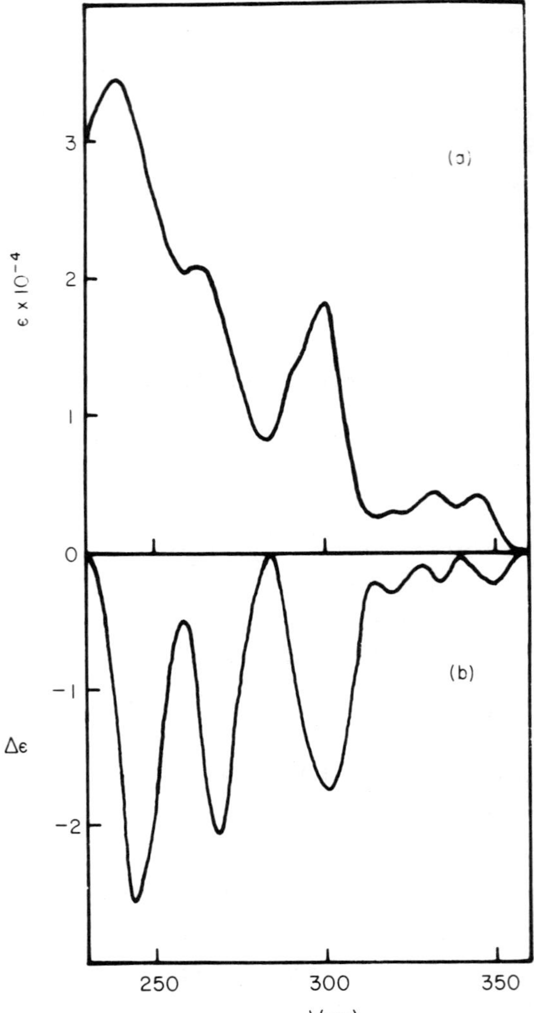

Fig. 8.25 (a) UV and (b) CD spectra of poly[(*S*)-9-(2-methylbutyl)-2-vinylcarbazole] obtained in the presence of Li-*n*-C_4H_9.

is not able to perturb the main chain conformation dissymmetrically [104]. Similarly, in copolymers of chiral MtA and MtMa with monomers having carbazole spaced from the polymerizable double bond, such as 2-(9-carbazolyl)ethyl methacrylate (9CEM) and 2-(9-carbazolylacetyloxy)ethyl methacrylate (9CAEM), there is a very low, if any,

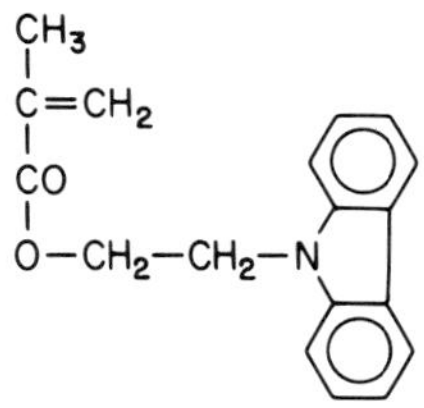

9CEM 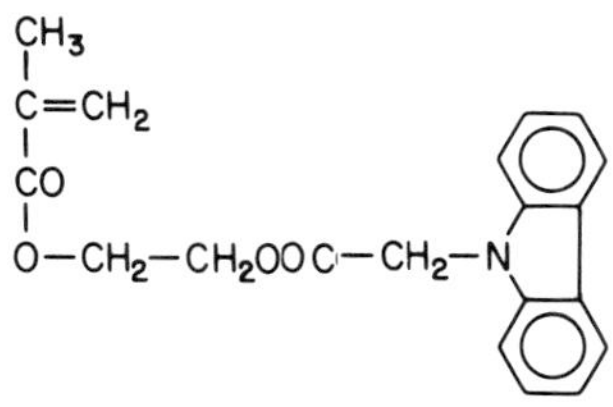 9CAEM

induced optical activity [41]. Apparently, the inherent flexibility of the spacer group enhances the conformational freedom of the carbazole nucleus, thus preventing both interactions among carbazole chromophores, as further evidenced by fluorescence emission properties, and the transmission of the chiral perturbation by the menthyl group. In accordance, similar copolymers from *m*-(9-carbazolyl) methyl styrene (*m*9CSt) and *p*-(9-carbazolyl)methyl styrene (*p*9CSt), where carbazole is removed from polymer backbone through the relatively rigid benzyl group, clearly show a significant induced CD [41].

In copolymers of NVC with various vinyl or vinylidenic optically active monomers it has been found that, independent of the nature of the chiral comonomer, composition and distribution of monomeric units, the carbazole moieties are always characterized by a more or less marked induced circular dichroism [35,38,42,43] (Figs. 8.26 and 8.27).

For all copolymers of NVC with optically active menthyl derivatives, the variation in induced CD absorption exhibits a maximum with increasing NVC content. For copolymers of NVC and MtVE, this maximum is very pronounced and occurs at approximately 20% NVC content [38]. Copolymers of NVC with MtA and MtMA show rather similar, but less pronounced maxima, at about 40 and 50% NVC contents, respectively [42]. Typical trends are shown in Fig. 8.28. The results suggest that a maximum in induced CD arises from the optimization of sequence lengths of chiral and prochiral units. In this respect the chiroptical properties of NVC-containing copolymers are rather similar to those of corresponding Ace copolymers.

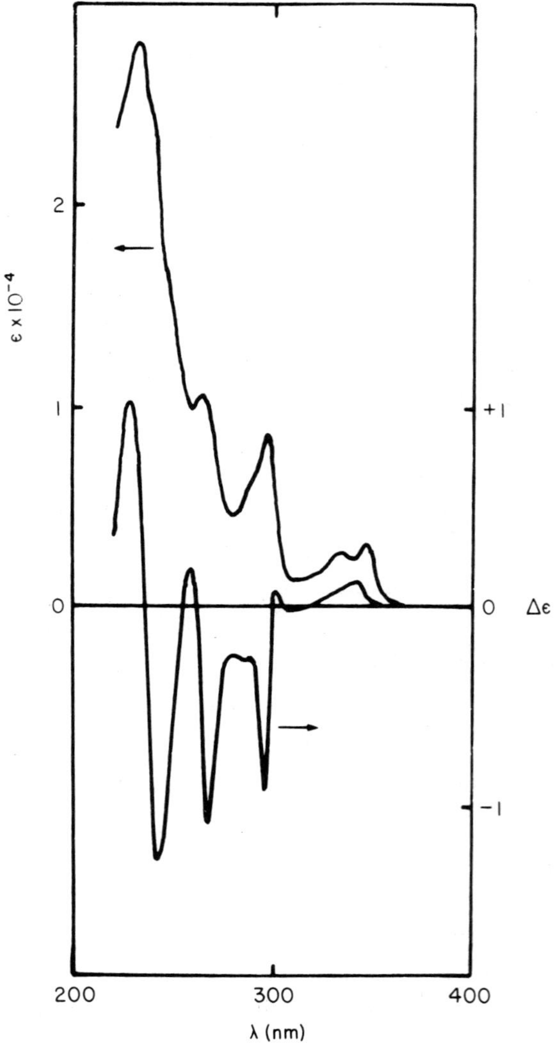

Fig. 8.26 UV and CD spectra of a copolymer of MtVE with NVC containing 17 mol% of NVC units.

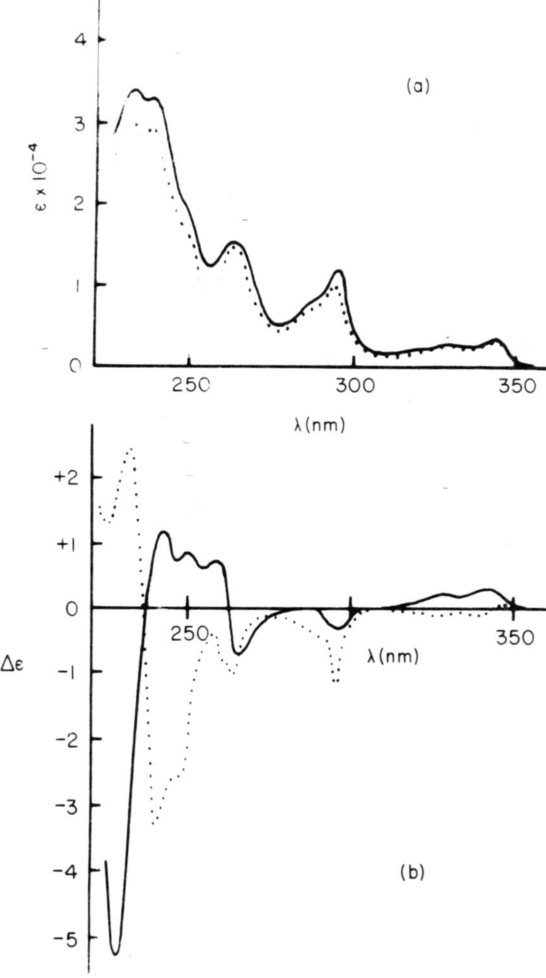

Fig. 8.27 (a) UV and (b) CD spectra of copolymers of NVC with (——) MtA and (···) MtMA containing about 65 mol% of NVC units.

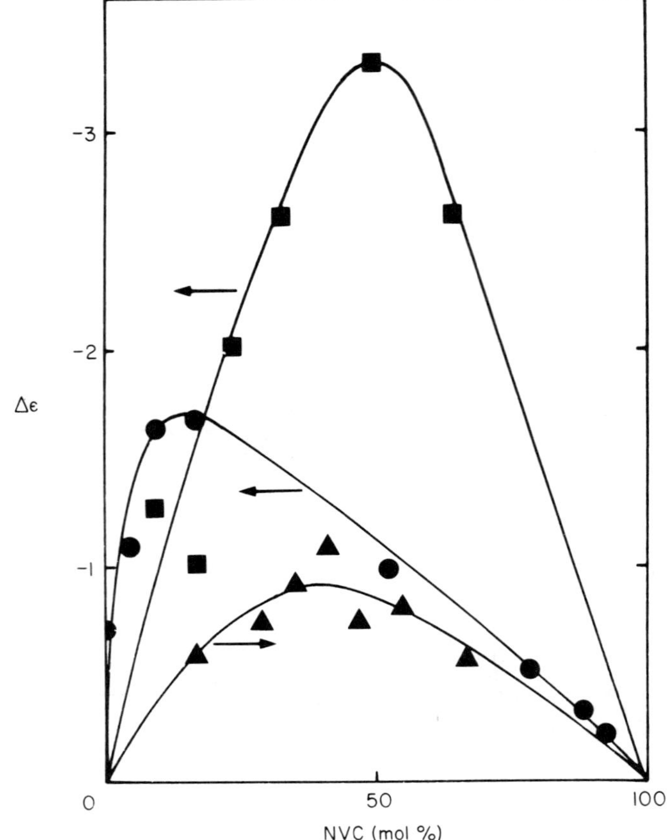

Fig. 8.28 Variation of the molar dichroic absorption coefficient (Δε) with composition in copolymers of NVC with ■ MtMA (240 nm), ▲ MtA (257 nm), and ● MtVE (265 nm).

Copolymers of NVC with MtVE have been shown to be block-like in nature [40] and as a consequence a significant amount of short sequences of NVC units is achievable only at very high MtVE contents. For free radically polymerized copolymers of NVC with MtA it is necessary to have an NVC content of approxiamtely 50% in order to obtain a significant contribution from NVC sequences of 2-3 units. This will automatically reduce the average sequence length for the chiral comonomer and hence will determine that optimum induced CD occurs at relatively high NVC contents.

Further support for this suggestion is provided by induced CD in copolymers of NVC and (-)-dimenthyl fumarate, in which case the natural tendency to alternation ensures that an optimum sequence distribution has not yet been observed even at 75% incorporation of NVC [43].

Examination of Figs. 8.26 and 8.27 shows that the precise shapes of the CD bands in the very high energy region may be attributable to an exciton splitting phenomenon, thus providing further evidence for the mutual interaction of aromatic chromophores in a chiral environment and in agreement with the well established tendency of carbazole units to form excimer-emitting sites in copolymers derived from NVC.

It is worth mentioning that the extent of induced CD on NVC units when they are inserted in copolymers with intrinsically chiral comonomers is of the same order of magnitude as that extrinsically induced by thermotropic cholesteric mesophases on achiral 9-ethylcarbazole when physically trapped within the mesophase [105]. Higher extrinsic CD effects are, by contrast, reported when lyotropic cholesteric mesophases, such as those obtained from poly(γ-benzyl-L-glutamate) in tetrachloroethane, are used as chiral traps for achiral carbazole derivatives and isoelectronic aromatic compounds [106,107].

Similarly, induced optical activity effects are observed when carbazole is bound, as a pendant substituent, to lyotropic poly(peptide)s [108-111]. The extent of the dissymmetric perturbation is generally high, and originates from the intrinsic chirality of polymer backbone in diluted solution, and, additionally, from the extrinsic chirality of the cholesteric phase when in concentrated solution.

8.3.4 Stilbene-containing polymers

CD spectra of copolymers of MtA with *trans*-4-acryloyloxystilbene exhibit a structured negative band centred at 310 nm corresponding to the first π→π* electronic transition of the stilbene chromophore [112] (Fig. 8.29). The intensity of this dichroic band monotonically increases with the content of chiral MtA units (Fig. 8.30). Independent of copolymer composition no exciton splitting is observed, possibly owing to the inherent flexibility of the ester spacer group connecting the stilbene nuclei to the main chain and reducing the backbone enforced interactions between neighbouring aromatic chromophores. In contrast, the CD spectra of copolymers of MtA with 4VS, where the stilbene group is tightly bound to the main chain, are characterized by the presence of a negative band at 285 nm and a positive one at 322 nm, attributed to exciton splitting (Fig. 8.29). The intensity of the induced CD is maximized at 50%

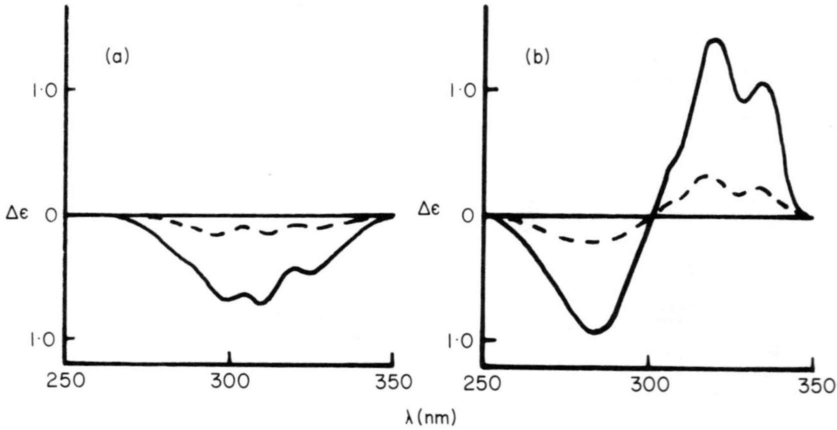

Fig. 8.29 CD spectra of copolymers of MtA with (a) 4AS and (b) 4VS containing about 60 mol% of MtA units: (——) all *trans* samples, (---) samples at the photostationary state after irradiation at 330 nm.

4VS units (Fig. 8.30), and it is interesting to observe that in copolymers of VS the optical rotation does not linearly depend on copolymer composition. In fact, the contribution from the vinyl aromatic chromophore to the optical rotation is maximized again in samples containing about 50% VC units (Fig. 8.31).

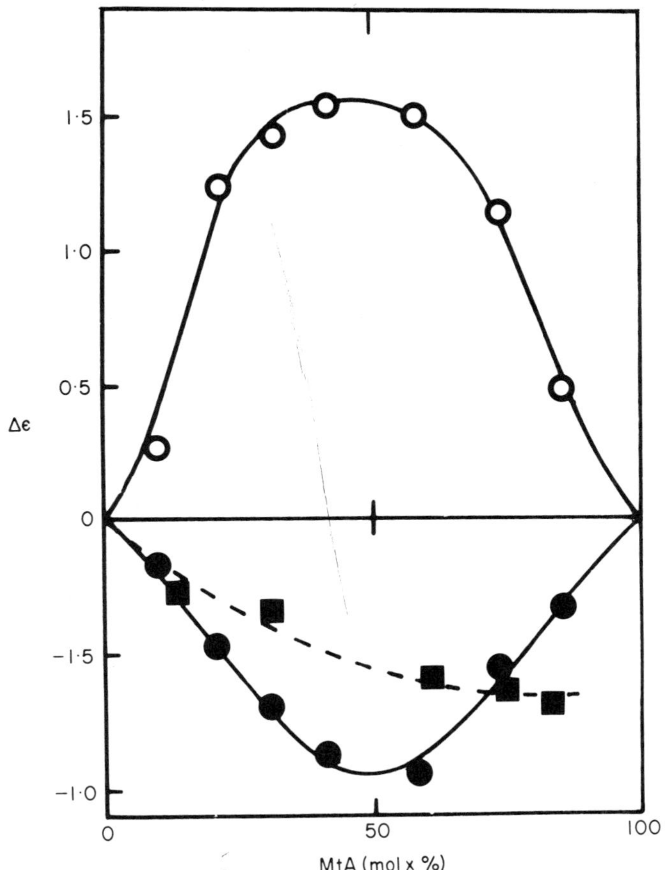

Fig. 8.30 Variation of the molar dichroic absorption coefficient (Δε) with the chemical composition in copolymers of MtA with *trans*-4AS (■, 313 nm), and *trans*-4VS (O , 322 nm; ● , 285 nm).

In copolymers of MtA with both 4AS and 4VS, *trans*→*cis* photoisomerization gives rise to a definite conformational rearrangement of the polymer backbone, as evidenced by the disappearence of the induced CD (Fig. 8.29), probably because accommodation of the bulky, non-planar *cis*-stilbene groups prohibits any local regular organization of polymer segments [52,89].

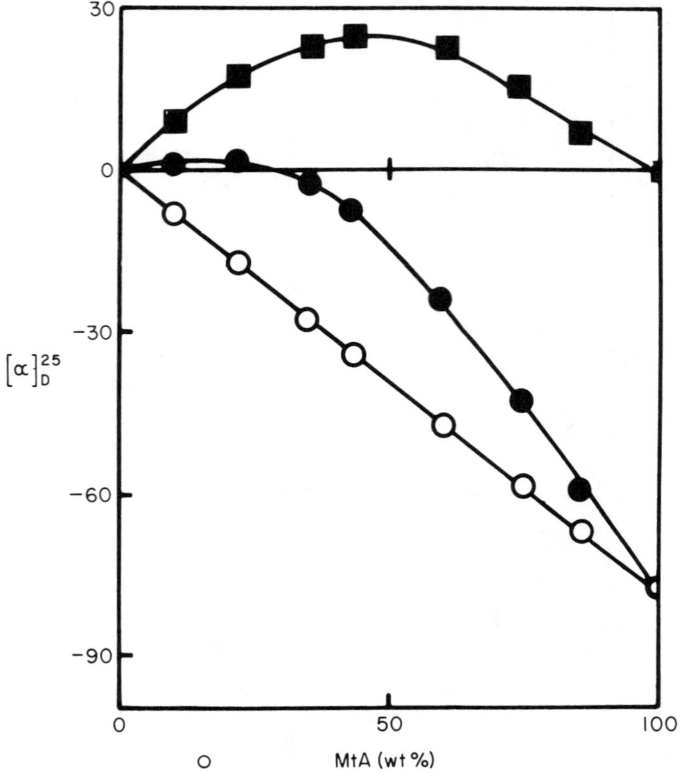

Fig. 8.31 Variation of specific optical rotation ($[\alpha]^{25}_{D}$) with chemical composition in (—●—) copolymers of MtA with 4VS and in (○) poly(MtA)/poly(4VS) mixtures, and (■) contribution to $[\alpha]^{25}_{D}$ from 4VS units.

8.4 CONCLUSIONS

The study of excimer fluorescence in polymers having pendant aromatic chromophores has been of interest and importance from different points of view. Generally excimer formation requires conformational eclipsing of aromatic chromophores and this can easily arise from adjacent monomer units in a polymer backbone, or via interactions of aromatic groups relatively separated within one polymer coil or even between two different macromolecules. Such studies are further complicated because the population of excimer-emitting sites will be greatly influenced by the extent of energy migration within, and between, macromolecular coils in solution and in the bulk phase. Poly(NVC) has been of special interest because its fluorescence spectrum gives clear evidence for both the anticipated conformationally eclipsed excimer, and a rather unique excimer involving only partial overlapping of adjacent carbazole units.

For other aromatic monomers covered in this review, excimer emission arises only via the conformationally eclipsed arrangements.

It is especially interesting that this conclusion may also be drawn for copolymers containing Ace units where the aromatic chromophore is incorporated directly into the backbone, and nearest-neighbour interaction is sterically prohibited.

The study of induced circular dichroism in polymers with optically active groups affords an additional technique for characterizing electronic transitions and, in favourable circumstances, permits the identification of interacting chromophores by observation of exciton splitting.

It is logical, therefore, to make joint studies of excimer emission and circular dichroism for appropriate copolymers in order to gain a better understanding of the configurational and conformational properties that influence the electronic interactions of aromatic groups in vinyl polymers.

For experimental convenience it has been necessary to utilize copolymers of suitable aromatic monomers with readily available optically active monomers derived from (-)-menthol, such as MtA and MtVE.

Generally the introduction of prochiral aromatic monomers into a copolymer backbone derived from optically active monomers gives rise to an induced circular dichroism in the aromatic chromophore, whose magnitude increases with increasing length of chiral unit sequences up to a saturation value at relatively short sequence lengths. Conversely the intrinsic polarizability characteristics of the different aromatic moieties appear determinant in affecting the magnitude of the induced CD with an increasing trend with increasing the polarizability [75]. This is the anticipated situation and reflects optically active substituent effects on isolated aromatic co-monomer units. In complete contrast, similar copolymers containing NVC, Ace and 4VS exhibit an optimized induced circular dichroism in samples in which, statistically, relatively short sequences of prochiral monomer units (2-4) are interposed between rather long sequences of chiral units (larger than 3). Significantly all these aromatic comonomers give copolymers that exhibit unusual excited-state properties as measured by fluorescence emission or depolarization.

A possible conclusion would be that polymer-chain induced conformational interactions of appended or in-chain chromophores, as reflected in excimer emission, should also determine an optimized copolymer structure for the maximization of induced optical activity. The latter phenomenon is particularly important in the design of copolymers that may be used as optically active reagents or supports for asymmetric syntheses, in which the aromatic co-unit is required to introduce the necessary reactive functionality.

In addition, the dependence of excimer properties in copolymers with aromatic units on the primary structure of the monomeric units, and the microstructure of the polymer backbone, offers the possibility to design polymeric systems with predictable photo-responses and energy transfer characteristics.

ACKNOWLEDGEMENTS

The authors are grateful to SERC and CNR for financial support.

REFERENCES

1. PEARSON, J.M. and STOLKA, M. (1981), *Poly*(N-*vinylcarbazole*), Gordon & Breach, New York.
2. OKAMOTO, K., OZEKI, M., ITAYA, A., KUSABAYASHI, S. and MIKAWA, A. (1975), *Bull. Chem. Soc. Jpn.* 48, 1362.
3. LANDMAN, U., LEDWITH, A., MARSH, D.G. and WILLIAMS, D.J. (1976), *Macromolecules* 9, 833.
4. PEARSON, J.M., TURNER, S.R. and LEDWITH, A. (1979), *Molecular Association*, Vol. 2, (ed. R. Foster), Academic Press, London.
5. SCHAFFERT, R.M. (1971), *IBM J. Res. Dev.* 15, 75.
6. PENWELL, R.C., GANGULY, B.N. and SMITH, T.W. (1978), *Macromol. Revs.* 13, 63.

7. CRYSTAL, R.C. (1971), *Macromolecules* 4, 379.
8. GRIFFITHS, C.H. (1975), *J. Polym. Sci., A2* 13, 1167.
9. OKAMOTO, K., YAMADA, M., ITAYA, A., KIMURA, T. and KUSABAYASHI, S. (1976), *Macromolecules* 9, 645.
10. KAWAMURA, T. and MATSUZAKI, M. (1978), *Makromol. Chem.* 179, 1003.
11. WILLIAMS, D.J. and FROIX, M.F. (1977), *ACS Polym. Prep.* 18(2), 445.
12. TERRELL D.R. and EVERS, F. (1982), *Makromol. Chem.* 183, 863.
13. TERRELL D.R. and EVERS, F. (1982), *J. Polym. Sci., Polym. Chem. Ed.* 20, 2529.
14. TERRELL, D.R. (1982), *Polymer* 23, 1045.
15. TERRELL, D.R., EVERS, F., SMOORENBURG, H. and VAN DER BOGAERT, H.M. (1982), *J. Polym. Sci., Polym. Phys. Ed.* 20, 1933.
16. OKAMOTO, S., AKANA, Y., KIMURA, A., HIRATA, H., KUSABAYASHI, S. and MIKAWA, H. (1969), *J. Chem. Soc., Chem. Commun.* 987.
17. WILLIAMS, D.J. (1970), *Macromolecules* 3, 602.
18. TSUCHIHASHI, N., HATANO, M. and SOHMA, J. (1976), *Makromol. Chem.* 177, 2739.
19. POWELL, R.C. and KIM, Q. (1973), *J. Luminescence* 6, 351.
20. JOHNSON, G.E. (1975), *J. Chem. Phys.* 62, 4697.
21. DESCHRYVER, F.C. VANDENDRIESSCHE, J., TOPPET, S., DEMEYER, K. and BOENS, N. (1982), *Macromolecules* 15, 406.
22. EVERS, F., KOBS, K., MENNING, R. and TERRELL, D.R. (1983), *J. Am. Chem. Soc.* 105, 5988.
23. ITAYA, A., OKAMOTO, K. and KUSABAYASHI, S. (1976), *Bull. Chem. Soc. Jpn.,* 42, 2082.
24. LEGRAND, M. (1973), in *Fundamental Aspects and Recent Developments in ORD and CD,* (ed. F. Ciardelli and P. Salvadori) Heyden, London. pp. 269 and 285.
25. CIARDELLI, F., AGLIETTO, M., CARLINI, C., CHIELLINI, E. and SOLARO, R. (1982), *Pure Appl. Chem.,* 54, 521.
26. MASON, S.F. (1979), in *Optical Activity and Chiral Discrimination* (ed. S.F. Mason), Reidel, Dordrecht, p.1.
27. HUG, W., CIARDELLI, F. and TINOCO, jr. I. (1974), *J. Am. Chem. Soc.* 96, 3407.
28. CIARDELLI, F., CHIELLINI, E. and CARLINI, C. (1979), in *Optically Active Polymers* (ed. E. Selegny), Reidel, Dordrecht, p.83.
29. SPASSKY, N., MOMTAZ, A. and SEPULCHRE, M. (1979), in *Preparation and Properties of Stereoregular Polymers* (ed. R.W. Lenz and F. Ciardelli), Reidel, Dordrecht, p.201.
30. OKAMOTO, Y., OHTA, K. and YUKI, H. (1978), *Macromolecules* 11, 724.
31. VAN BEIJNEN, A.J.M., NOLTE, R.J.M., DRENTH, W. and HEZEMANS, A.M.F. (1975), *Tetrahedron* 32, 2017.
32. PINO, P. (1965), *Adv. Polym. Sci.* 4, 393.
33. PINO, P., CIARDELLI, F. and ZANDOMENEGHI, M. (1970), *Ann. Rev. Phys. Chem.* 21, 561.
34. CIARDELLI, F., CARLINI, C., CHIELLINI, E., PIERONI, O., SALVADORI, P. and MENICAGLI, R. (1978), *J. Polym. Sci., Polym. Symp.* 62, 143.
35. CHIELLINI, E., SOLARO, R., PALMIERI, M. and LEDWITH, A. (1976), *Polymer* 17, 641.

36. CHIELLINI, E., SOLARO, R. and LEDWITH, A. (1977), *Makromol. Chem.* 178, 701.
37. CHIELLINI, E., SOLARO, R. and LEDWITH, A. (1978), *Makromol. Chem.* 179, 1929.
38. CHIELLINI, E., SOLARO, R., COLELLA, O. and LEDWITH, A. (1978), *Eur. Polym. J.* 14, 489.
39. HOUBEN, J.L., NATUCCI, B., SOLARO, R., COLELLA, O., CHIELLINI, E. and LEDWITH, A. (1978), *Polymer* 19, 811.
40. CHIELLINI, E., SOLARO, R., LEDWITH, A. and GALLI, G. (1980), *Eur. Polym. J.* 16, 875.
41. CHIELLINI, E., SOLARO, R., CIARDELLI, F., GALLI, G. and LEDWITH, A. (1980), *Polym. Bull.* 2, 577.
42. CHIELLINI, E., SOLARO, R., GALLI, G. and LEDWITH, A. (1980), *Macromolecules* 13, 1654.
43. GALLI, G., SOLARO, R., CHIELLINI, E. and LEDWITH, A. (1981), *Polymer* 22, 1088.
44. SOLARO, R., GALLI, G., MASI, F., LEDWITH, A. and CHIELLINI, E. (1983), *Eur. Polym. J.* 19, 433.
45. GHIGGINO, K.P., WRIGHT, R.D. and PHILLIPS, D. (1978), *Eur. Polym. J.* 14, 567.
46. TAGAWA, S., WASHIO, M. and TABATA, Y. (1979), *Chem. Phys. Lett.* 68, 276.
47. NG, D. and GUILLET, J.E. (1981), *Macromolecules* 14, 405.
48. ITO, M., TAZUKE, S. and OKAWARA, M. (1976), *Makromol. Chem.* 177, 621.
49. KEYANPOUR-RAD, M., LEDWITH, A., HALLAM, A., NORTH, A.M., BRETON, M., HOYLE, C. and GUILLET, J.E. (1978), *Macromolecules* 11, 1114.
50. LEDWITH, A., ROWLEY, N.J. and WALKER, S.M. (1981), *Polymer* 22, 435.
51. NAKAIRA, T., MARUYAMA, S., IVABUCHI, S. and KOJIMA, K. (1979), *Makromol. Chem.* 180, 1853.
52. ALTOMARE, A., CARLINI, C., CIARDELLI, F., SOLARO, R., HOUBEN, J.L. and ROSATO, N. (1983), *Polymer* 24, 95.
53. MERLE-AUBRY, L., HOLDEN, D.A., MERLE, Y. and GUILLET, J.E. (1980), *Macromolecules* 13, 1138.
54. HOLDEN, D.A. and GUILLET, J.E. (1980), in *Developments in Polymer Photochemistry*, Vol. 1 (ed. N.S. Allen), Applied Science, London, p.27.
55. HIRAYAMA, F. (1965), *J. Chem. Phys.* 42, 3163.
56. CHAPOY, L.L. and BIDDLE, D. (1983), *J. Polym. Sci., Polym. Lett. Ed.* 21, 621.
57. LIMBURG, W.W., YANUS, J.F., WILLIAMS, D.J., GOEDDE, A.O. and PEARSON, J.M. (1975), *J. Polym. Sci., Polym. Chem. Ed.* 13, 1133.
58. KEYANPOUR-RAD, M., LEDWITH, A. and JOHNSON, G.E. (1980), *Macromolecules* 13, 222.
59. LEDWITH, A., CHIELLINI, E. and SOLARO, R. (1979), *Macromolecules* 12, 240.
60. DELFINI, M., CIFANI, M., CONTI, F., PACI, M., SOLARO, R., GALLI, G. and CHIELLINI, E. (1980), Prep. IUPAC International Symposium on Macromolecules, Florence, Vol. 5, p.8.

61. YOKOYAMA, M., TAMAMURA, T., ATSUMI, M., YOSHIMURA, M., SHIROTA, Y. and MIKAWA, H. (1975), *Macromolecules* 8, 101.
62. SKILTON, P.F., SAKUROVS, R. and GHIGGINO, K.P. (1982), *Polym. Photochem.* 2, 409.
63. VALA, jr. M.T., HAEBIG, J. and RICE, S.A. (1965), *J. Chem. Phys.* 43, 886.
64. DAVID, C., LAVAREILLE, N.P. and GEUSKENS, G. (1974), *Eur. Polym. J.* 10, 617.
65. DAVID, C., LEMPEREUR, M. and GEUSKENS, G. (1973), *Eur. Polym. J.* 9, 1315.
66. SOLARO, R., RASPOLLI-GALLETTI, A.M. and CHIELLINI, E. (1984), *Macromolecules* 17, 2212.
67. SEGRE, A.L., DELFINI, M., SOLARO, R. and RASPOLLI-GALLETTI, A.M. (1985), *Macromolecules* in press.
68. MAJUMDAR, R.N., CARLINI, C., ROSATO, N. and HOUBEN, J.L. (1980), *Polymer* 21, 941.
69. FOX, F.B., PRICE, T.R., COZZENS, R.F. and McDONALD, J.R. (1972), *J. Chem. Phys.* 57, 534.
70. DAVID, C., PIENS, M. and GEUSKENS, G. (1976), *Eur. Polym. J.* 12, 621.
71. REID, R.F. and SOUTAR, I. (1977), *J. Polym. Sci., Polym. Lett. Ed.* 15, 153.
72. REID, R.F. and SOUTAR, I. (1978), *J. Polym. Sci., Polym. Phys. Ed.* 16, 231.
73. ALEXANDRU, L. and SOMERSALL, A.C. (1977), *J. Polym. Sci., Polym. Lett. Ed.* 15, 2013.
74. WEBBER, S.E., AVOTS-AVOTINS, P.E. and DEUMIE, M. (1981), *Macromolecules* 14, 105.
75. CHIELLINI, E., SOLARO, R. and CIARDELLI, F. (1982), *Makromol. Chem.* 183, 103.
76. McDONALD, J.R., ECHOLS, W.E., PRICE, T.R. and FOX, R.B. (1972), *J. Chem. Phys.* 57, 1746.
77. DAVID, C., LEMPEREUR, M. and GEUSKENS, G. (1972), *Eur. Polym. J.* 8, 417.
78. PHILLIPS, D., ROBERTS, A.J. and SOUTAR, I. (1980), *J. Polym. Sci. Polym. Lett. Ed.* 18, 123.
79. WANG, Y.C., MORAWETZ, H. (1975), *Makromol. Chem., Suppl.* 1, 283.
80. REID, R.F. and SOUTAR, I. (1980), *J. Polym. Sci. Polym. Phys. Ed.* 18, 457.
81. ANDERSON, R.A., REID, R.F. and SOUTAR, I. (1980), *Eur. Polym. J.* 16, 925.
82. DAVID, C., BAEYENS-VOLANT, D. and PIENS, M. (1980), *Eur. Polym. J.* 16, 413.
83. GALLI, G., SOLARO, R., CHIELLINI, E. and LEDWITH, A. (1983), *Macromolecules* 16, 497.
84. GALLI, G., SOLARO, R., CHIELLINI, E., FERNHYOUGH, A. and LEDWITH, A. (1983), *Macromolecules* 16, 502.
85. STORY, V.M., and CANTY, G. (1964), *J. Natl. Bur. Stan. A, Phys. Chem.* 68A, 165.
86. CIARDELLI, F., CARLINI, C., SOLARO, R., ALTOMARE, A., PIERONI, O., HOUBEN, J.L. and FISSI, A. (1984), *Pure Appl. Chem.* in press.

87. FISSI, A., HOUBEN, J.L., ROSATO, N., LOPES, L., PIERONI, O. and CIARDELLI, F. (1982), *Makromol. Chem., Rapid. Commun.* 3, 29.
88. ALTOMARE, A., CARLINI, C., PANATTONI, M. and SOLARO, R. (1984), *Macromolecules* 17, 2207.
89. ALTOMARE, A., CARLINI, C., CIARDELLI, F., PANATTONI, M. and SOLARO, R. (1983), Proceedings 6th Meeting of Italian Association of Macromolecules, Pisa, Vol.2, p.87.
90. PINO, P., CARLINI, C., CHIELLINI, E., CIARDELLI, F. and SALVADORI, P. (1968), *J. Am. Chem. Soc.* 90, 5025.
91. CARLINI, C. and CHIELLINI, E. (1975), *Makromol. Chem.* 176, 519.
92. CIARDELLI, F., CARLINI, C., CHIELLINI, E., SALVADORI, P., LARDICCI, L., MENICAGLI, R. and BERTUCCI, C. (1979), in *Preparation and Properties of Stereoregular Polymers* (ed. R.W. Lenz and F. Ciardelli), Reidel, Dordrecht, p.353.
93. CIARDELLI, F., SALVADORI, P., CARLINI, C. and CHIELLINI, E. (1972), *J. Am. Chem. Soc.* 94, 6536.
94. SALVADORI, P., LARDICCI, L., MENICAGLI, R. and BERTUCCI, C. (1972), *J. Am. Chem. Soc.* 94, 8598.
95. DE VOE, H. (1965), *J. Chem. Phys.* 43, 3199.
96. CHIELLINI, E., MAESTRINI, F., VERGAMINI, P. and CECCARELLI, G. (1975), *Chim. Ind. (Milan)* 57, 131.
97. CIARDELLI, F., SALVADORI, P., CARLINI, C., MENICAGLI, R. and LARDICCI, L. (1975) *Tetrahedron Lett.* 1779.
98. CIARDELLI, F., RIGHINI, C., ZANDOMENEGHI, M. and HUG, W. (1977), *J. Chem. Phys.* 81, 1948.
99. MAJUMDAR, R.N., CARLINI, C., NOCCI, R., CIARDELLI, F. and SCHULZ, R.C. (1976), *Makromol. Chem.* 177, 3619.
100. MAJUMDAR, R.N. and CARLINI, C. (1980), *Makromol. Chem.* 181, 201.
101. SCHULZ, R.C. and GUTHMANN, A. (1967), *J. Polym. Sci., Part B* 5, 1099.
102. BERTUCCI, C., CARLINI, C., CIARDELLI, F. and SALVADORI, P. (1979), Prep. IUPAC International Symposium on Macromolecules, Mainz, Vol. 1, p.273.
103. KONISHI, Y. and HATANO, M. (1976), *J. Polym. Sci. Polym. Lett. Ed.* 14, 351.
104. CIARDELLI, F., PIERONI, O., CARLINI, C. and MENICAGLI, R. (1972), *J. Polym. Sci., A1* 10, 809.
105. SAEVA, F.D. (1972), *J. Am. Chem. Soc.* 94, 5136.
106. SAEVA, F.D. and OLIN, G.R. (1973), *J. Am. Chem. Soc.* 95, 7882.
107. TSUCHIHASHI, N., NOMORI, H., HATANO, M. and MORI, S. (1975), *Bull. Chem. Soc. Jpn.* 48, 29.
108. NOMORI, H., ENAMOTO, T., YOSHIKAWA, M. and HATANO, M. (1976), *Makromol. Chem.* 177, 3077.
109. HATANO, M., NOMORI, H. and YOSHIKAWA, M. (1977), in *Peptides. Proceedings of the 5th American Peptides Symposium* (ed. M. Goodman and J. Meinhofer), Wiley, New York, p.390.
110. CHAPOY, L.L., BIDDLE, D., HALSTROM, J., KOVAKS, K., BRUNFELDT, K., QASIM, M.A. and CHRISTENSEN, T. (1983), *Macromolecules* 16, 181.
111. OSHIMA, R. and KUMAROTANI, J. (1979), *ACS Polym. Prep.* 20(1), 522.
112. ALTOMARE, A., CARLINI, C. and SOLARO, R. (1982), *Polymer* 23, 1355.

Index

NOTE: Figures indicated by underscored page numbers: footnotes by suffix n.